THE LAW OF
INTERNATIONAL WATERWAYS

THE LAW OF
INTERNATIONAL
WATERWAYS

With Particular Regard to Interoceanic Canals

by
R. R. BAXTER
with the research assistance of
JAN F. TRISKA

HARVARD UNIVERSITY PRESS
CAMBRIDGE, MASSACHUSETTS
1964

Distributed in Great Britain by Oxford University Press, London

Library of Congress Catalog Card Number 64-13420

PRINTED IN THE UNITED STATES OF AMERICA

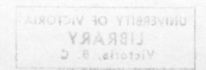

ACKNOWLEDGMENTS

In October 1954 the Harvard Law School undertook this study at the suggestion of the Suez Canal Company, which was then discharging functions of importance to the international community, and of the Carnegie Corporation of New York. Following the nationalization of the Suez Canal, the study was continued and completed.

Its financial support was drawn entirely from the Carnegie Corporation and funds available to the School under its program of International Legal Studies.

I wish to record my deep appreciation to the Harvard Law School and to the Carnegie Corporation of New York for this generous financial assistance, which has made possible the writing of this book.

In April 1955, the author spent a number of days in Egypt and at the offices of the Suez Canal Company in Paris for the purpose of viewing the Canal and observing the operations of the Company. He profited greatly from the assistance and information furnished him in generous measure by M. Georges-Picot, the General Manager of the Company, M. de Grièges, the Assistant General Manager, and the other officials and employees of the Company whom he met. Thanks are also owed to M. Boillot, the New York representative of the Company, for the documentation and information which he has provided.

During a further visit to the United Arab Republic in December 1960 and January 1961, I was graciously received by the Suez Canal Authority of the United Arab Republic and was given an opportunity to view the numerous improvements which have been made in the works and in the operations of the Canal since its nationalization. I am much indebted to the Suez Canal Authority for my reception at that time and for the information and documentation furnished me over the course of years.

It would be impossible to acknowledge the assistance of all of those who have contributed documentation, information, and other assistance which has been utilized in the writing of this study. I must, however, express my special gratitude to the following persons and agencies: Judge Jasper Y. Brinton, formerly President of the Court of Appeals of the Mixed Courts of Egypt, and presently Vice President of the Egyptian Society of International Law; the Embassy of the United Arab Republic in Washington;

v

Dr. Omar Z. Ghobashy, formerly of the Consulate General of the United Arab Republic in New York; the Suez Canal Authority; the Wasser- und Schiffahrtsdirektion Kiel; the Panama Canal Company; the Saint Lawrence Seaway Development Corporation; Mr. Charles M. Spofford; Dr. Jacob Robinson, formerly Counsellor, Permanent Delegation of Israel to the United Nations; Mr. Shabtai Rosenne, Legal Adviser, Ministry for Foreign Affairs of Israel; and the late Professor André Siegfried. The members of the International Rivers Committee of the International Law Association were good enough to furnish the author their observations on the Articles which appear in the Appendix, while they were still in draft form.

A portion of Chapter IV originally appeared in the *British Year Book of International Law* for 1954 and is reprinted here with the kind permission of the editor of the *Year Book* and of the Royal Institute of International Affairs.

It is a particular pleasure to record my thanks to Dr. Jan F. Triska of Stanford University for his research assistance in the writing of this volume. I am indebted to him not only for furnishing the basic materials for Chapter II, section E, and for research on a large number of other points but also for the important intellectual contribution he has made to this study.

I owe a heavy debt of gratitude to Judge (then Professor) Philip C. Jessup, Professors Milton Katz and Louis B. Sohn, and to the late Professor Edwin D. Dickinson, who patiently read the entire work in manuscript. I have benefited greatly from their criticism and suggestions. For any errors, omissions, or ambiguities which remain, the responsibility is, of course, mine alone.

It is fitting to conclude these acknowledgments with my warm thanks to Professor Milton Katz for his encouragement and wise counsel throughout the writing of this volume.

R. R. B.

Harvard Law School
Cambridge, Massachusetts
August 1, 1963

CONTENTS

vii

CONTENTS

THE LAW OF
INTERNATIONAL WATERWAYS

Abbreviations used in the notes

LE CANAL DE SUEZ — LE CANAL DE SUEZ: BULLETIN DE LA COMPAGNIE UNIVERSELLE DU CANAL MARITIME DE SUEZ

Compagnie Universelle — Compagnie Universelle du Canal Maritime de Suez

COMPAGNIE UNIVERSELLE, RECUEIL DES ACTES CONSTITUTIFS — RECUEIL CHRONOLOGIQUE ET ANNOTÉ DES ACTES CONSTITUTIFS DE LA COMPAGNIE UNIVERSELLE DU CANAL MARITIME DE SUEZ ET DES CONVENTIONS CONCLUES AVEC LE GOUVERNEMENT EGYPTIEN

MARTENS, N.R. — MARTENS, NOUVEAU RECUEIL DE TRAITÉS, 16 vols. (1817-1841)

MARTENS, N.R.G. — MARTENS, NOUVEAU RECUEIL GÉNÉRAL DE TRAITÉS, 20 vols. (1843-1875); 2d ser., 35 vols. (1876-1910); 3d ser., 40 vols. (1909-1943)

MARTENS, RECUEIL — MARTENS, RECUEIL DE TRAITÉS, 8 vols. (1817-1842)

REVUE DE LA NAVIGATION — REVUE DE LA NAVIGATION INTÉRIEURE ET RHÉNANE

CHAPTER I

INTRODUCTORY

T HE WORLD is "a planet dominated by its covering mantle of ocean, in which the continents are but transient intrusions of land above the surface of the all-encircling sea."[1] For all the immensity of the seas, there are places where the waters narrow and run between the land masses in straits, only to spread their expanse once more as the coasts are passed. Some oceans are accessible from the rest of the high seas only through these passages. The Mediterranean and the Aegean find their natural means of access to the oceans of the world through the Strait of Gibraltar, and the Black Sea opens to other waters through the straits which connect it with the Aegean and the Mediterranean. The waters of the Baltic Sea flow to the outer ocean through the Danish Straits, which consist not of one but of three passages, the Sound, the Great Belt, and the Little Belt. Not content with these channels, man has constructed a number of artificial straits, or canals, where an isthmus may be pierced to afford easy access between two portions of the high seas. While straits and the artificial waterways of Suez and Panama may serve the same purposes in terms of economic geography and navigation, the difference in their origins must inevitably place them in different categories for purposes of legal analysis. And finally, yet other important channels of commerce lead from the ocean highways deep into continents along those navigable rivers which carry the shipping both of the riparian states and of other nations as well.

Straits, canals, and rivers share the geographic characteristics of bringing maritime highways into close contiguity with land. This common use by shipping, whether merchant vessels or warships, sets the limits of this study, which concerns itself with problems of the navigation of international waterways. This is not to deny that these three types of waterways may be employed for a variety of other purposes, some of which may be inconsistent with the use of the waterway for navigation. The varying uses to which rivers may be subjected affords the clearest illustration of the pos-

[1] CARSON, THE SEA AROUND US 15 (1951).

1

sibility of competing demands for the benefits of such waterways.[2] The waters of a river may be diverted, most often through the instrumentality of a dam, which may block the waterway, for domestic or industrial consumption or for the irrigation of land. The waters of a river, again held back by a dam, may be used for the generation of power. The carrying away of waste may be yet another function for which the currents of a river may be employed. These uses, which may compete with one another and with navigation, may in some instances be reconciled with navigation of the waterway by shipping or may, on the contrary, make navigation an impossibility.[3] This additional complication of conflicting uses is far less characteristic of straits and canals, the real importance of which lies in their availability as highways of ocean shipping. Nevertheless, it is possible to conceive a situation in which the use of a strait for fisheries,[4] like other offshore areas of the seas,

---

[2] For a consideration of these varying uses, see United Nations, Economic Commission for Europe, Committee on Electric Power, Legal Aspects of the Hydro-electric Development of Rivers and Lakes of Common Interest 29-37 (U.N. Doc. No. E/ECE/136, E/ECE/EP/98 Rev. 1) (1952); BERBER, RIVERS IN INTERNATIONAL LAW 5-9 (1959); Sauser-Hall, *L'Utilisation industrielle des fleuves internationaux,* 83 HAGUE RECUEIL 465 (1953). The attempt to codify the main rules on the subject made in the Statute on the Régime of Navigable Waterways of International Concern, annexed to the Convention opened for signature at Barcelona, April 20, 1921, 7 L.N.T.S. 50, 1 HUDSON, INTERNATIONAL LEGISLATION 645 (1931), took into account, in paragraphs 1, 3, and 6 of article 10, the fact that navigable waterways might be devoted to other uses, consistent or inconsistent with their employment for navigation. H. A. Smith states that the conclusion is irresistible that "it is impossible to lay down any general rule as to the priority of interests upon all river systems." THE ECONOMIC USES OF INTERNATIONAL RIVERS 143 (1931).

[3] Treaty between the United States and Great Britain relating to Boundary Waters between the United States and Canada, signed at Washington, Jan. 11, 1909, 36 Stat. 2448, T.S. No. 548, contains in article VIII an order of precedence, which gives first priority to uses for domestic and sanitary purposes, second to uses for navigation, and third to uses for power and for irrigation purposes. No use is permitted which "tends materially to conflict with or restrain any other use which is given preference over it" in this listing.

In upholding the constitutionality of the Federal Water Power Act of 1920, the United States Supreme Court noted the inextricable connection between navigation and other uses of rivers. "In truth the authority of the United States is the regulation of commerce on its waters. Navigability, in the sense just stated, is but a part of this whole. Flood protection, watershed development, recovery of the cost of improvement through utilization of power are likewise parts of commerce control." United States v. Appalachian Electric Power Co., 311 U.S. 377, 426 (1940).

[4] Canada has, for example, been concerned about fishing rights in the Hecate Strait between the Queen Charlotte Islands and the mainland of British Columbia. See Canada, [1956] 7 H.C. DEB. 6700-03 (1956).

might enter into competition with the need of the users of the strait for free and unimpeded passage of the waterway. It is thus impossible to consider the navigation of international waterways without giving regard to the reconciliation of this use with the other, possibly conflicting, demands to which a watercourse may be subjected.

INTERNATIONAL WATERWAYS DEFINED

The answer to the question of which waterways—rivers, straits, and canals—are *international* waterways can at this stage be given only in factual terms. For purposes of analysis, international waterways must be considered to be those rivers, canals, and straits which are used to a substantial extent by the commercial shipping or warships belonging to states other than the riparian nation or nations.[5] In the case of rivers, it is necessary to add the additional qualification that the waterway run through or between two or more states and that the stretch of the river lying within the territory of any one state or bordering thereon be subject to a substantial amount of use by the vessels of other states. The international character of a waterway, thus conceived, rests upon factual considerations, for it is only by examination of the actualities of international usage that any conclusions about the requisites of international character become possible. To begin with a legal definition of such waterways would be inconsistent with the essentially inductive pattern of this study.

The geographic definitions of these three important types of international waterways unfortunately do not correspond with those which must be adopted for legal purposes.[6] In geographic terms, a strait is normally a narrow passage connecting two sections of the high seas,[7] but a precise de-

---

[5] This criterion must be distinguished from that employed in United States courts in determining whether a waterway is "navigable." Here capacity for navigation is the important element: ". . . a river having actual navigable capacity in its natural state and capable of carrying commerce among the States, is within the power of Congress to preserve for purposes of future transportation, even though it be not at present used for such commerce, and be incapable of such use according to present methods, either by reason of changed conditions or because of artificial obstructions." Economy Light & Power Co. v. United States, 256 U.S. 113, 123 (1921); *accord,* United States v. Appalachian Electric Power Co., *supra* note 3, at 407.

[6] Professor Hyde made clear the distinction by introducing his discussion of boundaries in straits with the statement, "There are straits and straits." 1 INTERNATIONAL LAW, CHIEFLY AS INTERPRETED AND APPLIED BY THE UNITED STATES 487 (2d rev. ed. 1945).

[7] Hall alludes to the difficulty of separating gulfs and straits from one another "in

limitation cannot be made of the width and length required of such a passage. The English Channel, which may be considered to be one great strait, ranges in width from 100 miles to a distance of 22 miles in the Strait of Dover. By way of contrast, the Bosporus, another strait of great international importance, is but 800 yards at its narrowest point. These extremes offer no real guidance for legal purposes. However, although a more precise legal definition has been attempted in the past, there has been such a diversity of opinion expressed that it remains difficult to formulate any definite rule. At the time when the breadth of the territorial sea was generally recognized to be three miles, the application of this rule to straits would have led to the conclusion that a strait is a natural channel six miles or less in width.[8] The territorial sea of each littoral state would extend out from land for three miles, and the area of the strait beyond these outer limits would constitute part of the high seas, open to free navigation by the ships of all nations. This view of the law, although attributed to the minority of the authorities,[9] was adopted by a subcommittee of the Hague Codification Conference of 1930[10] and seems indeed to have received the support of the majority of the governments which submitted comments.[11] An exception to this rule seemed necessary when both entrances to a strait were less than six miles in width but, between these two points, the strait broadened to a distance of more than six miles, leaving, by a rigid application of the three-mile rule, a pocket of the high sea completely surrounded by the territorial sea of the littoral states. The incongruity of the existence of such enclaves of the high sea led to the

principle." INTERNATIONAL LAW 195 (8th ed. Higgins 1924). The Strait of Juan de Fuca, to which he alludes, itself possesses some of the geographical characteristics of a gulf. In recent years, the problem of free navigation through the Gulf of Aqaba has been found to be inextricably linked with the question of passage through the straits which connect the Gulf with the Red Sea.

[8] This view was taken by the International Law Association in articles 14 and 15 of its draft convention on "Laws of Maritime Jurisdiction in Time of Peace." REPORT OF THE THIRTY-FOURTH CONFERENCE, VIENNA, 1926, at 44 and 103.

[9] 1 OPPENHEIM, INTERNATIONAL LAW 511 (8th ed. Lauterpacht 1955).

[10] LEAGUE OF NATIONS, ACTS OF THE CONFERENCE FOR THE CODIFICATION OF INTERNATIONAL LAW, VOL. II, REPORT OF THE SECOND COMMITTEE: TERRITORIAL SEA at 211 (Doc. No. C.351(b).M.145.1930.V). Because of failure to agree on the breadth of the territorial sea, the Second Committee on the Territorial Sea did not arrive at even a provisional decision on the subcommittee's report.

[11] LEAGUE OF NATIONS, CONFERENCE FOR THE CODIFICATION OF INTERNATIONAL LAW, BASES OF DISCUSSION, VOL. II, TERRITORIAL WATERS at 55-60 (Doc. No. C.74.M.39.1929.V).

conclusion that these areas might be incorporated in the territorial sea of the states concerned.[12]

There was, however, no general agreement in the teachings of the publicists that the method of delimiting the territorial sea, applicable elsewhere, had any relevance to determining the extent of the jurisdiction of the littoral states within a strait. It was thus maintained by some that the nations bordering on a strait might embrace the entire strait, without regard to the six-mile limitation, within their jurisdiction as part of their territorial sea.[13] Probably as a consequence of the long-standing misconception of the origin of the three-mile rule in the range of ancient cannon,[14] the extent to which the territorial sea might be broadened in straits was defined in terms of the range of artillery commanding the waters of the passage.[15] The practice of states has on occasion been in conformity with the doctrine that a littoral state may claim jurisdiction or sovereignty over a strait beyond the limits it normally claims for its territorial sea. The Protocol of March 10, 1873, between the United States and Great Britain fixing the Northwest Water Boundary between the United States and Canada, specified the boundary as a line running midway in the Strait of Juan de Fuca, which ranges between ten and fifteen miles in width.[16] The extension of the jurisdiction of the United

[12] LEAGUE OF NATIONS, ACTS OF THE CONFERENCE FOR THE CODIFICATION OF INTERNATIONAL LAW, VOL. II, *supra* note 10, at 220. The draft articles on the régime of the territorial sea which were prepared by the International Law Commission of the United Nations provided (art. 12, paras. 3 and 4) that "an area of sea not more than two miles across" which is enclosed by the territorial sea may be declared by the coastal states to be part of the territorial sea. *Report of the International Law Commission Covering the Work of Its Seventh Session, 2 May-8 July 1955,* U.N. GEN. ASS. OFF. REC. 10th Sess., Supp. No. 9, at 19 (A/2934) (1955). At the United Nations Conference on the Law of the Sea, article 12 of the International Law Commission's draft was deleted and replaced by a provision forbidding states having adjacent coasts to extend their territorial seas beyond the median line between them. Geneva Convention on the Territorial Sea and the Contiguous Zone, signed at Geneva, April 29, 1958, art. 12, 2 UNITED NATIONS CONFERENCE ON THE LAW OF THE SEA, OFF. REC. 133 (U.N. Doc. No. A/CONF.13/38) (1958); see *Report of the First Committee, id.* at 119.

[13] HALL, *op. cit. supra* note 7, at 193-95. Brüel states that the situation in which a strip of high sea runs through an entire strait "rather belongs to the litterature [*sic*] than to reality." 1 INTERNATIONAL STRAITS 41 (1947). This conclusion is based on the consideration that the three-mile limit for territorial waters is no longer generally recognized.

[14] Kent, *The Historical Origins of the Three-Mile Limit,* 48 AM. J. INT'L L. 537 (1954); Walker, *Territorial Waters: The Cannon Shot Rule,* 22 BRIT. Y.B. INT'L L. 210 (1945).

[15] 1 OPPENHEIM, *op. cit. supra* note 9, at 511.

[16] Protocol of a Conference at Washington, March 10, 1873, Respecting the North-

States beyond the three miles to which that country has traditionally adhered was subsequently justified on the basis of the lack of importance of the Strait to international shipping.[17] During World War I, Chile made even those portions of the Strait of Magellan which were more than three miles distant from shore part of its territorial sea for the purpose of defending its neutrality.[18]

In recent years, states have shown an increasing inclination to control activities off their shores by extending the limits of their territorial sea outward[19] to such an extent that it is no longer possible to say that the three-mile rule is recognized as placing bounds on the width of a nation's territorial sea. The International Law Commission, not without dissent,[20] recognized

---

west Water Boundary, T.S. No. 135, 63 BRITISH AND FOREIGN STATE PAPERS 354 (1872-73). The tracing of the boundary line in these terms was the result of an arbitral award rendered by the Emperor of Germany on October 21, 1872, 62 *id.* at 188 (1871-72). See [1872] 5 FOREIGN REL. U.S., pt. 2.

[17] On May 22, 1891, the Acting Secretary of State wrote to the Secretary of the Treasury that "the straits of Juan de Fuca are not a great natural thoroughfare or channel of navigation in an international sense . . ." 1 MOORE, DIGEST OF INTERNATIONAL LAW 658 (1906). The analogy which was drawn in the letter between Delaware Bay and the strait suggests a concept of a "historic strait."

When the interests of navigation were involved, the United States took the position that it would not recognize a claim to a territorial sea of more than a marine league. A Spanish claim of eight miles led Secretary of State Marcy to write with regard to "the thoroughfares of commerce between Cape St. Antonio and the Yucatan shore, or between the Keys of Florida and the Cuban coast," that "considering the vast amount of property transported over these thoroughfares it is of the greatest importance to the interests of commerce that the extent of the Spanish claim to jurisdiction in these two straits—for such they may be called—should be accurately understood." The Secretary of State to the United States Minister to Spain, July 7, 1855, in 11 MANNING, DIPLOMATIC CORRESPONDENCE OF THE UNITED STATES: INTER-AMERICAN AFFAIRS, 1831-1860, at 217 (1939).

[18] Decree of Dec. 15, 1914, 83 BOLETIN DE LAS LEYES I DECRETOS DEL GOBIERNO 1660 (1914), French translation in 23 REVUE GÉNÉRALE DE DROIT INTERNATIONAL PUBLIC, Documents at 13 (1916). It was stated *obiter* in The Bangor, [1916] P. 181, 185, that it was not inconsistent with the right of free passage of the Strait for commercial purposes that Chile should regard the passage "in whole or in part" as part of its territorial waters.

[19] For a listing, as of 1960, of the territorial seas and contiguous zones claimed under various national legislations, see the Synoptical Table prepared by the United Nations Secretariat, in SECOND UNITED NATIONS CONFERENCE ON THE LAW OF THE SEA, OFFICIAL RECORDS: SUMMARY RECORDS OF PLENARY MEETINGS AND OF MEETINGS OF THE COMMITTEE OF THE WHOLE 157 (U.N. Doc. No. A/CONF.19/18) (1960).

[20] Draft articles on the régime of the territorial sea, art. 3, *Report of the International Law Commission Covering the Work of Its Seventh Session, 2 May-8 July 1955,* U.N. GEN. ASS. OFF. REC. 10th Sess., Supp. No. 9, at 16 (A/2934) (1955).

that international practice in this regard is far from uniform and expressed the view that "international law does not justify an extension of the territorial sea beyond twelve miles," while maintaining at the same time that states are not required to recognize a breadth beyond three miles.[21] The Commission followed the recommendation made to the Hague Codification Conference by providing that the waters which may lie between the two belts of territorial sea within a strait form part of the high seas.[22]

No success in resolving the question of the permissible width of a state's marginal sea was achieved by the Geneva Conference on the Law of the Sea of 1958. Disagreement on the point at that conference was the most important single reason for the convening of the Second Conference on the Law of the Sea, which met at Geneva in 1960.[23] Not least among the objections to a territorial sea of 12 miles was the consideration that narrow waterways of a breadth of 24 miles or less, such as the English Channel and the Red Sea, would become, for legal as well as geographic purposes, international straits.[24] In view of the widely shared sentiment in favor of the recognition of a territorial sea of more than three miles' width, there can no longer be any justification for an exceptional principle of international law which would authorize the further widening, beyond a six, nine, or 12 mile limit, of the territorial sea within a strait. Had the Geneva Conference of 1958 desired to concede that states may lay greater claims within straits than elsewhere, it would have said so. The fact is that it did not.

The application of the usual rules on the territorial sea and the high seas would suggest that international law need not make any special pro-

---

[21] Five members of the Commission, including those from the United States and the Soviet Union, expressed disagreement with this formulation. The reasons why they did so differed widely. *Id.* at 15.

[22] Draft articles, art. 12, *id.* at 19.

[23] Resolution VIII, April 27, 1958, 2 UNITED NATIONS CONFERENCE ON THE LAW OF THE SEA, OFF. REC. 145 (U.N. Doc. No. A/CONF.13/38) (1958).

[24] See the remarks of the Delegate of the United Kingdom in the First Committee, April 18, 1958, 3 *id.* at 163; Sorenson, *Law of the Sea,* [1958] INTERNATIONAL CONCILIATION 193, 245. The effect of recognition of a territorial sea of six or 12 miles was described in the following terms by Mr. Arthur H. Dean, Chairman of the United States Delegation to the Geneva Conference of 1958: "There are approximately 116 important international straits in the world which could be affected by the choice of a limit for territorial seas. *All* would become subject to national sovereignties if a 12-mile rule were established. Fifty-two would become subject to national sovereignties if a 6-mile rule were adopted." Statement made before the Senate Foreign Relations Committee, Jan. 20, 1960, 42 DEP'T STATE BULL. 251, 260 (1960).

vision for navigation in geographic straits in which the overlapping territorial seas of the littoral nations do not, as it were, block the strait. However, as the result of the liberation of certain areas of the high seas from the control of littoral states over past years,[25] a number of international agreements recognize freedom of passage through straits whose width far exceeds that of the sum of the belts of territorial sea running through it. An instance of this process of opening an international waterway of considerable proportions to free use by all nations was the recognition by Japan of the Strait of Formosa as a "great sea highway of the nations," beyond the exclusive control or appropriation of that country.[26] The stipulation of the Anglo-French Declaration of 1904[27] which forbids the fortification of the Moroccan coast "in order to secure the free passage of the Straits of Gibraltar" is illustrative of yet another basis of international interest in such wider waterways.[28] The strategic or geopolitical significance of waterways affording access between two areas of the high seas[29] is such that control of the littoral territory is of capital importance. Finally, the channels used in the navigation of even such broad straits may require shipping to pass through the territorial waters of the littoral states.[30]

How much use, or what necessity of use, by nonlittoral states is necessary to give an international complexion to a strait is a question which must be answered in less definite terms than can be applied to the delimitation of national jurisdiction in these waterways. The criteria sometimes applied in the past were the essentiality of the waterway to passage between two sections

[25] See Attorney-General for the Province of British Columbia v. Attorney-General for the Dominion of Canada, [1914] A.C. 153, 174 (P.C.), in which it was pointed out that the "narrow seas" doctrine discussed by the older authorities is "a principle which may safely be said to be now obsolete."

[26] Identic Note of the French, German, and Russian Ministers to the Japanese Minister for Foreign Affairs regarding Retrocession of the Liaotung Peninsula, Oct. 18, 1895, 1 MacMurray, Treaties and Agreements With and Concerning China, 1894-1919, at 53 (1921).

[27] Declaration between Great Britain and France respecting Egypt and Morocco, together with the Secret Articles, signed at London, April 8, 1904, art. VII, 101 British and Foreign State Papers 1053 (1912).

[28] See 2 Brüel, *op. cit. supra* note 13, at 139-56, concerning the political background of the Declaration of 1904; and Treaty between France and Spain Stipulating the Respective Situations of the Two Countries with regard to the Sheriffean Empire, signed at Madrid, Nov. 27, 1912, art. 6, 7 Martens, Nouveau Recueil général de traités, 3d ser. 323, 326 (1913).

[29] See Eliot, *The World's Strategic Waterways*, 1 United Nations World, Sept. 1947, p. 30.

[30] 1 Brüel, *op. cit. supra* note 13, at 42.

of the high seas and its use by considerable numbers of foreign ships.[31] It was substantially this position which was taken by the Albanian Government in the *Corfu Channel Case*[32] before the International Court of Justice. The Court, however, denied the correctness of this standard, stating in its opinion:

It may be asked whether the test is to be found in the volume of traffic passing through the Strait or in its greater or lesser importance for international navigation. But in the opinion of the Court the decisive criterion is rather its geographical situation as connecting two parts of the high seas and the fact of its being used for international navigation. Nor can it be decisive that this Strait is not a necessary route between two parts of the high seas, but only an alternative passage between the Aegean and the Adriatic Seas.[33]

It is sufficient that the waterway be "a useful route for international maritime traffic." It is impossible to answer in the abstract how many straits meet this requirement of being "useful" for international navigation, for the test applied by the Court lays more emphasis on the practices of shipping than on geographic necessities.[34] A listing which would be confined to those straits which are of great strategic interest would, however, represent no more than a minority of those straits with which international law concerns itself.[35]

[31] This, for example, is the view of Hyde. 1 INTERNATIONAL LAW, CHIEFLY AS INTERPRETED AND APPLIED BY THE UNITED STATES 488 (2d rev. ed. 1945).

[32] Judgment of April 9, 1949, [1949] I.C.J. Rep. 4.

[33] *Id.* at 28.

[34] *Id.* at 29.

[35] An interesting listing of strategic waterways, including both straits and canals, has been attempted by Major George Fielding Eliot. In the category of maritime highways between two of the great oceans he placed the Mediterranean system, including Gibraltar, the Sicilian Straits, the Suez Canal, the Red Sea, and the Strait of Bab-el-Mandep; the Panama Canal; and the waterways linking the Indian and Pacific oceans. Among the important passages giving access to enclosed seas, he listed the Turkish Straits, the Danish Straits (with the Kiel Canal as an alternative route), the various entrances to the Sea of Japan, the Strait of Ormuz, and St. George's Channel and the Irish Channel. *The World's Strategic Waterways,* 1 UNITED NATIONS WORLD, Sept. 1947, p. 30 at 31-34.

In his argument in the *Corfu Channel Case,* Sir Eric Beckett stated that, "It is scarcely surprising that the works of jurists do not contain comprehensive lists of international straits, since they are concerned to expound principles and not to provide catalogues." There follows an ironic reference to an exhaustive list of straits, cited by Professor Cot, counsel for Albania; something was thought to be wrong with the copy of the volume in the Peace Palace Library, since no such list was found in it. 4 CORFU CHANNEL CASE—PLEADINGS, ORAL ARGUMENTS, DOCUMENTS 549-50 (I.C.J. 1949).

For an extensive listing of international straits, see Kennedy, *A Brief Geographical and Hydrological Study of Straits which Constitute Routes for International Traffic,*

It may at this point be objected that if a strait consists either entirely of belts of territorial sea or of territorial waters separated by a strip of high seas, the law applicable to such waterways should be the usual law applying to the jurisdiction of states on the high seas or in the territorial sea, as the case may be. This conclusion would suggest that there is no reason for the establishment of any special body of law applicable to straits alone. The existence, however, of a variety of conventional arrangements having application to individual waterways and of a body of customary law dedicated to this particular type of waterway makes it impossible to start from this easy assumption. It is also not without significance that the International Court of Justice, in considering the right of Great Britain to send its warships through the Corfu Strait, based its conclusions upon the law of straits without considering the more general question, which had been much debated by the parties, whether states have under international law a right to send warships through the territorial sea not included in a strait in time of peace.[36] It follows that here, as elsewhere, logic must yield to experience. To what extent the law governing straits may be assimilated to that concerning the territorial or high seas must accordingly await a consideration of the practice of states, judicial decisions, and treaties bearing on the subject.[37]

If the number of straits which are of international concern is large, the list of *interoceanic* canals which are of importance to the international shipping community goes to the opposite extreme of brevity. The definition of international canals in terms of interoceanic waterways naturally has the effect of excluding those canals which are part of the international river systems of Europe and other continents. The régime of these artificial inland waterways is often assimilated to that of the rivers upon which they are dependent or which they link. The provisions of the Treaty of Versailles applicable to the Elbe, the Oder, the Niemen, and the Danube[38] were also given application to lateral canals constructed to duplicate or improve the naturally navigable sections of these "river systems." Similar stipulations, imposing obligations on Germany, governed the artificial waterways forming

---

Preparatory Document No. 6, 1 UNITED NATIONS CONFERENCE ON THE LAW OF THE SEA, OFF. REC. 114 (U.N. Doc. No. A/CONF.13/38) (1958).

[36] Corfu Channel Case, [1949] I.C.J. Rep. 4, 30.

[37] See p. 159 *infra.*

[38] Treaty of Peace between the Allied and Associated Powers, signed at Versailles, June 28, 1919, art. 331, 112 BRITISH AND FOREIGN STATE PAPERS 1 (1919).

part of the Rhine and Moselle systems.[39] Such is the extent to which rivers have been made navigable by the construction of lateral canals and passages connecting two navigable portions of these waterways and by the improvement of the river itself that rivers are often spoken of as "canalized." There may even come a stage in the improvement of a river for navigation when it becomes more accurate to refer to it in terms of an artificial waterway or canal than in terms of the natural watercourse it once was. Of such nature is the St. Lawrence Seaway. In addition, the régime of international rivers has been extended in a number of instances to canals which were not constructed with a view to improving the navigability of an international river.[40] The facts that these internal canal systems resemble rivers in their geography and that natural waterways have to an increasing extent become artificial as a consequence of improvements which have been made in their courses point to considering the two types of water highways as different aspects of systems of inland, fluvial navigation, differing essentially in degree of artificiality.[41]

As inland canals may be considered to be artificial forms of rivers, so may interoceanic canals be regarded in a geographic sense as artificial straits.[42] Some canals, it must be emphasized, are not of international concern, even though they may in theory connect two seas. A clear instance is the Baltic-White Sea Canal, which can take ships of 3000 tons or less and this only during the 165 days of the year that the waterway is usable.[43] While the Gota Canal runs from Goteberg on the North Sea to Stockholm on the

---

[39] Article 362. Germany and the other parties to the Treaty also agreed in article 361 that the projected "deep-draught Rhine-Meuse navigable waterway" should be placed under the same administrative régime as the Rhine itself.

[40] E.g., the Grand Canal d'Alsace. Commission Centrale pour la Navigation du Rhin, 1st sess., 1952, *Communiqué du Secrétariat,* 24 REVUE DE LA NAVIGATION INTÉRIEURE ET RHÉNANE 387 (1952).

[41] See H. A. SMITH, THE ECONOMIC USES OF INTERNATIONAL RIVERS 10 (1931).

[42] This is not to suggest that the geographic similarity necessarily entails an identity of law bearing on the two types of waterways. The statement of the Permanent Court of International Justice that even the passage of belligerent warships does not compromise the neutrality of the state under whose jurisdiction either a strait or a canal lies cannot be read as extending this far. The S.S. Wimbledon, P.C.I.J., ser. A, No. 1, at 28 (1923).

The analogy of a canal to a strait connecting two parts of the high seas also fails when the canal offers access to an inland port. The obvious example is the Manchester Ship Canal. See WOYTINSKY AND WOYTINSKY, WORLD COMMERCE AND GOVERNMENTS: TRENDS AND OUTLOOK 489-90 (1955).

[43] Villard, *La Navigation intérieure en U.R.S.S.,* 24 REVUE DE LA NAVIGATION 268 (1952).

Baltic, its capacity is limited to extremely small vessels, and the traffic which it carries seems to be essentially local.[44] Another Russian waterway, the Volga-Don Canal, which was completed in 1952, was to permit oceangoing vessels to proceed from the Baltic Sea to the Black Sea upon the completion of certain improvements in the course of the Volga,[45] but the combined watercourse does not appear to be of international significance. The Corinth Canal comes close to the line between those canals which are of international importance and those which are of purely national concern. The waterway is short and narrow, and, in terms of number of transits, its use is largely by Greek vessels.[46] However, the annual tonnage figures for Greek and foreign vessels are approximately equal.[47] This circumstance, when viewed in combination with the substantial number of foreign transits, makes it impossible to say that the waterway is without international importance.[48]

The three interoceanic canals which are indubitably of international concern are the Suez, Panama, and Kiel[49] canals. Although the extent to which such waterways have been the subject of international negotiations and agreement must be taken into account in determining the depth of the interest of the nonlittoral states, the decisive criterion is the amount of use of the waterway by the shipping of foreign nations. It is possible, by the application of this test, to arrive at a mathematical assessment of the international stake in the passage, which partially explains, although it may not justify, the attitude of foreign nations to the riparian state or the agency responsible for the operation of the waterway. The international significance of the canals must, of course, also be sought in the interests of those nations which rely on the passages for certain essential products or for the export of their raw materials and manufactured products and of those countries whose

[44] The Canal is 24 meters wide and only four meters deep. The size of the locks is such that they can accommodate ships of 400 tons or less. Bonét-Maury, *Navigation intérieure suédoise: Le Göta Canal,* 21 NAVIGATION DU RHIN 503 (1949).

[45] WOYTINSKY AND WOYTINSKY, *op. cit. supra* note 42, at 490, citing Scandinavian Shipping Gazette (Copenhagen), Feb. 1952, p. 182.

[46] According to information supplied by the Bank of Greece, Athens, 6006 vessels passed through the Canal during a typical year (1953); 4889 of these were under the Greek flag.

[47] In round figures, for 1953, 986,000 tons of Greek shipping and 928,000 tons of foreign shipping.

[48] No international agreement or declaration, however, holds the waterway open to free usage by the vessels of all nations. Concerning the early history of the Corinth Canal, see GERSTER, L'ISTHME DE CORINTHE ET SON PERCEMENT (Budapest, 1896).

[49] In German, the Nord-Ostsee-Kanal. The name Kaiser-Wilhelm-Kanal has not been used for some years.

nationals actually own the fleets which use the canals.[50] The extent of their interest cannot, however, be translated into mathematical terms with the same readiness as statistics on the use of the waterways by ships of various flags.

About one sixth of the world's seaborne commerce has in recent years passed through the Suez Canal.[51] The reliance of foreign nations on this waterway is immensely greater than the dependence of other states on the Panama and Kiel canals. During the first quarter of 1963, for example, ships flying the flag of the United Arab Republic (excluding small vessels) accounted for only 55 out of a total of 4,626 transits of the Suez Canal.[52] In terms of net tonnage, ships of United Arab Republic registry ranked twenty-second among the users of the waterway. The ten largest users, again in terms of net tonnage, were during 1962 Great Britain, Liberia, Norway, France, Italy, Greece, the Netherlands, Germany, Sweden, and Panama.[53] The current tendency to place vessels under flags of convenience provided by certain smaller countries in order to avoid the impact of tax and labor laws[54] means

---

[50] See p. 22 *infra*.

[51] Statement of Secretary of State Dulles at the Suez Conference, Aug. 16, 1956, in THE SUEZ CANAL PROBLEM: JULY 26—SEPTEMBER 22, 1956, at 73 (Dep't of State Pub. 6392) (1956).

[52] UNITED ARAB REPUBLIC, SUEZ CANAL AUTHORITY, MONTHLY REPORT, March 1963, at 47.

Pursuant to an exchange of notes between the Chairman of the Board of the Suez Canal Company and the Egyptian Minister of Commerce and Industry of March 7, 1949, Egyptian vessels of 300 tons or less not carrying passengers were exempted from tolls. See RECUEIL CHRONOLOGIQUE ET ANNOTÉ DES ACTES CONSTITUTIFS DE LA COMPAGNIE UNIVERSELLE DU CANAL MARITIME DE SUEZ ET DES CONVENTIONS CONCLUES AVEC LE GOUVERNEMENT EGYPTIEN 241 (3d ed. 1950). The United Arab Republic has not modified the toll structure established when the Canal was being operated by the Suez Canal Company.

[53] UNITED ARAB REPUBLIC, SUEZ CANAL AUTHORITY, SUEZ CANAL REPORT, 1962, at 110. The measurement of tonnage for the purpose of computing tolls for passage through the Suez Canal is accomplished according to the Regulations for the Measurement of Tonnage recommended by the International Tonnage Commission assembled at Constantinople, 1873, MINUTES OF PROCEEDINGS at xxi, app. II. The rules are reprinted in COMPAGNIE UNIVERSELLE DU CANAL MARITIME DE SUEZ, RÈGLEMENT DE NAVIGATION 83 (1953 ed.), which has been maintained in force by the Suez Canal Authority of the United Arab Republic.

[54] See BOCZEK, FLAGS OF CONVENIENCE: AN INTERNATIONAL LEGAL STUDY (1962); McDougal, Burke, and Vlasic, *Maintenance of Public Order at Sea and the Nationality of Ships*, 54 AM. J. INT'L L. 25 (1960); Comment, *The Effect of United States Labor Legislation on the Flag-of-Convenience Fleet: Regulation of Shipboard Labor Relations and Remedies against Shoreside Picketing*, 69 YALE L.J. 498 (1960).

that the criterion of flag does not give an altogether accurate picture of the beneficiaries of the right of passage. The artificiality of statistics based on the flag of vessels does not, however, conceal the overwhelming preponderance of foreign interest in the Suez Canal, as contrasted with that of the United Arab Republic.

The application of the same test of tonnage to the Panama Canal indicates a closer balancing of the interests of the United States, which operates the Canal and exercises jurisdiction over the Canal Zone, and those of foreign nations. United States flag vessels are the largest users of the waterway, accounting for about one sixth of the total net tonnage during the fiscal year 1962 and for the largest number of transits.[55] The nine flags making the next greatest use of the Canal were Norway, Great Britain, Germany, Liberia, Japan, Greece, the Netherlands, Panama, and Sweden.[56]

The relationship of domestic and foreign use of the Kiel Canal bears a close resemblance to that prevailing in the case of the Panama Canal. Germany is the largest user of the waterway, which is operated by the German Government. The German share of the total tonnage has of late been somewhat more than one quarter, and 34 other nations have been represented among the users of the Canal.[57] In terms of tonnage, some recent figures indicate that Germany is followed by Sweden, Finland, the Netherlands, Great Britain, Norway, Poland, the Soviet Union, Denmark, and Liberia.[58] The size of the ships using the Canal differs markedly from that of the vessels passing through the Suez and Panama canals. The majority of transits through the Kiel Canal are performed by vessels of less than 4,000 tons.[59] By way of comparison, the average net tonnage of all types of ships using the Suez Canal is now well over 10,000 tons.[60] The registered net tonnage of

[55] 10,900,843 net vessel tons, Panama Canal measurement, out of a total of 65,378,945 tons. The figures on the Panama Canal are based on the fiscal year, running from July 1 of one year to June 30 of the year next following. ANNUAL REPORT OF THE PANAMA CANAL COMPANY AND THE CANAL ZONE GOVERNMENT FOR THE FISCAL YEAR 1962, at 57 (1962).

[56] *Id.* at 10.

[57] *The Traffic in the Third Quarter of the Calendar Year 1962*, [1962] NORD-OSTSEE-KANAL, Nos. 3/4, 7 at 10 and 13. The NORD-OSTSEE-KANAL: KIEL CANAL, published by the Wasser- und Schiffahrtsdirektion Kiel on behalf of the Federal Ministry of Transport, appears in both German and English; articles are cited herein by their titles in English.

[58] *Ibid.*

[59] *The Traffic of Merchant Vessels Classified according to Size, id.* at 21, 23-24.

[60] UNITED ARAB REPUBLIC, SUEZ CANAL AUTHORITY, MONTHLY REPORT, Feb. 1963, at 4.

ships transiting the Panama Canal averaged 5,864 tons during fiscal year 1962.[61]

The importance of these waterways to the shipping trade and to those who depend on the waterways for the transport of their imports and exports is also illustrated by the average number of transits per day on each of the three waterways. The current figure for the Suez Canal is 54 ships per day.[62] Seven years ago, in the last normal month prior to nationalization the Canal was operating at what was then top capacity with over 44 transits per day.[63] The daily average has increased steadily as the amount of petroleum and its by-products shipped through the Canal has mounted. The substantial increase in the number and size of ships using the Canal has necessitated a succession of long-term programs of improvement of the waterway.[64] The average number of daily transits through the Panama Canal has risen to more than 30.[65] Improvements which will increase the capacity have likewise been necessitated by the continuing high level of traffic.[66] Recent statistics show that an average of 244 ships per day are passing through the Kiel Canal,[67] but it must be borne in mind that many of these are of small tonnage[68] and

[61] ANNUAL REPORT OF THE PANAMA CANAL COMPANY AND THE CANAL ZONE GOVERNMENT FOR THE FISCAL YEAR 1962, at 11 (1962).

[62] UNITED ARAB REPUBLIC, SUEZ CANAL AUTHORITY, MONTHLY REPORT, March 1963, at 3.

[63] LE CANAL DE SUEZ: BULLETIN DE LA COMPAGNIE UNIVERSELLE DU CANAL MARITIME DE SUEZ, No. 2,324 (Aug. 15, 1956), at 1. The accommodation of the ships utilizing the Canal required the institution of a system of daily convoys, which permits ships proceeding north and those proceeding south to pass each other in the bypasses provided for this purpose. Formerly mooring of larger vessels to the sides of the Canal was necessary in order to permit passing in the watercourse.

[64] An eighth major program of improvement had been undertaken by the Company prior to the nationalization of the Canal. This program called for the construction of bypasses at Port Said and at Kabret in the Grand Lac Amer, as well as for the deepening and widening of the Canal in order to permit the passage of supertankers at normal speed and without excessive erosion of the banks. The Suez Canal Authority of the United Arab Republic, after the nationalization of the Canal, decided to undertake an even more ambitious program of improvements, only the first stage of which was to be the completion of the eighth program of the Suez Canal Company. The "Nasser Project" looks to the eventual doubling of the entire Canal and its deepening to permit the transit of the largest tankers in the world. UNITED ARAB REPUBLIC, SUEZ CANAL AUTHORITY, SUEZ CANAL REPORT, 1958, at 64.

[65] ANNUAL REPORT OF THE PANAMA CANAL COMPANY AND THE CANAL ZONE GOVERNMENT FOR THE FISCAL YEAR 1962, at 5 (1962).

[66] *Id.* at 15-17.

[67] Based on figures contained in *The Traffic in the Third Quarter of the Calendar Year 1962, supra* note 57, at 8.

[68] *60 Years of Traffic on the Kiel Canal,* [1955] NORD-OSTSEE-KANAL, No. 1, 23 at 30.

that different methods of statistical reporting make difficult any precise comparisons between the three canals in terms of tonnage or transits. The statistics furnished by the operator of each of the three waterways do, however, lend emphasis to the importance which these passages hold for shipping in general, as well as for the fleets registered under the flags of nonlittoral states.

Rivers, as has been observed,[69] acquire international significance because of competing national demands for the use of their waters for a variety of purposes, whether by way of their appropriation for the generation of power, the irrigation of land, or industrial and domestic consumption, or by the use of the river as a highway of commerce or as a convenient open sewer. Those states through which a river capable of these uses flows must inevitably give thought to the adjustment of their interests in the waterway, and so the river becomes of international concern. International character, in the sense of a river's being of consequence to more than one nation, is thus a coefficient of physical situation and actual or potential use. For purposes of navigation, geography and trade together determine the international character of rivers[70] which flow through the territories of two or more states. In a more literal sense than is true of straits and canals such rivers are highways rather than mere passages. Often, as in Europe, they form part of a complex system of internal waterways, partially natural and partially artificial in nature. Such systems, and individual international rivers as well, compete with systems of road and rail transportation on land, and serve many of the same functions. This very analogy of function between fluvial, rail, and road transport lies at the basis of the economic problems which competing forms of transportation have brought to European inland navigation.[71] Differing geographic

[69] See p. 1 *supra.*

[70] The use of this term is not intended to constitute a prejudgment of the issue whether foreign vessels are entitled under customary international law to free use of the waterway.

[71] The Economic Conference on Rhine Navigation was held at Strasbourg and Basel in February, April, and May 1952, in order to consider means of preventing crises in the economy of the Rhine. Commission Centrale pour la Navigation du Rhin, 2d sess., 1952, *Communiqué du Secrétariat,* 24 REVUE DE LA NAVIGATION 728 (1952). The economic health of the Rhine has also been of concern to the European Conference of Transport Ministers. Commission Centrale pour la Navigation du Rhin, sess. d'automne 1954, *Communiqué du Secrétariat,* 26 *id.* at 717 (1954); CONFÉRENCE EUROPÉENNE DES MINISTRES DES TRANSPORTS, CINQUIÈME RAPPORT ANNUEL 31 (1959).

In connection with the establishment of a common market for coal and steel, the European Coal and Steel Community has been forced to devote attention to freight rates for these commodities and in particular to the establishment of international rail-

situations may give a variety of aspects to the exploitation of rivers for the purposes of navigation and commerce. To some nations, isolated from the sea by neighboring states, a river may offer a means of access to the high seas and to the markets of the world, but it is by no means clear that customary international law recognizes any right of way or servitude in favor of the landlocked state by reason of this circumstance alone.[72] Bolivia, isolated from the seacoast as the result of its conquest by Chile, thus relies on the Amazon River for its access to the Atlantic Ocean through the territory of Brazil.[73] The Rhine River serves a like function for Switzerland. The international significance of a river may also arise from the fact that it constitutes the boundary between two states and is needed for navigation by the vessels of the two nations it separates. Of such nature is the St. Lawrence River, which has been made navigable from the Great Lakes to the Atlantic Ocean by

---

way through rates. The introduction of such through rates led France to establish a compensation fund for water-transport rates in order to counterbalance the competitive advantage which would have been enjoyed by rail transport. The High Authority of the Community decided that this scheme constituted an import duty and discriminated between producers, and it ordered the French Government to bring an end to the plan by early 1955. EUROPEAN COAL AND STEEL COMMUNITY HIGH AUTHORITY, MONTHLY REPORT, 3rd year, No. 1, at II,3,1 (Jan./Feb. 1955). An agreement concerning freights and conditions of transport for coal and steel on the Rhine has been concluded by the members of the Community and entered into force in 1958. CONFÉRENCE EUROPÉENNE, CINQUIÈME RAPPORT, *supra* at 31. These arrangements demonstrate that it is quite impossible to view the economic situation of inland navigation in isolation from land transport and from the economy of the area as a whole.

See also *Recent Trends in the Use of Some European Inland Waterways for the Transport of Goods,* 3 TRANSPORT AND COMMUNICATIONS REVIEW, No. 1, at 3 (1950).

[72] The Faber Case (Germany v. Venezuela), RALSTON, VENEZUELAN ARBITRATIONS OF 1903, 600 at 630 (1904) (dictum).

The Geneva Convention on the High Seas calls for the conclusion of agreements whereby landlocked states would be accorded "free transit" through the territory of the nations which separate them from the sea. Such agreements would be in implementation of the principle that "States having no sea-coast *should have* free access to the sea" (emphasis supplied). Geneva Convention on the High Seas, signed at Geneva, April 29, 1958, art. 3, 2 UNITED NATIONS CONFERENCE ON THE LAW OF THE SEA, OFF. REC. 136 (U.N. Doc. No. A/CONF.13/38) (1958); see Boas, *Landlocked Countries and the Law of the Sea,* 4 AMERICAN BAR ASSOCIATION, SECTION OF INTERNATIONAL AND COMPARATIVE LAW BULLETIN, No. 1, at 22 (1959).

[73] The river was opened to free navigation by virtue of the Treaty of Commerce and Fluvial Navigation between Bolivia and Brazil, signed at Rio de Janeiro, Aug. 12, 1910, 7 MARTENS, N.R.G. 3d ser. 632 (1913), in implementation of article 5 of the Treaty of Delimitation between Bolivia and Brazil, signed at Petropolis, Nov. 17, 1903, 3 *id.* 3d ser. 62, 65 (1910).

ships of considerable size through improvements undertaken on both the Canadian and United States sides of the international boundary.[74] The navigation of boundary waters is, because of the multiplicity of its instances, not an uncommon subject for conventional arrangements.[75] Finally, a river may offer a convenient means of transport between various nations which are situated along its course. These geographic bases for the concern of states in inland waterways are not, of course, mutually exclusive, and the significance of a river may arise from any combination of these factors, which are seldom found to exist in isolation. The Rhine itself fulfills a threefold function—as Switzerland's link with the sea, as a navigable boundary between France and Germany, and as a fluvial highway of concern not only to these nations but to the Netherlands and to Belgium as well. Although rivers in these categories may become of international concern because of the fact that they flow through or border on the territory of more than one state, this is not to say that their international significance begins and ends with the interests of the littoral states. The very importance of these waterways as arteries of international trade makes them of concern to all those nations who employ them in connection with their trading activities and, through them, to the international community as a whole.

It may be objected at this point that rivers are fundamentally different from straits and canals in that they are navigated by shipping of an entirely different character from the vessels which employ passages connecting stretches of the high seas. In the majority of instances, cargo which has been

[74] The Canadian St. Lawrence Seaway Authority was established by The St. Lawrence Seaway Authority Act, 15 & 16 Geo. 6, c. 24, and the United States Saint Lawrence Seaway Development Corporation by the St. Lawrence Seaway Act, 68 Stat. 92, as amended, 33 U.S.C. §§ 981-90. The works constructed on the St. Lawrence by the two nations are described in Senate Committee on Foreign Relations, *St. Lawrence Seaway Manual,* S. Doc. No. 165, 83d Cong., 2d Sess. 29 (1955).

[75] E.g., Treaty between the United States and Great Britain relating to Boundary Waters between the United States and Canada, signed at Washington, Jan. 11, 1909, 36 Stat. 2448, T.S. No. 548; Convention between Finland and Russia concerning the Maintenance of River Channels and the Regulation of Fishing on Water Courses Forming Part of the Frontier between Finland and Russia, signed at Helsingfors, Oct. 28, 1922, 19 L.N.T.S. 184, revived, March 13, 1948, 67 U.N.T.S. 157; Convention between Hungary and Czechoslovakia concerning the Settlement of Technical and Economic Questions on the Hungarian-Czechoslovak Frontier Section of the Danube and on that of the Tisza below the Confluence of the Szamos, signed at Budapest, Aug. 24, 1937, 189 L.N.T.S. 404, revived, Feb. 27, 1948, 26 U.N.T.S. 119; Treaty between the United States of America and Mexico relating to the Utilization of the Waters of the Colorado and Tijuana Rivers and of the Rio Grande, signed at Washington, Feb. 3, 1944, 59 Stat. 1219, T.S. No. 994.

carried across the sea must be transshipped to river craft if it is to be transported up the river. It is to this circumstance that the great ports of Antwerp, Rotterdam, and Amsterdam, which serve as the ocean ports for the Rhine, owe their importance. The possibility of a difference in legal principles, depending on whether a river is navigable from the sea by oceangoing vessels or is capable of use only by river vessels was suggested in the *Faber* case,[76] although its precise *ratio decidendi* is far from clear. The Umpire characterized as "very different matter[s]" the navigation of the Catatumbo and Zulia rivers by oceangoing vessels in order to carry goods through Venezuela to Colombia and the transshipment of such cargo to river craft within Venezuela, which would not only have required the use of shore facilities but also would have, in his view, extended "the claim of free navigation of rivers to a new case."[77] The opinion is of interest in suggesting that free navigation may entail something more than the unfettered right to navigate a ship between the shores of a river without regard to problems of transshipment.

Some rivers, such as the Amazon,[78] the St. Lawrence, and the Danube, are navigable to some extent by oceangoing vessels, and thus clearly constitute, for the purposes both of geography and of navigation, branches of the high seas. The very word "seaway" is a dramatic way of stating that the St. Lawrence River has been made navigable by oceangoing vessels.[79] In the case of the Danube, the two maritime ports of Galatz and Braila mark the point at which the river ceases to be navigable from the Black Sea.[80] The distinction between the fluvial and the maritime Danube is reflected in the different régimes which historically have been established over the two sections of the river. The Treaty of Paris of 1856, which proclaimed the Danube open to the vessels of all nations, established the European Commission of the Danube for the maritime portion of the river and the Riparian Countries Commission for the fluvial portion.[81] Through a combination of circumstances, the European Commission, which had originally been intended to

[76] (Germany v. Venezuela), *supra* note 42, 600 at 625.

[77] *Ibid.*

[78] Oceangoing vessels can proceed 1000 miles up the river. International traffic, however, is not substantial. WOYTINSKY AND WOYTINSKY, *op. cit. supra* note 42, at 488.

[79] The number of ocean vessels which may use the Seaway is limited by its controlling depth of 27 feet. St. Lawrence Seaway Act, § 3(a), 68 Stat. 93, 33 U.S.C. § 983(a).

[80] See Benoist, *La Conférence de Belgrade sur le statut du Danube,* 20 NAVIGATION DU RHIN 409, 410 (1948); Gorove, *Internationalization of the Danube: A Lesson in History,* 8 J. PUB. L. 125, 150 (1959).

[81] General Treaty of Peace between Austria, France, Great Britain, Prussia, Russia,

be a temporary and technical body, was, during the period preceding the First World War, the only effective international agency dealing with the waterway.[82] Between the two World Wars, the European Commission continued to carry on its functions with regard to the maritime section of the Danube, while an international commission concerned itself with the remainder of the river.[83] The Convention regarding the Régime of Navigation on the Danube, which was signed at Belgrade on August 18, 1948, perpetuates this distinction by creating a special administration for the lower Danube as well as for the Iron Gates.[84] The fact that this administration would permit control of access to the Danube to rest in the hands of two states was one of the principal objections raised to the new arrangement by those states which were not members of the Soviet bloc.[85]

The few attempts which have been made to define international rivers in multilateral treaties reflect the multiplicity of elements which go into the definition of these waterways—geographic position, essentiality for commerce, existing use for commerce, navigability in the technical sense, and capacity for improvement so as to be made navigable, to name some of the more important ones. The Statute on the Régime of Navigable Waterways of International Concern, which was drawn up at Barcelona in 1921,[86] defined as navigable waterways of international concern "all parts which are naturally navigable to and from the sea of a waterway which in its course, naturally navigable to and from the sea, separates or traverses different States, and also any part of any other waterway naturally navigable to and from the sea, which connects with the sea a waterway naturally navigable which separates or traverses different States."[87] Navigability may mean either that the waterway is "now used for ordinary commercial navigation," that is, currently an artery of commerce, or that it is "capable by reason of its natural con-

Sardinia, and the Ottoman Porte, signed at Paris, March 30, 1856, arts. 16 and 17, 15 MARTENS, N.R.G. 770, 776-77 (1857), 46 BRITISH AND FOREIGN STATE PAPERS 8 (1865).

[82] CHAMBERLAIN, THE REGIME OF THE INTERNATIONAL RIVERS: DANUBE AND RHINE 57, 126 (1923).

[83] Under the Convention Instituting the Definitive Statute of the Danube, signed at Paris, July 23, 1921, 26 L.N.T.S. 173.

[84] Article 23, 33 U.N.T.S. 196.

[85] Statements by Ambassador Cannon, Chairman of the U.S. Delegation at the Belgrade Conference, on Aug. 9, 1948, 19 DEP'T STATE BULL. 219, 220 (1948), and on Aug. 13, 1948, *id*. at 284.

[86] 7 L.N.T.S. 50, 1 HUDSON, INTERNATIONAL LEGISLATION 645 (1931).

[87] Article 1, para. 1.

ditions of being so used . . ."[88] The fact that the Statute has not been ratified by or acceded to by the majority of the important fluvial states tends to cast doubt on whether it establishes an acceptable definition of international waterways for the purposes of international law. The Treaty of Versailles specifically declared certain rivers within specified limits to be international and added to this enumeration, which included the Elbe, Oder, Niemen, and Danube, "all navigable parts of these river systems which naturally provide more than one State with access to the sea, with or without transshipment from one vessel to another," together with the lateral canals and channels constructed in order to improve the navigability of the waterway.[89]

For present purposes international rivers—that is, those waterways which are not within the exclusive jurisdiction of a single state—must be regarded as being those navigable rivers which connect two or more states, which flow into or connect with the sea, and which are in fact employed or are capable of being employed for navigation. The very fact that such waterways connect two or more states and furnish access to the free highways of the seas is sufficient to make them of concern to more than one nation. To what extent the riparian and nonriparian states may on this basis assert a voice in the control of the waterway is a matter which must await later discussion.[90]

Types of Interest in International Waterways

The geographic situation of international waterways and the use made of them combine to produce various categories of interests in these maritime passages. These interests may be those of governments or they may be those of natural or juridical persons or groups of such persons. However, since international relations are so organized that states speak for the interests of their nationals, one may speak of these interests as being those of states for the purposes of international law. The several types of interest may be classified, in an admittedly rough and ready way, as being those of the user of the waterway, those of the sovereign exercising jurisdiction over the territory through which the passage flows, and those of the operator or supervisor of the waterway. To these there may perhaps be added a fourth group consisting of those states which have a political or strategic interest in the water-

---

[88] Article 1, subpara. 1(*a*).
[89] Article 331.
[90] See Chapter II, Section E *infra*.

way, even though they may not be vitally concerned with it in any of these preceding respects. These categories obviously require further explanation.

The *users* of an international waterway must themselves be broken down into a number of subclasses. The most obvious of these covers the nations whose ships carry goods through the waterway. The identification of nations falling in this category is made difficult by the widespread practice of registering vessels under "flags of convenience"[91] in order to avoid taxation and more rigid laws concerning labor standards and safety. Statistics on the use of international waterways are usually maintained in terms of national use as determined by the registry of vessels employing the waterway. These figures can give a wholly distorted picture of the concern of the shipping trade of various countries in a particular waterway. Sandwiched in among the important maritime powers which are the heaviest users of the Suez Canal are Liberia and Panama,[92] both well known as international refuges for shipping. The same two countries appear among the top ten users of the Panama Canal, according to the criterion of number of passages by flag of registry.[93] A more accurate view of the extent of the interest of the shipping trade of various countries could be obtained from figures showing the number of transits by, and tonnage of vessels owned by, nationals of particular countries or by corporations having the national character of such nations.[94]

Those who make their livelihood from the carriage of goods through a waterway have in a sense less dependence on it than does another category of users—those who rely on the waterway for the transport of their imports or exports. The dependence of Europe on the Suez Canal for the transport of oil from the Middle East is the most striking instance of the significance of a waterway for the economic life of a user. In 1956, Europe drew approximately 70 per cent of its oil supplies from the Middle East and Far

[91] The subject is comprehensively and learnedly discussed in BOCZEK, FLAGS OF CONVENIENCE: AN INTERNATIONAL LEGAL STUDY (1962), and McDougal, Burke, and Vlasic, *Maintenance of Public Order at Sea and the Nationality of Ships,* 54 AM. J. INT'L L. 25 (1960).

[92] See p. 13 *supra.*

[93] ANNUAL REPORT OF THE PANAMA CANAL COMPANY AND THE CANAL ZONE GOVERNMENT FOR THE FISCAL YEAR 1962, at 10 (1962).

[94] This is not to underestimate the difficulty of ascertaining the national character of a corporation.

The continuing relevance of the flag of a vessel for international purposes is borne out by the Advisory Opinion on the Constitution of the Maritime Safety Committee of the Inter-Governmental Maritime Consultative Organization, [1960] I.C.J. Rep. 150.

East. The effect of the Suez crisis of that year was to sever the route over which was transported about two thirds of the 2,165,000 barrels per day which Europe had to import from those areas in order to maintain its economy.[95] In terms of tonnage transported, oil continues to be the single most important commodity shipped through the Suez,[96] Panama,[97] and Kiel[98] canals. So long as petroleum and petroleum products maintain their profound significance for the economy of the world, so long will the avenue through which they travel be of concern to the whole international community.[99]

The significance of canals for the transport of oil should not be allowed to obscure the fact that these waterways remain important avenues of commerce for dry cargoes as well. Nearly a quarter of Great Britain's import and export trade in raw materials and in manufactured products is carried on with nations to which Suez offers access.[100] Even the United States, remote from the Canal as it is, uses the passage for the transportation of substantial portions of its imports of such strategic materials as rubber, tin, and manganese, in addition to petroleum.[101]

The dependence of states on particular waterways is seldom an absolute one in the sense that alternative means or routes of transport do not in fact exist. In some cases, a competing method of transport may be substituted. Water transport along the Rhine has undergone severe competition from

[95] UNITED STATES, DEPARTMENT OF THE INTERIOR, OFFICE OF OIL AND GAS, REPORT TO THE SECRETARY OF THE INTERIOR FROM THE DIRECTOR OF THE VOLUNTARY AGREEMENT RELATING TO FOREIGN PETROLEUM SUPPLY, AS AMENDED MAY 8, 1956; CONCERNING THE ACTIVITIES OF THE FOREIGN PETROLEUM SUPPLY COMMITTEE UNDER THE VOLUNTARY AGREEMENT AND THE ACTIVITIES OF THE MIDDLE EAST EMERGENCY COMMITTEE AND ITS SUBCOMMITTEES UNDER THE PLAN OF ACTION, FOR THE PERIOD APRIL 1, 1956 THROUGH JUNE 30, 1957, vol. I at 5 (1957); cf. the somewhat different figures presented in ORGANISATION FOR EUROPEAN ECONOMIC CO-OPERATION, EUROPE'S NEED FOR OIL: IMPLICATIONS AND LESSONS OF THE SUEZ CRISIS 11 (1958).

[96] UNITED ARAB REPUBLIC, SUEZ CANAL AUTHORITY, SUEZ CANAL REPORT, 1962, at 117.

[97] ANNUAL REPORT OF THE PANAMA CANAL COMPANY AND THE CANAL ZONE GOVERNMENT FOR THE FISCAL YEAR 1962, at 65-66 (1962).

[98] *The Traffic in the Third Quarter of the Calendar Year 1962*, [1962] NORD-OSTSEE-KANAL, Nos. 3/4, 7 at 12.

[99] While it is possible that nuclear-generated power may eventually to some extent supplant oil as an energy source, atomic power presently makes only a modest contribution to energy supplies. See Cavers, *The Development of an International Atomic Economy*, 5 CONFLUENCE 29 (1956).

[100] *Via Suez*, 180 THE ECONOMIST 419, 420 (1956).

[101] Wall Street Journal, July 30, 1956, p. 1, col. 1. For example, approximately 60 per cent of United States imports of rubber come from Malaya and are shipped through the Suez Canal.

railroads,[102] which would gladly assume the transport of goods carried on the inland-waterways system of Europe, of which the Rhine forms an important part. The relatively small saving in distance, and hence in operating cost, secured through the use of the Kiel Canal is such, when balanced against tolls, that a number of ships find it more profitable not to use the waterway.[103] Alternatives to the shipment of petroleum and other important supplies through the Suez Canal exist in the form of pipelines[104] and of the route about the Cape of Good Hope.

The existence of theoretical alternative routes must not be allowed to obscure the fact that their use would be attended with grave, if not insuperable, difficulties. The first of these is necessarily cost. Even though the route may exist and the extra cost can be borne, there still exists the possibility that the organization of shipping has been geared to the existence of the waterway. During the Suez crisis of 1956-57, tankers were in such short supply that it was physically impossible to move over the much longer route about the Cape of Good Hope the quantity of oil normally imported from the Near and Far East through the Suez Canal. As a report of the Organisation for European Economic Co-operation put it, "To maintain full supplies for Europe, simply by re-routing round The Cape oil previously carried through the Canal and shipped from the pipeline terminals of Banias and Tripoli would have required an 80 per cent increase in the tanker capacity normally serving Europe. Such additional tanker capacity did not exist; indeed, the total number of tankers laid up or 'mothballed' was less than twenty."[105] The inevitable consequences were severe restrictions on the use of oil, cooperative measures to secure allocation of existing supplies to the users, greatly increased costs, and a temporary alteration in the pattern of supply of petroleum to Europe. The blocking of the Canal stimulated renewed consideration of the practicability in economic terms of the use on

[102] *Recent Trends in the Use of Some European Inland Waterways for the Transport of Goods, supra* note 71, 3 at 5.

[103] ARNOLD, DIE GRUNDLAGEN DER TARIFPOLITIK FÜR DEN NORDOSTSEEKANAL (No. 20 in Kieler Studien) at 10-11 (1951).

[104] The charges demanded by countries through which pipelines run from the Persian Gulf to the Mediterranean bear a close relationship to the tolls for the use of the Suez Canal, and an increase in the latter could be expected to be reflected in demands for higher payments in the former instance. See *Pipeline versus Tanker,* 176 THE ECONOMIST 1037 (1955).

[105] ORGANISATION FOR EUROPEAN ECONOMIC CO-OPERATION, EUROPE'S NEED FOR OIL: IMPLICATIONS AND LESSONS OF THE SUEZ CRISIS 23 (1958).

the Cape route of supertankers larger in capacity than those able to pass through the Suez Canal.[106]

In time of war or other armed conflict, the use of a longer exposed sea route and the necessity of using a great number of tankers and other cargo vessels could affect the quantity of oil supplied to Europe in such a way as to determine the outcome of a war. It is impossible to overlook the fact that both the Panama[107] and Kiel[108] canals were constructed with a specific view to their significance for the naval power of the nations controlling them. The strategic significance of straits for warships and for cargo vessels in time of war is reflected in the efforts which were made during the Second World War to prevent Spain and Turkey from assisting the Axis powers through their control of the Strait of Gibraltar and the Turkish Straits.

In a more abstract sense, the severing of communications between states which are normally maintained through an international waterway may cause the countries concerned to drift apart. New alliances may even be sought if a nation can no longer rely on prompt and efficient communications through a waterway which has hitherto been part of its ocean highways to other lands.[109]

The *territorial sovereign* may be a single state through which the entirety of an international waterway flows or it may be one of several riparian or littoral states. It is convenient to include within this category of states the belligerent occupant, although admittedly such an occupant does not exercise sovereignty. It was upon this basis, among others, that the United States claimed a right to participate in the Belgrade Conference to Consider the

---

[106] *Id.* at 45; N.Y. Times, July 28, 1956, p. 3, col. 5.

[107] S. REP. No. 1, 57th Cong., 1st Sess. 7 (1901). The voyage of the *Oregon* around Cape Horn in 1898 underlined the importance of a transisthmian canal to the defense of the United States. See DuVAL, CADIZ TO CATHAY: THE STORY OF THE LONG DIPLOMATIC STRUGGLE FOR THE PANAMA CANAL 87 (2d ed. 1947).

There has recently been some questioning of the continued strategic importance of the Panama Canal. See Travis and Watkins, *Control of the Panama Canal: An Obsolete Shibboleth?*, 37 FOREIGN AFFAIRS 407 (1959).

[108] Bismarck wrote that he had attached great importance to the building of the Kiel Canal "in the interest of German sea-power." 2 GEDANKEN UND ERINNERUNGEN 29 (1898); see also THE KIEL CANAL AND HELIGOLAND (Foreign Office, Peace Handbook No. 41) at 1-5 (1920); CAI SCHAFFALITZKY DE MUCKADELL, THE KIEL CANAL 4-6 (1947).

[109] Doubt about the utility of the Suez Canal in the event of a future war had led South Africa, even before the Suez crisis, to reorganize her sea defense and to engage in joint exercises with French forces based at Madagascar. N.Y. Times, Sept. 19, 1954, § 1, p. 12, col. 3.

Free Navigation of the Danube, which was held in 1948.[110] In other respects, the identification of the territorial sovereign presents no difficulties.

A watercourse which must be maintained or improved to make it capable of navigation requires the establishment of an *operating agency* to perform this function. In practice, a considerable variety of solutions has been found to the problem of assuring an efficient instrument for the discharge of this responsibility. An artificial international waterway flowing through the territory of a state may be administered by an agency of the government of that nation. The Kiel Canal, which falls in this category, is the responsibility of the Wasser- und Schiffahrtsdirektion Kiel. This agency is accountable to the Ministry of Transport of the Federal Republic of Germany, and it is the federal budget which finances the Canal.[111] An autonomous agency, the Suez Canal Authority, was created to operate the waterway after the Egyptian nationalization of the Canal.[112] The New Corinth Canal Corporation[113] operates the canal of that name.[114] A canal through foreign territory may be operated and maintained by a foreign company, as was once true in the case of Suez,[115] or by a foreign government, as in the case of the Panama Canal. The construction and operation of an international waterway may, on the other hand, be undertaken by two nations, each of which performs and finances its own portion of the construction under a coordinated plan. This

[110] At the opening of the conference, Ambassador Cannon stated that "the United States has the responsibility of participating directly in the problems of the Danube by reason of what is still a provisional situation as regards the treaty with Austria and by reason of the American occupation of that zone of Germany through which the navigable Danube flows." 19 Dep't State Bull. 198 (1948). The same reason dictated the membership of the United States in the Central Commission for the Navigation of the Rhine after World War II. See Arrangement providing for Participation by the United States of America in the Central Commission of the Rhine, signed at London, Oct. 4 and 29, and Nov. 5, 1945, 60 Stat. 1932, T.I.A.S. No. 1571.

[111] Lorenzen, *The Administration of the Kiel Canal,* [1953] Nord-Ostsee-Kanal, No. 2, 20 at 21.

[112] *Décret-loi du Président de la République, No. 146 de 1957 relatif aux Statuts de l'Organisme du Canal de Suez,* Journal officiel du Gouvernement Egyptien No. 53*bis* "C," July 13, 1957.

[113] Statutes of the New Corinth Canal Corporation (Athens, 1933). The Corporation replaced a bankrupt company; the majority of its stock was turned over to the National Bank of Greece.

[114] See Gerster, L'Isthme de Corinthe et son percement (1896), regarding the history of the Corinth Canal.

[115] The characterization of the Compagnie Universelle du Canal Maritime de Suez as a foreign corporation is based solely on its administration and finances and does not take into account the difficult legal question of the national character of the Company and of its actual *siège social.*

device of separate but cooperative activities has been resorted to in connection with work on the St. Lawrence Seaway.[116] The St. Lawrence Seaway Authority was established by Canada, and the Saint Lawrence Seaway Development Corporation by the United States, each of which was made responsible for certain construction and operating functions within the jurisdiction of the country concerned.[117]

International commissions have frequently been employed for the supervision or coordination of activities on the three major categories of international waterways. In the case of rivers, the most noteworthy of these are the Central Commission for the Navigation of the Rhine and the various Danube commissions. The necessity of assuring freedom of passage for ships through international straits, which, being natural waterways, do not call for extensive work of maintenance and improvement, may of itself be a sufficient reason for the establishment of an international supervisory commission. This was the role filled by the now defunct Turkish Straits Commission, established by article 10 of the Convention of Lausanne.[118] These procedural devices for maintaining the effectiveness of international passages exhibit varying degrees of national or international participation and of jurisdiction and power.

The Interests Defined

Once it is established that there are in general three categories of nations having an interest in an international waterway, whether it be a river, canal, or strait, it is necessary to inquire whether there is any perceptible identity of interests within each class, and if so, what these shared interests are. For example, if the user of an international river assesses the waterway and its use in the same terms as the user of an interoceanic canal or a strait, and it can further be demonstrated that there are similar common concerns in the cases of the operator or supervisor and of the territorial sovereign as well, a substantial foundation will have been laid for a comparison of the legal problems occasioned by the three types of waterway.

[116] Pursuant to Agreements between the United States of America and Canada regarding the Saint Lawrence Seaway, signed at Ottawa, Aug. 17, 1954, and at Washington, June 30, 1952, 5 U.S.T. 1784, T.I.A.S. No. 3053.

[117] Canada by The St. Lawrence Seaway Authority Act, 15 & 16 Geo. 6, c. 24; the United States by the St. Lawrence Seaway Act, 68 Stat. 92, 33 U.S.C. §§ 981-90.

[118] Convention relating to the Régime of the Straits, signed at Lausanne, July 24, 1923, 28 L.N.T.S. 115.

Without attempting to establish the order of importance of various interests, one may nevertheless assert that the user of any type of international waterway is vitally concerned with free access to the waterway and with free and unimpeded passage through the waterway. A user desires to keep to a minimum the political and strategic complications produced by war, and it therefore demands that passage be as free in time of war as in time of peace. The impediments to passage which the user fears may be a prohibition on use of the waterway or restrictions of that use short of outright prohibition. These prohibitions or restrictions may be either overt or intentional, as in the case of express restrictions laid on passage by the ships of certain nations, or may be the consequence of the operation of local law which indirectly restricts use of the waterway. In the latter category belong customs laws and regulations or sanitary regulations which may interfere with expeditious passage. The law or regulation in question may actually have been adopted with the purpose of limiting use of the waterway. An unnecessarily burdensome law or heavy-handed application of the law may achieve the same result, even though neither the law nor its enforcement had as its purpose the hampering of navigation.

Closely related to this concern with freedom from legal restraints is the interest of the user in facility of passage through the waterway in a physical sense. If the channel is not dredged by the state through the territory of which the river, canal, or strait runs, navigation through it may be rendered impossible. In general, the user will wish the waterway to accommodate the shipping it is economically feasible and practicable to send through the waterway. Users recognize that this requirement must be held in bound by the dictates of reason, especially if the waterway is an artificial one. No actual or potential user of the St. Lawrence Seaway has suggested that the United States and Canada have an obligation to go beyond the 27-foot depth now set for the waterway in order that it may accommodate any variety of seagoing vessel. On the other hand, there has been legitimate pressure from the shipping trade directed to the deepening of the Suez Canal in order to accommodate supertankers, which have in the past been required to limit their cargo in order to permit safe navigation through the Canal.[119] The ancillary

---

[119] The *World Glory,* a supertanker owned by Niarchos, Ltd., draws 37½ feet when laden. The owner in the past sent it through the Suez Canal with less than its full capacity of oil, since the Canal, even after the completion of the revised eighth program of improvement, will not accommodate ships drawing more than 37 feet. Wall Street Journal, April 27, 1955, p. 6, col. 4; UNITED ARAB REPUBLIC, SUEZ CANAL AUTHORITY, SUEZ CANAL REPORT, 1959, at 60-62; *id.,* 1962, at 10.

facilities which are needed to permit use of the waterway are numerous. Aids to navigation, such as buoys, lights, and shore markings are indispensable; pilotage and facilities for towing may likewise be needed. In addition, the user must be assured not only that navigation through the waterway will be assisted but also that physical impediments to passage are not created. A bridge which is too low or, as has happened in the history of the Suez Canal, a bridge which was closed at the wrong time[120] or a dam and hydroelectric installations on the waterway may have the effect of closing it down.

Finally, the user of a natural or artificial waterway desires that passage be as cheap as may be consistent with proper maintenance and operation of the waterway. Pressure of this sort by a user was responsible for the abolition of the Danish Sound dues,[121] and the shipping trade has exercised an important influence on the level of tolls for the Suez and Panama canals. Ideally, the user would wish to see some correlation between the amount of tolls and the services actually rendered to shipping. The user can understand the imposition of reasonable tolls for the use of artificial waterways or even natural waterways which have been improved, but it rightly demands that the right to passage through a natural watercourse not be conditioned on the payment of a tax.

The operator or supervisor of a waterway may, as previously indicated, be a government department, a private company, a public corporation, or an international organization. Whatever form its organization takes, it plays, in its operating or supervisory capacity, the role of an entrepreneur. Expressed in the most general terms, its object is to provide the best possible service while, depending on the requirements laid upon it by the agency to which it is responsible, generating a reasonable rate of return on invested capital or at least operating without loss to its superiors. As a business, like a railroad or a shipping company or an airline, it has only a minimal concern with politics. In more precise terms, the operator or supervisor shares the interest of the user in free access to the waterway in time of war as in time

[120] Despite the existence of an agreement between the former Suez Canal Company and the Egyptian State Railways concerning the hours during which the railroad bridge at El Ferdan would be closed, there were numerous occasions on which convoys were held up by the closing of the bridge during periods when, under the terms of the agreement, it should have been open.

[121] By the Treaty between Great Britain, Austria, Belgium, France, Hanover, Mecklenburg-Schwerin, Oldenburg, The Netherlands, Prussia, Russia, Sweden and Norway, and the Hanse Towns, on the one part, and Denmark, on the other part, for the Redemption of the Sound Dues, signed at Copenhagen, March 14, 1857, 16 MARTENS, N.R.G. pt. 2, at 345 (1858), 47 BRITISH AND FOREIGN STATE PAPERS 24 (1866).

of peace because it desires to see maximum utilization of the waterway. Because discrimination among flags will also drive away some of its potential customers, it has a significant interest in resisting attempts by the territorial sovereign to exploit the waterway for political, economic, or strategic purposes. The obvious instances of the operator's position in this respect have been the contention of the former Suez Canal Company that it was above politics,[122] and the view of the Security Council in its resolution of October 13, 1956, that "the operation of the Canal should be insulated from the politics of any country."[123]

Because the operator or supervisor is dedicated to the principle of maximum use of the waterway, it will be at pains to provide a service which is in every respect efficient and meets the demands of those who use the waterway. Its zeal in this respect may be expected to be roughly proportionate to the amount of competition offered by alternative routes and methods of transport. The program of improvement which was undertaken by the Suez Canal Company prior to its nationalization by Egypt was induced, in part at least, by the possibility that heavily laden supertankers might have to utilize the route about the Cape of Good Hope if the Canal was not improved in order to accommodate them.[124] If it is to satisfy the user, the operator or supervisor must ensure that the waterway is properly maintained, that necessary ancillary facilities such as pilots and towage are available, and that the administration and actual conduct of transits are effected with speed, accuracy, and a minimum of burden to the user.

[122] At the annual meeting of the shareholders held on June 9, 1953, the Chairman of the Conseil d'Administration, M. Charles-Roux, stated: "Les rapports internationaux peuvent connaître des vicissitudes; votre Compagnie y reste étrangère. L'accomplissement du devoir qui lui est propre limite le champ de son activité; il est assez vaste pour qu'elle s'y renferme. Sa politique consiste essentiellement à n'en pas faire . . . Ce qu'il y a encore de plus certain, en pareille matière, c'est qu'il importe qu'en tous temps et en tous circonstances les bateaux passent de la Méditerranée à la Mer Rouge et vice versa, et que l'on en est aussi convaincu sur place qu'au dehors." LE CANAL DE SUEZ, No. 2,286 (June 15, 1953), at 9518.

[123] U.N. Doc. No. S/3675. The quoted phrase is the third of the so-called "Six Principles" for the settlement of the Suez question. President Nasser had already agreed in principle that the Canal should be "insulated from politics." Letter from President Nasser to Prime Minister Menzies, Sept. 9, 1956, in THE SUEZ CANAL PROBLEM, JULY 26-SEPTEMBER 22, 1956, 317 at 322 (Dep't of State Pub. 6392) (1956). The legal obligations undertaken by Egypt are set forth in the Egyptian Declaration of April 24, 1957, annexed to letter to the Secretary-General of the United Nations from the Egyptian Minister for Foreign Affairs of the same date (U.N. Doc. No. A/3576, S/3818) (1957).

[124] See *Note sur le 8ème programme des travaux d'amélioration du Canal de Suez*, Supplement to LE CANAL DE SUEZ, No. 2,306 (Feb. 15, 1955).

In return for these services, the operator or supervisor will, if charged with financial responsibilities, seek a rate of return which will at a minimum make the waterway self-sustaining. High operating cost or competition from other routes or means of transportation may render it difficult to achieve even this level.[125] On the other hand, the operator of a waterway like the Suez Canal can expect to derive a very substantial financial advantage through the possession of a waterway vital to the movement of important commodities.[126] So long as competing means of transport compute their rates on the basis of Canal tolls and the alternative route imposes higher financial burdens on the shipper, so long can a high rate of return on invested capital be justified in business terms, if not in those of facilitating international commerce.

The nation through the territory of which an international waterway flows may, of course, have an interest in the maritime passage in the aspect of a user or of an operator of the waterway. If these interests be excluded for analytic purposes, there remain a number of national concerns of which account must be taken. In the first place, the territorial sovereign must have regard for the protection of its national interests. One of these is that no portion of its territory should be exempt from its jurisdiction. In all places which are comprehended within its territory or territorial sea, its law must be permitted to operate. Crimes must not go unpunished, in its view, even though their situs and consequences bear some relation to the ships using the waterway. Transiting vessels must not be allowed to spread disease through the littoral state. The state must likewise be able to protect itself against violation of its customs laws and against the entry of undesirable persons. The state must be even more preoccupied with the danger to its security that the waterway may present. Not only must it oppose any possible use of the waterway for an actual hostile attack against it, but it must also ensure that the waterway is not used for the economic sustenance of an enemy or even of a potential enemy. A right of free passage through the waterway must not, in its view, be extended so far as to make the waterway a strategic and eco-

---

[125] The Kiel Canal, for example, has operated at a deficit during a majority of the years of its existence. ARNOLD, *op. cit. supra* note 103, at 33. The difficulty of operating the Canal at a profit may be attributable, in part at least, to the circumstance that the saving in distance secured by use of the Canal is in no wise comparable to that gained by the use of the Suez or Panama canals.

[126] This is not to suggest that the profits of the Suez Canal Company were, or those of the Suez Canal Authority are, excessive. Attempts to secure a reduction in tolls have, however, normally been supported by the contention that the operator of the waterway had been deriving exorbitant profits at the expense of the shipping trade.

nomic asset to that very nation with which it may be engaged in hostilities.[127]

In a positive sense, the territorial sovereign will, in all probability, attempt to derive some strategic or political advantage from its control of the channel. Its aim may be to secure some financial advantage from the control of the waterway—that very purpose which was asserted to be the basis for the nationalization by Egypt of the Suez Canal Company.[128] Beyond the financial prize involved, the control of a waterway may have profound strategic significance. Turkey, as the "Guardian of the Straits," exercises power over the whole of the Black Sea through its position athwart the Bosporus and Dardanelles. The other great Black Sea power, Russia, has historically attempted to extend its sway over the entry into the Sea. Its last major effort in this respect was the diplomatic campaign waged after the conclusion of the Second World War. It may be expected that the possibility of revision of the Montreux Convention may, at a propitious time, inspire further demands by the Soviet Union.[129] Great Britain has, for similar reasons, clung to Gibraltar as the entrance to the Mediterranean, part of the great waterway which links that country with important sources of raw materials, a major portion of the Commonwealth, and the whole East. And Egypt, not unmindful of the tremendous strategic and political import of the Isthmus of Suez, first secured the withdrawal of British forces from that area[130] and then attempted to destroy the last vestiges of Anglo-French control of the Isthmus by seizing the Suez Canal and the assets of the Suez Canal Company in Egypt.[131]

The segregation and identification of the various national interests which influence the control and operation of rivers, canals, and straits must not be

[127] This view is the essence of the case of the United Arab Republic against the unrestricted use of the Suez Canal by Israeli vessels and by non-Israeli vessels carrying goods to and from that country during the undeclared war between Israel and the Arab States. See p. 221 *infra.*

[128] See Speech by President Nasser, Alexandria, July 26, 1956, in THE SUEZ CANAL PROBLEM, JULY 26-SEPTEMBER 22, 1956, at 25 (Dep't of State Pub. 6392) (1956).

[129] Article 28 of the Convention concerning the Régime of the Straits, signed at Montreux, July 20, 1936, 173 L.N.T.S. 213, 7 HUDSON, INTERNATIONAL LEGISLATION 386 (1941), provides that the Convention is to remain in force for twenty years, that is, until 1956, and that thereafter the treaty may be denounced on two years' notice by notification given to the French Government.

[130] By the Agreement between the Government of Great Britain and the Egyptian Government regarding the Suez Canal Base, signed at Cairo, Oct. 19, 1954, GREAT BRITAIN T.S. No. 67 (1955).

[131] "The Universal Company of the Maritime Suez Canal has been the key to occupation." REPUBLIC OF EGYPT, MINISTRY FOR FOREIGN AFFAIRS, WHITE PAPER ON THE NATIONALISATION OF THE SUEZ MARITIME CANAL COMPANY 10 (1956).

allowed to obscure the fact that the interests of the government of a state and of that state's nationals may be antithetic. As the owner of 44 per cent of the shares of the Suez Canal Company, the British Government had a legitimate concern with the dividends which were paid on these shares. British shipping companies, which were and continue to be among the largest users of the waterway, were equally devoted to the principle that tolls for the use of the Canal should be kept as low as possible. When the Governor-General of Australia protested in 1906 that immigration and trade were being discouraged through the deterrent effect of tolls on shipping, the British directors of the Company replied that to lower tolls would be to effect a subsidy for British shipping at the expense of the large revenues then accruing to the Exchequer from the dividends of the Company. The response of the Earl of Elgin to the Governor-General emphasized the point that "due regard" would have to be paid to "the interests of those who have a purely financial concern in the affairs of the Suez Canal."[132] The battle over the level of tolls, which has been carried on throughout the life of the Canal, again came to a head in 1931, when a severe decline in shipping led to renewed demands on the British Government for a reduction in the level of tolls.[133]

The United States, as operator of the Panama Canal, has had similar disputes with American shipping companies, which are, in national terms, the largest user of the waterway.[134] The conflict is dramatically illustrated by the unsuccessful suit for $27,000,000 filed by a group of shipping companies for alleged overcharges on tolls.[135] The fleets of barges which ply the inland waterways of Europe are owned by those very littoral states which, as territorial sovereigns, may attempt to place limitations on the free use of the waterway by the vessels of other states. So long as international river commissions, such as the Central Commission for the Navigation of the Rhine, served only the purpose of an organ of consultation—a diplomatic conference

[132] WILSON, THE SUEZ CANAL: ITS PAST, PRESENT, AND FUTURE 211-15 (1933).

[133] *Id.* at 152-58.

[134] ANNUAL REPORT OF THE PANAMA CANAL COMPANY AND THE CANAL ZONE GOVERNMENT FOR THE FISCAL YEAR 1958, at 66 (1959).

[135] Panama Canal Co. v. Grace Line, Inc., 356 U.S. 309 (1958). The decision in that case, adverse to the interests of the shipping industry, led to renewed demands for revision of the accounting practices of the Panama Canal Company and for judicial review of tolls under the Administrative Procedure Act. Bills to this effect were introduced in the 86th Congress, 2d Session. See *Hearings on H.R. 8983 and H.R. 10968 before the Subcommittee on Panama Canal of the House Committee on Merchant Marine and Fisheries,* 86th Cong., 2d Sess. (1960).

on a permanent footing[136]—so long did the individual states, members of the Commission, combine in one person the interests of a territorial sovereign, a user, and an operator of the river watercourse. With the increase in the power and the functions of such a commission, it tended to assume a personality of its own,[137] and the Commission became an operator or supervisor of the waterway with interests not necessarily identical with those of the individual states represented within it.

International waterways are the cause of international disputes when it proves impossible to harmonize the interests of those states or agencies which are concerned with the watercourse as users, operators or supervisors, and territorial sovereigns. One of these conflicts of interest—that occasioned by the nationalization of the Suez Canal and the subsequent invasion of Egypt by forces from Israel, Great Britain, and France—has been the cause of one of the greatest world crises since the Second World War, comparable in importance to the blockade of Berlin and the Korean conflict. The three types of interests to which reference has been made were clearly identifiable during the conflict. Egypt, as the territorial sovereign, had seized for itself the operation and supervision of the waterway which ran through its territory. The interests of the former operator, the Compagnie Universelle du Canal Maritime de Suez, were represented on the international plane by Great Britain and France, which, through public and private shareholdings, held the largest beneficial interest in the Company. The third interested group were the users of the Canal and those who relied upon it for the transport of merchandise, a group of nations in which Great Britain and France held an extremely important place. Within these two countries, there was no necessary harmony of views between the Suez Canal Company, on the one hand, and shipping lines, trading companies, business enterprises, and consumers, on the other. Ultimately the needs of the economies of the United Kingdom and France exerted a more powerful influence on the position taken by those two governments than the demands of the Company itself, with its natural preoccupation with the restoration of the Canal concession or proper compensation.[138]

[136] Van Eysinga, La Commission centrale pour la Navigation du Rhin 16-17 (1935).

[137] *Id.* at 120.

[138] See, e.g., personal letter from M. Charles-Roux, Chairman of the Suez Canal Company, to Mr. Dillon, Under-Secretary of State of the United States, April 19, 1957, protesting the payment of tolls to Egypt and reiterating the concern of the Company with the question of compensation, and Mr. Dillon's reply, May 6, 1957, in Compagnie

The necessity of taking account of the existence of a tripartite distribution of national interests in the Canal points to the futility of such a device as the Suez Canal Users Association to provide a solution for the problem of Suez.[139] In so far as it sought to "facilitate any steps which may lead to a final or provisional solution of the Suez Canal problem and to assist the members in the exercise of their rights as users of the Suez Canal in consonance with the 1888 Convention, with due regard for the rights of Egypt,"[140] it was a legitimate organization of users, which might have performed a useful negotiating function with Egypt. But its contemplated mobilization of pilots for the passage through a canal which was already manned and operated by Egypt[141] and the collection of tolls for services which it could not provide and passages which it could not secure[142] might have been effective negotiating tactics but could not be considered to be any solution of the Canal prob-

---

UNIVERSELLE, THE SUEZ CANAL COMPANY AND THE DECISION TAKEN BY THE EGYPTIAN GOVERNMENT ON 26TH JULY 1956, SECOND PART, at 65-66 (1957).

[139] Under the Declaration Providing for the Establishment of a Suez Canal Users Association, Sept. 21, 1956, *op. cit. supra* note 128, at 365; see also Resolution on the Organization of the Suez Canal Users Association, adopted on Oct. 4, 1956, by the Council of the Association, 35 DEP'T STATE BULL. 580 (1956).

Considerably before the establishment of the Suez Canal Users Association, the users of the Panama Canal had been considering the desirability of an organization of users which might consult with the Panama Canal Company. *Hearings before the House Committee on Merchant Marine and Fisheries on the Study of the Operations of the Panama Canal Company and Canal Zone Government,* 84th Cong., 1st Sess. 53 (1955).

[140] Article II, para. (1), of the Declaration. The wording of the Declaration is in a neutral form which would permit its interpretation as meaning either that the Suez Canal Users Association was to function as an organ of consultation or that it was to serve as an operating agency. An example of the latter is paragraph (2) of article II, which states that one of the purposes of the Users Association is "to promote safe, orderly, efficient and economical transit of the Canal by vessels of any member nation . . ."

[141] Statement by Secretary Dulles, Sept. 19, 1956, at the Second Suez Conference at London, in THE SUEZ CANAL PROBLEM, JULY 26-SEPTEMBER 22, 1956, at 353, 355 (Dept. of State Pub. 6392) (1956). It was thought that the presence of qualified pilots aboard the ships would make it harder for the Egyptian authorities to deny passage to vessels which might otherwise be held up through the shortage of pilots available to the Egyptian Suez Canal Administration. Extemporaneous Statement by Secretary Dulles, Sept. 19, 1956, *id.* at 359.

[142] Declaration Providing for the Establishment of a Suez Canal Users Association, Sept. 21, 1956, art. II, para. (4), *id.* at 365. It was contemplated by Secretary Dulles that Canal tolls would be paid by United States vessels using the Canal to the Users Association, which would in turn make "certain payments" to Egypt. Press Conference of Sept. 26, 1956, 35 DEP'T STATE BULL. 543, 545-46 (1956).

lem itself. If the Canal was once more to be operated as an international waterway, some harmonization of the interests of the users, the operator, and the territorial sovereign was required, whether by way of creating an institution for that purpose or by agreement on the régime of the Canal. That harmonization of interests could not be secured through the medium of a body representing one or two of the contending factions opposing Egyptian operation of the Canal.

Stability in the administration of an international waterway is more likely to be a consequence of its political geography and its use than of any legal formula which may be evolved. International disputes are less likely to arise if one state operates the waterway and administers the territory through which it runs and is in addition a substantial user of the river, canal, or strait. Germany both operates the Kiel Canal and is its principal user. It can therefore take no measure affecting the users of the Canal without affecting its own interests and the interests of its nationals. As the operator of the Panama Canal, the United States must likewise reckon with the fact that its nationals are the largest users of the waterway. To the extent the United States exercises in the Canal Zone the powers which it would have if it were sovereign,[143] it is accordingly called upon to respect the needs of its own operating agency, the Panama Canal Company, and of the users of the Canal, many of whom are Americans. While conflicts of interests will arise, they will in the main be matters of domestic concern and will not necessarily give rise to international controversy. Although the states operating these two important international waterways furnish a larger volume of use than any other single state, the restraint of a stake in both operation and use would be felt even if these nations lagged behind others in terms of use but still furnished a substantial portion of the shipping passing through the waterway.

Similarly, the international river commissions of Europe give representation to states through whose territory the river runs and whose boatmen navigate upon the river. If measures are taken on a nondiscriminatory basis,

[143] Under the face-saving formula of the Convention between the United States of America and the Republic of Panama for the Construction of a Ship Canal to Connect the Waters of the Atlantic and Pacific Oceans, signed at Washington, Nov. 18, 1903, art. III, 33 Stat. 2234, T.S. No. 431. The position of the United States is, of course, complicated by the fact that the slender ribbon of land and surrounding waters which make up the Canal Zone lie within Panamanian territory. The United States is thus under constant pressure to increase its payments to Panama and to allow increased Panamanian jurisdiction within the Canal Zone.

impediments to navigation will equally affect their own nationals and aliens. If, on the contrary, an attempt is made to discriminate against the shipping and boatmen of other riparian states, it may be expected that those other states will be in a position to take like measures against the shipping and boatmen of the state which first threw up the barriers. A strong reciprocity of interest and the possibility of retaliation in kind thus exercise a restraint, which may admittedly vary in degree, upon the state which seeks to extend its dominion over the river and its traffic.

Unfortunately, there is no such unity or reciprocity of interest in the case of the Suez Canal. The interest of the United Arab Republic in the capacity of a user is slight; it ranked twenty-second among the users in terms of net tonnage of shipping moving through the Canal during 1962.[144] The 204 transits made by vessels registered in the United Arab Republic stand in stark contrast with 4,072 by ships sailing under the British flag.[145] Its limited degree of dependence upon the Canal in terms of use means that it views the Canal primarily from the standpoint of the economic and strategic profit to be derived from it rather than from the point of view of a nation which also has an important stake in its usage. This circumstance, when seen in light of the tremendous economic and political stakes involved in Suez, goes far to explain why the problems of that waterway are more vexatious than those of Panama and Kiel, which, together with Suez, make up the Big Three of interoceanic canals.

Analogies Between International Waterways

It may well be asked at this point whether the existence of three types of interests in international waterways *ratione personae* necessarily indicates that conflicts between these interests may be subjected to a uniform solution, regardless of the type of waterway—river, canal, or strait. The geographic facts, the extent of the littoral state's interest, and the degree to which the waterway is subjected to international use may vary from case to case. A river running through a country presents a far different problem from a strait upon which a state may border. A strait, on the other hand, is a natural waterway, while a canal is by nature the work of man.

[144] United Arab Republic, Suez Canal Authority, Suez Canal Report, 1962, at 156.

[145] *Ibid.* It should be noted that vessels of under 300 tons gross Suez measurement are not required to pay tonnage dues.

To these objections the answer must be given that analogies may be drawn between the legal problems of different international waterways and solutions found on the basis of these analogies only to the extent that there exist resemblances between the factual situations presented and between the competing demands which are made with respect to the waterway. As regards international waterways of the same type, it is reasonable to assume that a solution found to a vexing problem regarding one may have application to the same problem arising with respect to other waterways of the same type. The draftsmen of the Hay-Pauncefote Treaty of 1901[146] slipped naturally into the language of the convention governing navigation of the Suez Canal when they dealt with the question of freedom of transit in connection with the projected canal across the Isthmus of Panama. Article III of the Treaty provides: "The United States adopts, as the basis of the neutralization of such ship canal, the following Rules, substantially as embodied in the Convention of Constantinople, signed the 28th October, 1888, for the free navigation of the Suez Canal . . ." The provisions of the Act of Navigation of the Congo of 1885[147] were in a corresponding fashion drawn from similar stipulations in treaties regulating rivers in Europe and in Latin America, including the Final Act of the Congress of Vienna, the Treaty of Paris of 1856, the Act of Navigation of the Danube of 1857, the Public Act relative to the Navigation of the Mouths of the Danube of 1865, and identical treaties concluded in 1853 between France, Great Britain, and the United States, on the one hand, and Argentina on the other, concerning the free navigation of the Parana and Uruguay rivers.[148] Identity of problems and

---

[146] Treaty between the United States and Great Britain to Facilitate the Construction of a Ship Canal, signed at Washington, Nov. 18, 1901, 32 Stat. 1903, T.S. No. 401. Article XVIII of the Convention between the United States of America and the Republic of Panama for the Construction of a Ship Canal to Connect the Waters of the Atlantic and Pacific Oceans, signed at Washington, Nov. 18, 1903, 33 Stat. 2234, T.S. No. 431, provided that the Canal should "be opened . . . in conformity with all the stipulations" of the Hay-Pauncefote Treaty and thus adopted by reference the language of the earlier agreement. Bunau-Varilla later wrote that it was his and Hay's desire that the Convention of Constantinople "should become, in a permanent way, the directing principle of the operation of the Panama Canal." PANAMA: THE CREATION, DESTRUCTION, AND RESURRECTION 373 (1920).

[147] Chapter IV of the General Act of the Conference of Berlin, signed at Berlin, Feb. 26, 1885, 10 MARTENS, N.R.G. 2d ser. 414, 420 (1885-86), 76 BRITISH AND FOREIGN STATE PAPERS 4 (1884-85).

[148] Protocol No. 1 (Meeting of Nov. 15, 1844), Conference of Berlin on West Africa, 10 MARTENS, N.R.G. 2d ser. 199 (1885-86); FRANCE, DOCUMENTS DIPLOMATIQUES, AFFAIRES DU CONGO ET DE L'AFRIQUE OCCIDENTALE 55 at 59 (1885).

solutions cannot be universal, since political considerations, geography, and degree of international concern may vary from case to case. But it is not too much to expect that formulas worked out with respect to one waterway of a given type may offer some guidance in the solution of problems occasioned by another watercourse of the same type.

A further objection which might be raised to any attempt to give orderly development to a general body of law relating to interoceanic canals in particular is that each of the three major canals is already regulated by a treaty, establishing a separate régime for the particular waterway. The existence of different conventions governing the canals is not, however, an obstacle to the drafting of general principles on the subject and may actually be an additional reason for making such an attempt. If the provisions of the treaties do incorporate the same legal right or duty in different language, it should be the task of the international lawyer to bring that principle to light. More than this, a common interpretation of provisions of these treaties has grown up in a manner which makes it difficult to say whether these interpretations follow by logical implication from articles of the treaties or have an independent existence as rules of customary international law. Finally, there are certain matters, not dealt with in the treaties, which are susceptible of harmonious resolution through principles of general application.

Recourse to analogies between different types of international waterways, as for example between straits and rivers, is more difficult but by no means unrewarding. The analogies are never complete ones, but they may be found with respect to individual aspects of the operation and control of differing varieties of channels of maritime communication. To take an obvious example, some agency must be provided to operate and maintain an interoceanic canal. In an analogous manner, international rivers must likewise be improved and maintained and have for this reason frequently been placed under international administrations charged with responsibility for carrying on these activities or for supervising or coordinating such work by the littoral states. The helpfulness of the analogy in connection with an operating or supervising agency is emphasized by the fact that rivers which have been made navigable through dredging, the installation of locks, and other measures are often spoken of as "canalized." On the other hand, international straits require no construction or continuing maintenance or administration, except for such comparatively minor matters as marking and the furnishing of pilots. To the question of what sort of régime may be established for the operation of an interoceanic canal, such as Suez, experience derived both from other canals and from international rivers may afford some answer, while, on the contrary, the practice relating to straits affords little assistance.

In another respect, interoceanic canals and straits are not dissimilar, for both share the characteristic of providing access between two stretches of open sea. The great proportion of the traffic passing through them is in transit and does not engage in trade with ports situated directly on the strait or canal. By contrast, international rivers are important not so much for the purpose of through transit as for the access they afford to the markets of the littoral states. From the standpoint of geopolitics, geography makes the position of rivers on the one hand and canals and straits on the other quite different in time of war. To the question of how to assure free passage through international canals, experience derived from international straits is thus more likely to give an answer than are conclusions drawn from international rivers.

Significant analogies for the solution of problems relating to rivers, canals, or straits do not stop short with those furnished by these waterways alone. A guarantee of free navigation on a river where two or more states are riparians is not dissimilar to the securing of freedom of navigation on a lake which is similarly situated.[149] Indeed, in the Treaty relating to Boundary Waters between the United States and Canada,[150] rivers, lakes, and canals connecting boundary waters are treated as a common category of "boundary waters." Ports may have to be established as ancillary facilities for international canals, and here precedents for the international administration of ports may prove to be of value.[151] If interoceanic canals or straits are to be made subject to international control or supervision, with corresponding obligations on the users to bear the costs of maintenance or improvement,[152] it may be necessary

---

[149] Revision of the International Regulations for Navigation of Lake Constance, signed at Stuttgart, Dec. 29, 1909, 7 MARTENS, N.R.G. 3d ser. 435 (1913).

[150] Treaty between the United States and Great Britain relating to Boundary Waters between the United States and Canada, signed at Washington, Jan. 11, 1909, 36 Stat. 2448, T.S. No. 548. See in particular the preliminary article and article I.

[151] See Convention on the International Régime of Maritime Ports, signed at Geneva, Dec. 9, 1923, 58 L.N.T.S. 285, 2 HUDSON, INTERNATIONAL LEGISLATION 1156 (1931); Convention concerning the Territory of Memel, signed at Paris, May 8, 1924, Annex II (Port of Memel), 29 L.N.T.S. 85 at 109.

[152] In presenting the results of the First London Conference to President Nasser, Prime Minister Menzies pointed out that one of the advantages of an international body to operate the Canal would be that "the future financial burdens involved in such maintenance and improvement would be carried out and handled by the new body and therefore Egypt would in fact be relieved of them." Letter of Sept. 7, 1956, in THE SUEZ

to borrow from experience gained in the international control and financing of lighthouses and other aids to navigation,[153] of weather stations,[154] and even of air navigation services, such as those necessary for air traffic across the North Atlantic.[155] As in the case of other types of waterways, the analogies are limited ones and must be applied to specific, limited questions where a borrowing may prove profitable.

The applicability to interoceanic canals of precedents and practices which have their source in waterways of the same category and even in watercourses of a different nature is borne out by the case of *The S.S. Wimbledon*.[156] In that case, the Permanent Court of International Justice stated:

> The precedents therefore afforded by the Suez and Panama Canals invalidate in advance the argument that Germany's neutrality would have necessarily been imperilled if her authorities had allowed the passage of the "Wimbledon" through the Kiel Canal, because that vessel was carrying contraband of war consigned to a State then engaged in an armed conflict. Moreover they are merely illustrations of the general opinion according to which when an artificial waterway connecting two open seas has been permanently dedicated to the use of the whole world, such waterway is assimilated to natural straits in the sense that even the passage of a belligerent man-of-war does not compromise the neutrality of the sovereign State under whose jurisdiction the waters in question lie.[157]

CANAL PROBLEM, JULY 26-SEPTEMBER 22, 1956, 309 at 315 (Dep't of State Pub. 6392) (1956).

[153] Convention as to Cape Spartel Light-House, signed at Tangier, May 31, 1865, 14 Stat. 679, T.S. No. 245, 1 MALLOY, TREATIES, CONVENTIONS, INTERNATIONAL ACTS, PROTOCOLS AND AGREEMENTS BETWEEN THE UNITED STATES OF AMERICA AND OTHER POWERS, 1776-1909, at 1217 (1910); International Convention concerning the Maintenance of Certain Lights in the Red Sea, signed at London, Dec. 16, 1930, 5 HUDSON, INTERNATIONAL LEGISLATION 853 (1936) (did not enter into force); International Agreement regarding the Maintenance of Certain Lights in the Red Sea, signed at London, Feb. 20, 1962, TRACTATENBLAD VAN HET KONINKRIJK DER NEDERLANDEN, 1962, No. 128.

[154] Agreement on North Atlantic Ocean Stations, signed at Paris, Feb. 25, 1954, 6 U.S.T. 515, T.I.A.S. No. 3186, 215 U.N.T.S. 268.

[155] Agreement on the Joint Financing of Certain Air Navigation Services in Iceland, signed at Geneva, Sept. 25, 1956, 9 U.S.T. 711, T.I.A.S. No. 4048; Agreement on the Joint Financing of Certain Air Navigation Services in Greenland and the Faroe Islands, signed at Geneva, Sept. 25, 1956, 9 U.S.T. 795, T.I.A.S. No. 4049.

[156] P.C.I.J., ser. A, No. 1 (1923).

[157] *Id.* at 28. The decision of the Court demonstrates what growth had taken place in international law in the course of less than half a century, for in 1885 Lawrence had written: "Now, since there are no precedents to guide us, and since the analogies applicable to the case are many and diverse, we are forced to the purely negative conclusion that, agreement apart, International Law does not afford any rules for the conduct of states in reference to the Suez Canal . . . If we ask for precedents, none are forth-

On its face the analogy which is made is not a general one but is limited to the passage through a canal of belligerent merchant vessels and warships and vessels carrying cargo to the belligerents, while the littoral state is at peace. Since in factual terms the analogy between straits and canals appears to have equal force with respect to passage through canals in time of peace and when the littoral state is at war, the principle suggested in *The S.S. Wimbledon* would appear to be applicable in these circumstances as well.

In the practice of states, analogies have been drawn with great frequency between individual watercourses of the same general type and between various categories of waterways. Reference has already been made to the incorporation of the language of the Convention of Constantinople of 1888 in the Hay-Pauncefote Treaty regarding the Panama Canal.[158] It has even been suggested that the adoption of the régime of Suez for Panama has of itself given rise to "un droit commun des canaux internationaux."[159] Diplomats have been quick to seize upon analogies between the Suez and Panama canals. An attempt in 1947 by the Egyptian Prime Minister to secure a United States military mission to assist Egypt in anticipation of the time when Egypt would become solely responsible for the defense of the Suez Canal was reported to have been rejected for the reason, among others, that it would create an undesirable precedent for the Panama Canal.[160] During the debate in the Security Council in 1954 concerning the measures taken by Egypt to prevent the shipment of supplies to Israel, the Colombian representative referred to the traditional concern of his country with the free navigation of international canals by reason of Colombia's dependence on Panama for its trade and also alluded in that connection to the principles governing the navigation of rivers in South America.[161]

---

coming; if we reason from analogy, we are lost in confusion; if we search for an agreement, we cannot find one." *The Suez Canal in International Law,* in ESSAYS ON SOME DISPUTED QUESTIONS IN MODERN INTERNATIONAL LAW 56-57 (2d rev. ed. 1885).

[158] See p. 38 *supra.*

[159] Guillien, *Un Cas de dédoublement fonctionnel et de législation de fait internationale: Le Statut du Canal de Suez,* in 2 LA TECHNIQUE ET LES PRINCIPES DU DROIT PUBLIC: ETUDES EN L'HONNEUR DE GEORGES SCELLE 735, 743 (1950).

[160] HOSKINS, THE MIDDLE EAST: PROBLEM AREA IN WORLD POLITICS 67 (1954).

[161] U.N. SECURITY COUNCIL OFF. REC. 9th year, 664th meeting 5-7 (S/PV.664) (1954). In the later discussions of the detention of the *Bat Galim* by Egypt, the Belgian Representative said of the Suez Canal: "It is an artificial waterway, and therefore not governed in all respects by the rules of ordinary international law concerning natural straits." U.N. SECURITY COUNCIL OFF. REC. 10th year, 688th meeting 1 (S/PV.688) (1955).

The possibility of parallels between the Suez and Panama canals did not long escape the attention of parties and bystanders to the dispute concerning the nationalization of the Suez Canal Company by Egypt. The Panamanian Minister to Egypt upheld in August 1956 the action taken by Egypt and stated that Panama would never allow the Panama Canal to be placed under international control. Concurrently Panama protested to Great Britain its exclusion from the First London Conference on the Suez Canal, attendance at which, it contended, was justified by the quantity of Panamanian flag shipping using the Canal.[162] In the debate in the Security Council in October, Mr. Shepilov, the Representative of the Soviet Union, inquired why internationalization was considered necessary for the Suez Canal when other important maritime canals of an international character, like the Panama Canal, were controlled and operated by only one state.[163] These references to the analogy between the two great canals fell on receptive ears in Egypt. Parallels were drawn between unilateral control of Suez by Egypt and the position of the Turkish Straits, Gibraltar, and Panama as waterways controlled by single nations.[164] One writer concluded, on the basis of these precedents, that international control "is an accidental, unnatural arrangement, contrary to the sound legal statutes of the waterways and straits and impairing a nation's territorial sovereignty."[165]

In his memoirs, Sir Anthony Eden has recorded that the Americans were "nervous about the Panama Canal" in the discussions in London following the nationalization of the Suez Canal.[166] The Suez Canal, so the United States position ran, was an international matter, whereas the Panama Canal was an "American and not an international waterway" as a consequence of the fact that its juridical status was governed solely by a treaty between Panama and the United States. The Americans were therefore determined to keep the Panama Canal out of the discussions and to have the case on the Suez Canal grounded on the Convention of Constantinople of 1888.[167] The same theme was taken up in public statements made at the time by President

---

[162] N.Y. Times, Aug. 21, 1956, p. 4, col. 5.

[163] U.N. SECURITY COUNCIL OFF. REC. 11th year, 736th meeting 16 (S/PV.736) (1956).

[164] N.Y. Times, Aug. 21, 1956, p. 4, col. 5; Abel Amer, *The Suez Canal & Other Waterways,* in the SUEZ CANAL: FACTS AND DOCUMENTS (Selected Studies No. 5) 109 at 111-15 (Cairo, 1956).

[165] Abel Amer, *supra* note 164, at 115.

[166] THE MEMOIRS OF ANTHONY EDEN: FULL CIRCLE 485 (1960).

[167] *Id.* at 487.

Eisenhower[168] and Secretary Dulles. Suez had been internationalized, whereas in the case of the canal under United States administration there is no treaty giving other nations any rights, except, in the words of Mr. Dulles, "for a treaty with the United Kingdom which provides that it has the right to have the same tolls for its vessels as for ours."[169] It is safe to conclude that there was no meeting of the minds between Sir Anthony Eden and Mr. Dulles concerning the dimensions of the treaty relationships of the United States. Moreover, Mr. Dulles' reference to the Hay-Pauncefote Treaty of 1901 overlooks the wider scope of that agreement, which lays the Panama Canal "free and open to the vessels of commerce and of war of all nations observing these Rules."[170] The invocation of the maxim *pacta tertiis nec nocent nec prosunt* in this context may justify the position of the United States with respect to the use of the Panama Canal by ships of third states but would at the same time mean that the United States and other nations which are not parties to the Convention of Constantinople of 1888 have no legal right to use the Suez Canal.

Both the Panamanian Government and Latin American scholars were alive to the purposes of the United States in asserting a legal basis for separate treatment of the questions of Suez and Panama. In a Declaration on the Suez and Panama canals, the Panamanian Academy of International Law spoke of "fundamental analogies"—although not complete ones—between the two waterways.[171] The same term was employed by a spokesman for the Panamanian Government.[172] These national positions taken during the negotiations of 1956 are, at the least, illustrative of the possibility of reasoning by analogy from one canal to another. The inaccuracy of the observations made about the Panama Canal by the President and Secretary of State of the United States support the belief that the Suez and Panama canals bore a closer legal kinship than was made to appear at that time.

[168] News Conference Statement by President Eisenhower, Aug. 8, 1956, *op. cit. supra* note 152, at 45.

[169] News Conference Statement by Secretary Dulles, Aug. 28, 1956, *op. cit. supra* note 152, at 301.

[170] Treaty between the United States and Great Britain to Facilitate the Construction of a Ship Canal, signed at Washington, Nov. 18, 1901, art. III, 32 Stat. 1903, T.S. No. 401.

[171] Declaration of the Panamanian Academy of International Law on the Suez and Panama Canals, Jan. 31, 1957; see also ALFARO ET AL., LOS CANALES INTERNACIONALES at xxxii-xxxv (1957).

[172] Letter by Sr. Arias, Panamanian Ambassador to Great Britain, The Times (London), May 1, 1957, p. 11, col. 5.

The Convention of Constantinople of 1888 has also exerted an influence on the Turkish Straits. European governments attempted to apply its principles to the Bosporus and Dardanelles when the provisions of the Convention seemed to be to their advantage.[173] The articles of the unratified Treaty of Sèvres dealing with the Turkish Straits[174] would have established for that waterway a régime akin to that of the Suez Canal.[175] The deep-rooted Russian policy of seeking a strong hand in the control of the Bosporus and the Dardanelles was more recently reflected in Molotov's reaction to the proposal made at Potsdam that the Straits be "internationalized." He inquired of Mr. Churchill whether Suez was operated under the principle of an international guarantee and of international control, to which Mr. Churchill replied that "the British had an arrangement with which they were satisfied and under which they had operated for some seventy years without complaints."[176]

The Final Act of the Congress of Vienna, from which the modern development of river law must be traced, purported in its Articles Concerning the Navigation of Rivers which in their Navigable Course Separate or Traverse Different States to give general application to the following principle: "La navigation dans tout le cours des rivières indiquées dans l'article précédent [those navigable rivers traversing or separating several states], du point où chacune d'elles devient navigable jusqu'à son embouchure, sera entièrement libre, et ne pourra, sous le rapport de commerce, être interdite à personne . . ."[177] As previously indicated,[178] the language of the Final Act of the Congress of Vienna has been imitated frequently in treaties relating to other international rivers, but it is doubtful whether the foregoing principle has gained general acceptance in customary international law. Indeed, a number of authorities have suggested that the wide variety of possible geographic factors and the influences of economics and politics render any general law relating to international rivers undesirable and impossible of

---

[173] HOSKINS, *op. cit. supra* note 160, at 24.

[174] Treaty of Peace between the British Empire and Allied Powers and Turkey, signed at Sèvres, Aug. 10, 1920, arts. 37-61, 113 BRITISH AND FOREIGN STATE PAPERS 652, 661 (1923).

[175] HOSKINS, *op. cit. supra* note 160, at 26; Tchirkovitch, *La Question de la révision de la Convention de Montreux concernant le régime des Détroits Turcs: Bosphore et Dardanelles,* 56 REVUE GÉNÉRALE DE DROIT INTERNATIONAL PUBLIC 189, 190 (1952).

[176] TRUMAN, YEAR OF DECISIONS 385-86 (1955).

[177] Signed June 9, 1815, 2 MARTENS, NOUVEAU RECUEIL DE TRAITÉS 361, 414 (1818).

[178] See p. 38 *supra*.

attainment.[179] However, the very admission that the establishment of an international commission for each international river may be a more effective means of adjusting conflicting interests in the river than any general international river law[180] points of itself to a significant institutional analogy in the administration of international rivers. The establishment of some common rules in an area where the tendency has hitherto been to emphasize differences may, it is hoped, be facilitated by the statement of principles, of general application to international rivers, which has been adopted by the International Law Association as the basis for further studies.[181]

An attempt to draw parallels between an international river and a strait was made in the correspondence which was conducted between the United States and Great Britain with regard to the St. Lawrence during the period 1824 to 1826. The right of United States citizens to navigate that river to and from the sea was upheld by the United States on the theory that the St. Lawrence was in effect a strait connecting the Gulf of St. Lawrence and the high seas with the Great Lakes.[182] Other correspondence from the United States referred to the accepted tolls for the use of European international rivers and to the Danish Sound dues,[183] but the British reply rejected the American contention and referred to "the marked difference in principle between Rivers and Straits."[184] The lack of favor with which the argument made by the United States has been received[185] may be attributable, in part

[179] Brierly, The Outlook for International Law 43 (1944); Quint, *Nouvelles Tendances dans le droit fluvial international,* 12 Revue de droit international et de législation comparée, 3d ser. 325, 340 (1931). Cf. the view of H. A. Smith in The Economic Uses of International Rivers (1931), who draws up a number of principles applicable to international rivers (at 150-53) but warns against premature codification of this body of practice and precedents about international rivers (at 158).

[180] Brierly, *op. cit. supra* note 179, at 43; Kaeckenbeeck, International Rivers 15-16 (1918).

[181] International Law Association, New York Conference, *Resolution on the Uses of the Waters of International Rivers,* in Report of the Forty-Eighth Conference held at New York at viii (1959).

[182] Letter from Mr. Clay to Mr. Gallatin, June 19, 1826, 6 American State Papers, Foreign Relations, 2d ser. 762 (1859); 19 British and Foreign State Papers 1089 (1834).

[183] Argument of the American Plenipotentiary annexed to the Protocol of the 18th Conference, London, June 19, 1824, 6 American State Papers, Foreign Relations, 2d ser. 769-71 (1859); 19 British and Foreign State Papers 1067-75 (1834).

[184] Reply of the British Plenipotentiaries annexed to the Protocol of the 24th Conference, London, July 19, 1824, 6 American State Papers, Foreign Relations, 2d ser. 772-75 (1859); 19 British and Foreign State Papers 1075, 1082-83 (1834).

[185] Kaeckenbeeck, *op. cit. supra* note 180, at 16; Brierly, *op. cit. supra* note 179, at 43.

at least, to the fact that the analogy was too broadly stated. The difficulties to which an attempt to establish a complete parallel between rivers and straits would give rise do not necessarily militate against recourse to more limited analogies, for example, in respect to supervisory bodies.

The most ambitious attempt that has been made to throw open a large number of international waterways to free and unrestricted use by all the nations of the world was a proposal by President Truman at the Berlin (Potsdam) Conference in 1945. The exact scope of the suggestion is not altogether clear. In a radio address made at the time, Mr. Truman said:

> One of the persistent causes for wars in Europe in the last two centuries has been the selfish control of the waterways of Europe. I mean the Danube, the Black Sea Straits, the Rhine, the Kiel Canal, and all the inland waterways of Europe which border on two or more states.
>
> The United States proposed at Berlin that there be free and unrestricted navigation of these inland waterways. We think this is important to the future peace and security of the world. We proposed that regulations for such navigation be provided by international authorities.[186]

He later wrote in his memoirs that he had also included the Suez Canal and the Panama Canal in the list of watercourses which were to be made "free waterways for the passage of freight and passengers of all countries."[187] When the proposal by Mr. Truman was recalled at the time of the Suez Canal crisis, a question was raised whether he had indeed suggested the internationalization of the Panama Canal. Secretary Dulles replied at a press conference that a careful search had been made of the records of the Department of State and that no such offer had come to light.[188] In any case, the proposal is significant for its bracketing of rivers, straits, and a canal as ripe for a régime which would at the least be a common one as regards freedom of transit and the establishment of international supervisory agencies. Unfortunately the plan came to naught at the subsequent meeting of the Council of Foreign Ministers.[189]

Soviet authorities on international law tend to particularize in dealing with legal problems of international waterways, as they do with respect to

---

[186] Radio address of Aug. 9, 1945, 13 Dep't State Bull. 208, 212 (1945).

[187] Truman, Year of Decisions 377 (1955).

[188] News Conference Statement by Secretary Dulles, Aug. 28, 1956, in The Suez Canal Problem, July 26-September 22, 1956, at 295, 298 (Dep't of State Pub. 6392) (1956).

[189] Statement by Mr. Bevin, Secretary of State for Foreign Affairs, in the House of Commons, Oct. 9, 1945, 414 H.C. Deb. (5th ser.) 40 (1945).

other areas of the law. They are consequently unwilling to concede that any general law exists with respect even to straits in general or canals in general. In dissenting in the *Corfu Channel Case,* Judge Krylov stated that, "Contrary to the opinion of the majority of the judges, I consider that there is no such thing as a *common* regulation of the legal régime of straits. Every strait is regulated individually."[190] It is not surprising that the Bosporus and the Dardanelles should be the first straits which he uses to illustrate his contention. The standard Russian text on international law distinguishes three categories of straits—national, lying entirely within the jurisdiction of a single state; straits affording access between the open sea and a closed sea, such as the Baltic or the Black Sea; and straits connecting two open seas. Only the last should be open to free navigation by all states, while straits affording access to closed seas should be open to all merchant vessels but closed to the warships of nonriparians.[191] According to another Russian authority, writing long before the Suez crisis, the Dardanelles or the Danube River are of concern only to the states to which these waterways provide access, while the Suez and Panama canals are true international waterways which ought to be placed under international supervision.[192]

It must in conclusion be emphasized that the theoretical possibility of analogies between rivers, canals, and straits and the fact that on repeated occasions states have drawn these analogies in the conduct of their international relations are not intended to suggest that there is necessarily any general body of law common to all three categories. The development of any customary international law has been handicapped rather than aided by the fact that the status of each individual maritime or fluvial channel of navigation has normally been fixed on a conventional basis. However, as the jurisprudence of the Permanent Court of International Justice and the International Court of Justice in the case of *The S.S. Wimbledon* and the *Corfu Channel Case* has indicated, this circumstance has not prevented the establishment of certain principles of international law having common application to a particular category of international waterways or to two such types of watercourse. For the moment, however, we must content ourselves with ex-

[190] [1949] I.C.J. Rep. 74-75.

[191] KOROVIN ET AL., MEZHDUNARODNOE PRAVO 308-10 (1951); see Kulski, *Les Tendances contemporaines dans le droit international soviétique,* 31 REVUE DE DROIT INTERNATIONAL, DE SCIENCES DIPLOMATIQUES ET POLITIQUES 272, 279-80 (1953). Concerning the Russian view of the Turkish Straits, see DRANOV, CHERNOMORSKIE PROLIVY (1948).

[192] Serezhin, *The World's Sea Routes and International Relations,* NEW TIMES, Jan. 9, 1947, p. 28 at 31.

tracting such general wisdom as can be derived from experience gained in connection with the shared problems of international waterways. Some of this may present itself in the form of customary law, some as a workable compromise of conflicting interests which might be the basis for the solution of a dispute submitted to an international tribunal. Other legal experience may serve as a guide in suggesting what matters should be taken into account in the establishment of the status of an individual waterway by treaty and what pitfalls should be avoided. For these purposes it may not always be necessary to deal with the vexing and perhaps unanswerable questions as to when usage becomes custom and custom becomes law, or how a law-making treaty or a settled pattern of conventional arrangements is transmuted into customary international law binding on all states.

CHAPTER II

THE OPERATING OR SUPERVISORY AGENCY

W ITH ONE relatively unimportant exception,[1] no strait has ever been placed under the supervision of an agency of an international character. On the other hand, river commissions, while not to be found on every international river, have proved to be useful adjuncts to the maintenance of free navigation, and interoceanic canals have in all instances been subjected to the administration of a special operating agency, whether international or national. The differing treatment which has been accorded the various types of waterways makes it appropriate to ask whether this circumstance is a consequence of history alone or has some basis in principle.

A strait is, in the first place, a natural waterway, which means that it requires a minimum of improvement and maintenance, aside from such matters as the marking of channels and the installation of aids to navigation like lighthouses or buoys. Possibly more important for analytic purposes is the fact that a strait is, by comparison with rivers and canals, functionally little related to the land masses through which it flows. There is relatively little communication between the vessel transiting a strait and the shores it passes and reciprocally little control exercised by the authorities ashore over the vessel.[2] The strait, even though it constitutes territorial sea, is in fact treated like an arm of the sea itself. Only exceptionally does this type of water highway have any economic relation to the territory through which it passes, in the sense of being a channel of trade between that territory and other lands. On the other hand, infinitely more potential obstacles to free passage, in both the legal and technical sense, exist in the case of international rivers and interoceanic canals. A river must be improved for navigation and at

---

[1] The Straits Commission created for the Bosporus and Dardanelles by the Convention relating to the Régime of the Straits, signed at Lausanne, July 24, 1923, 28 L.N.T.S. 115.

[2] In this respect, passage through the territorial waters of a strait differs little from "innocent passage" through territorial seas not forming a strait.

the same time be kept free of man-made obstacles to that navigation. A canal is, by its very nature, the work of man, a work which must be constantly improved and repaired, a work which, like a piece of machinery, must have an operator. The channels must be dredged, the locks operated. Geography, which makes a strait an arm of the sea and a river or canal a fluid highway across the territory of a nation, is productive of another important distinction in the legal régime of the two categories of waterways. The river or canal and the vessels upon it are necessarily in intimate contact with the shore. Whether it be a highway serving the littoral nations or an avenue of commerce of slight importance to the riparian state, the artificiality, the proximity to land, and the length of a canal or river link it closely to the economy of the country through which it flows. In terms of security, closer control must be maintained over such instruments of force as warships within a narrow waterway crossing an extended area of land than is true of these same vessels passing through a strait. And since an interoceanic canal or an international river traverses territory in the same way that a road or a rail line passes over land, it is only natural that there should be a greater reaching out of legal controls to such a waterway than to the open waters of a channel linking the high seas. The very ease with which restraints may be placed over passage through these narrow waterways, as distinguished from straits, encourages the exercise of such control.

The conflicting demands of freedom of commerce and of free passage by the user of the waterway and of security, the maintenance of public order, the desire for political prestige and influence, the protection of the economy, and financial profit exerted by the territorial sovereign clash abruptly over international rivers and canals. International institutions or national agencies which may carry on technical, engineering, and regulatory functions for the waterway have been devices by which the reconciliation of these conflicting demands has been sought. The degree of control exercised by these bodies is of great variety, but the two polar concepts are perhaps best expressed in the terms "operation" and "supervision," the latter implying a limited function of coordination, investigation, and recommendation. The effect of the bringing into existence of such an agency may be that the controversy between user and territorial sovereign may be carried on both outside the new organ, as heretofore, and within the new agency of coordination and control. The creation of an entity to supervise or operate can, it follows, merely create a new field of contest, but one which may be somewhat removed from politics. To the extent that the agency charged with this responsibility of operation or supervision acquires autonomy and assumes an existence of its own, what had initially been a struggle between the user and the territorial sovereign may become a tripartite conflict in which the operator or supervisor may find itself a third force capable of aligning itself with either of the two other antagonists. More often than not, the operator or supervisor has taken sides with the user.

The principal forms of operating or supervising organizations are the private corporation (as in the case of Suez before 1956), the agency of a state operating a canal through foreign territory (as in the case of Panama), the agency of a state operating a canal through the territory of that state (as in the cases of Suez and Kiel today), and the international commission. These will be discussed in turn.

SECTION A

ADMINISTRATION BY A PRIVATE COMPANY
(THE SUEZ CANAL)

Prior to the nationalization of the Suez Canal Company by the Egyptian Government on July 26, 1956, the Suez Canal was operated by the Compagnie Universelle du Canal Maritime de Suez, a private corporation, under concessions granted by Egypt in 1854 and 1856.[3] An enterprise which would in a modern world be undertaken only by a governmental agency or by an international organization was in the mid-nineteenth century characteristically left to private initiative. The eighteenth and nineteenth centuries had witnessed numerous instances in which trading and development companies, supported by private capital, had been endowed with powers scarcely less sweeping than those of a government for the purpose of economic exploitation of overseas areas by the European powers.[4] When in 1854 Ferdinand

[3] First Act of Concession, Nov. 30, 1854, Compagnie Universelle, Recueil des actes constitutifs 2; Firman de concession et cahier des charges pour la construction et l'exploitation du Grand Canal Maritime de Suez et dépendances, Jan. 5, 1856, *id.* at 6. The texts of the concessions are also to be found in 55 British and Foreign State Papers 970 (1864-65), and in English translation in The Suez Canal Problem, July 26-September 22, 1956, at 1 (Dep't of State Pub. 6392) (1956), and in *The Suez Canal: A Selection of Documents relating to the International Status of the Suez Canal and the Position of the Suez Canal Company,* 5 Int'l & Comp. L.Q., Sp. Supp. 1 (1956).

[4] See 1 Hyde, International Law, Chiefly as Interpreted and Applied by the United States 24 (2d rev. ed. 1945).

de Lesseps, the builder of the Canal, first approached the new Viceroy of Egypt, Said Pasha, on the question of building a waterway across the Isthmus of Suez, the world was still in such a state that a work of this nature was capable of achievement through the boundless energy and imagination of one man,[5] aided by funds made available by private persons. There was probably no thought at this time that the construction and operation of the Canal by a private company would insulate the waterway from politics, for De Lesseps and his canal were embroiled in international politics from the very inception of the project. But by 1888, when agreement was finally reached on the neutralization of the Canal and the maintenance of free passage through it, the Company had already demonstrated that it was, if not altogether neutral, at least somewhat more remote from the pressures of international rivalry than the governments which were concerned with the future of the waterway. The signatories to the Convention of Constantinople were actually less concerned with the administration of the Canal by the Suez Canal Company than they were with securing a right of free passage while Egypt remained under a temporary British occupation which appeared more and more like a permanent one with each passing year.[6]

A word should be said about the organization and structure of the former Compagnie Universelle du Canal Maritime de Suez before proceeding to an examination of the juridical status of the Company and of its relation to the régime of the Canal. The Company was a *société anonyme* organized to construct and operate "un canal maritime de grande navigation" through

---

One of the functions of the Royal Niger Company was the performance, observation, and undertaking of "all the obligations and stipulations relating to the River Niger" to which Great Britain had agreed in the Act of Navigation of the Niger, General Act of the Conference of Berlin, signed at Berlin, Feb. 26, 1885, 10 MARTENS, N.R.G. 2d ser. 414, 424 (1885-86); see Charter granted to the National African Company (Limited), July 10, 1886, art. 15, 77 BRITISH AND FOREIGN STATE PAPERS 1022, 1027 (1885-86); PILLIAS, LA NAVIGATION INTERNATIONALE DU CONGO ET DU NIGER 129 (1900).

[5] The colorful circumstances of the granting in principle of the Concession of 1854 are described in De Lesseps' words in ORIGINES DU CANAL DE SUEZ 37-40 (1890?); see also EDGAR-BONNET, FERDINAND DE LESSEPS: LE DIPLOMATE, LE CRÉATEUR DE SUEZ 190-203 (1951).

[6] HALLBERG, THE SUEZ CANAL: ITS HISTORY AND DIPLOMATIC IMPORTANCE 278-91 (1931).

the Isthmus of Suez.[7] Of the shares of the company, 44.15 per cent were owned by the Government of the United Kingdom,[8] which acquired them from the financially embarrassed Khedive of Egypt in 1875 in a dramatic purchase conceived by Prime Minister Disraeli.[9] The remainder of the shares, considered before the nationalization to be gilt-edged securities, were widely owned by private investors in France and in other countries. The administration of the Suez Canal Company rested in a *conseil d'administration*, consisting, according to the Statutes of the Company, of 32 directors representing the principal nationalities interested in the enterprise.[10] Sixteen directors of the Company were, as of the date of the purported nationalization of the Company by the Egyptian Government, French, nine British, five Egyptian, one Dutch, and one American.[11] The British directors comprised three representatives of the Government in its capacity as a shareholder in the Canal Company and six individuals representing British shipowners,[12] who have over the course of the years been the largest users of the waterway. The presence of five Egyptian directors on the Board was a consequence of Egyptian demands, to which recognition was given by the convention concluded with that government in 1949,[13] for increased participation by that nation in the affairs of the Company prior to the anticipated termination of the concession in 1968. An executive committee (*comité de direction*), composed of seven French directors and two British, was charged with the day-to-day direction of the affairs of the Suez Canal Company.[14]

The management of the Suez Canal Company was strongly centralized in Paris, its *domicile administratif*,[15] where a French staff carried on its

[7] COMPAGNIE UNIVERSELLE, STATUTS, arts. 1 and 2 (1949); the Statutes of the Company are also set forth in RECUEIL DES ACTES CONSTITUTIFS 12-28.

[8] COMPAGNIE UNIVERSELLE, RENSEIGNEMENTS CONCERNANT LES TITRES DE LA COMPAGNIE 12 (1953).

[9] Correspondence respecting the Purchase by Her Majesty's Government of the Suez Canal Shares belonging to the Egyptian Government, GREAT BRITAIN, EGYPT No. 1 (1876) (C. 1391).

[10] STATUTS, art. 24.

[11] COMPAGNIE UNIVERSELLE, THE SUEZ CANAL COMPANY AND THE DECISION TAKEN BY THE EGYPTIAN GOVERNMENT ON 26TH JULY 1956 (26TH JULY-15TH SEPTEMBER 1956) 38 (1956).

[12] 279 H.C. DEB. (5th ser.) 1650 (1932-33).

[13] Exchange of notes between M. Charles-Roux, President of the Company, and Riaz Bey, Egyptian Minister of Commerce and Industry, Cairo, March 7, 1949, COMPAGNIE UNIVERSELLE, RECUEIL DES ACTES CONSTITUTIFS 241.

[14] STATUTS, arts. 24, 37-42; COMPAGNIE UNIVERSELLE, RENSEIGNEMENTS GÉNÉRAUX 7 (1955).

[15] STATUTS, art. 3.

activities under the direction of the general manager. An agency was also maintained in London and a delegation of the Company in New York.[16] The operation and maintenance of the Canal in Egypt was not supervised by a single manager at the offices of the Company in Ismailia, but by a triumvirate of officials—a chief of administrative services, a chief of transit, and a chief engineer. This tripartite division of control was maintained throughout the lower echelons of the Company in the Suez Canal area.[17] One may surmise that the purpose of this scheme of organization was not alone to keep control of the activities of the Company centered in Paris but also to emphasize the essentially international and non-Egyptian character claimed for the enterprise. Relations with the Egyptian Government were carried on through an *agent supérieur* of the Company in Cairo, who fulfilled functions which might be described in a rough and ready way as those of an ambassador.[18] In 1956, when the control of the Canal was taken from the Company, a majority of the staff in Egypt was European, with French, British, and Greek nationals preponderating.[19] Although the Suez Canal Company had substantially increased the number of Egyptians working for the company, until at the end they comprised over 40 per cent of the personnel of the Company,[20] the higher ranking staff was overwhelmingly French. The highest post occupied by an Egyptian when the writer visited the offices of the Company in 1955 was a deputy to the *agent principal* for transit at Port Tewfik, one of the two terminals of the Canal. Despite the presence of substantial numbers of non-French Europeans in the operating sections of the organization—British, for example, in the pilotage service, and Greeks in the shops and on the Company's ships—the organization had an unmistakably French cast, which must have had much to do with persuading the Egyptian populace that the Company was an instrument of French "colonialism."[21]

In two significant instances the affairs of the Suez Canal Company have been subjected to the scrutiny of international commissions. The first of

[16] Compagnie Universelle, Renseignements généraux 27-28 (1955).

[17] *Id.* at 30-49.

[18] Statuts, art. 42; Firman de concession et cahier des charges pour la construction et l'exploitation du Grand Canal Maritime de Suez et dépendances, Jan. 5, 1856, art. 9, Compagnie Universelle, Recueil des actes constitutifs 7.

[19] Compagnie Universelle, *op cit. supra* note 11, at 39.

[20] *Ibid.*

[21] El-Hefnaoui, Les Problèmes contemporains posés par le Canal de Suez 275 (1951).

these was the consequence of objections which had been raised to the method of computation of tonnage employed by the Company in order to fix tolls.[22] The Ottoman Porte convened an international conference to deal with the matter in 1873.[23] The Moorsom system recommended by the conference was unacceptable to De Lesseps, and the Porte persuaded him to put the method in operation only by sending a force of 10,000 men to take possession of the Canal.[24] The level of tolls, although frequently the subject of protests to the Company, has not subsequently been the occasion for any international conference. In 1883, De Lesseps signed an agreement with British shippers concerning improvements in the operation of the Canal and more favorable financial terms.[25] Among other things, the agreement provided for the establishment of the International Consultative Commission on Suez Canal Works, which continued in existence until the nationalization of the Company. The Commission, which consisted of a number of eminent engineers of various nationalities, was only an adviser to the Company and in that capacity made recommendations to the Company and reviewed the various programs of improvements.[26] The existence of such a body, limited in powers as it was, was not only useful in an affirmative sense but also, in negative terms, protected the Company from criticism regarding the maintenance and increase in capacity of the Canal.

The Suez Canal Company, as has already been pointed out, owes its right to exploit the Canal to the definitive concession which was granted it in 1856.[27] From that date, the Company followed the policy of placing its relations with Egypt, both with respect to property and facilities over which the Company had jurisdiction and with regard to Egyptian control

[22] HALLBERG, *op. cit. supra* note 6, at 224-25.

[23] See Correspondence relative to the question of the Suez Canal dues together with the Procès-verbaux of the meetings held by the International Commission at Constantinople (1874), GREAT BRITAIN, COMMERCIAL No. 19 (1874) (C. 1075). An extract from the Regulations for the Measurement of Tonnage recommended by the International Tonnage Commission appears in COMPAGNIE UNIVERSELLE, RÈGLEMENT DE NAVIGATION 82 (1953).

[24] HALLBERG, *op. cit. supra* note 6, at 227.

[25] Reymond, *Histoire de la navigation dans le Canal de Suez*, 3 MÉMOIRES DE LA SOCIÉTÉ D'ETUDES HISTORIQUES ET GÉOGRAPHIQUES DE L'ISTHME DE SUEZ 90-91 (1956).

[26] See, e.g., *Note sur le 8ème programme des travaux d'amélioration du Canal de Suez*, Supplement to LE CANAL DE SUEZ, No. 2,306 (Feb. 15, 1955), at 6, concerning the examination of the eighth program of improvements by the Commission.

[27] Firman de concession et cahier des charges pour la construction et l'exploitation du Grand Canal Maritime de Suez et dépendances, Jan. 5, 1856, COMPAGNIE UNIVERSELLE, RECUEIL DES ACTES CONSTITUTIFS 6.

over important aspects of its operations, on a consensual basis through the conclusion of agreements which, at least in form, strongly resemble international agreements. The conventions and other agreements which were concluded between Egypt and the Company over the 90 years following the definitive concession of 1856 were concerned with a wide variety of matters, of which exemption from customs duties, tolls for Egyptian vessels, the establishment of free ports, expropriation for the needs of the Company, the retrocession of properties and facilities to the Egyptian Government, and the employment of Egyptians by the Company may be cited as typical.[28] In 1947, this established pattern of dealings, which the Company had found to be an important weapon in defense of its independence, was challenged by the enactment by Egypt of a new law, having application to all companies, which required that at least 40 per cent of the directors be of Egyptian nationality and that 75 per cent of the personnel employed by corporations be Egyptian.[29] The Suez Canal Company qualified on neither score. To have hired this number of Egyptian employees would have required a radical change in the staffing of the Company, while the appointment of a total of 13 Egyptian directors would have given that country a powerful voice in the conduct of the affairs of the Company. The juridical basis for the application of the law to the Company was, in the Egyptian view, a simple one. According to article 16 of the Convention of 1866 between Egypt and the Company, "La Compagnie Universelle du Canal Maritime de Suez étant Egyptienne, elle est régie par les lois et usages du pays." But the article goes on to say: "Toutefois, en ce qui regarde sa constitution comme Société et les rapports des associés entre eux, elle est, par une convention spéciale, réglée par les lois qui, en France, régissent les Sociétés anonymes."[30] The Statutes of the Company provided that the Company would be governed by the principles applicable to French corporations, since it had been established by analogy to these.[31] Thus there existed a substantial basis for the application to the Company of Egyptian laws relating to employment. However, the composition of the *conseil d'administration* was a matter of the internal regulation of the Company and was governed by French law and not by Egyptian,

[28] These agreements are collected in the RECUEIL DES ACTES CONSTITUTIFS, the "treaty series" of the Company.

[29] Loi No. 138 de 1947 édictant certaines dispositions relatives aux sociétés anonymes, Supplément au JOURNAL OFFICIEL DU GOUVERNEMENT EGYPTIEN, No. 74, Aug. 11, 1947.

[30] Convention du 22 février 1866, COMPAGNIE UNIVERSELLE, RECUEIL DES ACTES CONSTITUTIFS 41, 46.

[31] STATUTS, art. 73.

unless Egyptian law should be regarded as prevailing over the concession in the event of conflict between the two. The Company relied heavily on the fact that its relations with Egypt had in the past been governed by conventions and that its "universal" and "international" character precluded the extension to it of Egyptian law.[32] The controversy was resolved by the conclusion of a new convention and exchange of notes between Egypt and the Company.[33] This reaffirmation of the conventional basis of the relations between the two was bought at a substantial price. The demands of Egypt for the Egyptianization of the staff of the Company were scaled down, and a compromise was reached whereby five Egyptian directors would be appointed to the *conseil d'administration*. But the Company was forced to agree to turn over to the Egyptian Government 7 per cent of its gross revenues from the operation of the Canal, to transfer certain facilities and functions to the Egyptian Government, and to exempt vessels of Egyptian registry of less than 300 tons from the payment of tolls.[34]

As late as May 30, 1956, shortly before the nationalization of the Company by Egypt, another matter, which had been in an unsettled state for the previous nine years, was cleared up in an agreement between Egypt and the Company. The handsome revenues of the Company, largely paid in hard currencies in New York, London, and Paris, were an attractive prospect to an essentially poor country. In 1947, the National Bank of Egypt, acting in accordance with the Egyptian Exchange Control Law, made certain proposals to the Company regarding the repatriation of its earnings and the control of its foreign exchange assets. The Egyptian Foreign Exchange Law would actually have required repatriation of all of the Company's earnings, but the Bank was prepared to establish a special régime for

[32] COMPAGNIE UNIVERSELLE, REPORT, 1949, at 6. The Egyptian case is presented in EL-HEFNAOUI, *op. cit. supra* note 21, at 257-60.

[33] Exchange of notes between the President of the Suez Canal Company and the Egyptian Minister of Commerce and Industry, Cairo, March 7, 1949, and Convention, signed at Cairo, March 7, 1949, COMPAGNIE UNIVERSELLE, RECUEIL DES ACTES CONSTITUTIFS 241, 245.

[34] When the agreement was submitted to the Egyptian Parliament for ratification, it was severely criticized by the opposition, which maintained that it was altogether favorable to the Company and that it failed to give recognition to the subjection of the Company to the sovereignty and laws of Egypt. EL-HEFNAOUI, *op. cit. supra* note 21, at 265-68. In defense of the agreement, the president of the Council of Ministers said: "[L]a question se borne à une application de la loi sur les sociétés, et ne touche en rien l'assujettissement à la législation." In his contention, the agreement could be justified on the basis of the dislocation of the Suez Canal Company's activities which would be occasioned by the application of the Egyptian companies law to it. *Id.* at 268-69.

the Company by reason of its special status and function.[35] The Company replied that the requirements established by the Bank as to the form in which tolls must be paid would effect a discriminatory treatment of vessels using the Canal, in violation of articles 15 and 17 of the Firman of Concession of 1856, and raised certain other objections to the Egyptian proposals.[36] A compromise was worked out, which remained in effect until 1956, when the demands of Egypt for the repatriation of the earnings of the Company were renewed. An agreement concluded in that year provided for the transfer to Egypt, there to be invested, of a progressively increasing quantity of sterling funds, which the Company was to be allowed to repatriate upon the termination of the concession in 1968.[37]

When the Egyptian Government purportedly nationalized the Company in July, the Company and Egypt had concluded over one hundred such conventions and other agreements governing their relations.[38][What juridical significance can be attached to this settled practice is not clear, for it cannot be determined whether, from a legal point of view, the various agreements constituted acts of grace by Egypt or whether Egypt was under an obligation to treat with the Company in this fashion.]Nor does the mere existence of these agreements answer the question whether these were merely contracts susceptible of modification by the sovereign which was a party to them. The arbitral award rendered by Napoleon II in 1864 spoke of the concessions as creating for the signatories "des engagements réciproques, de l'exécution desquels il ne leur a pas été permis de s'affranchir" and that their stipulations presented in sum "un véritable contrat."[39] If these agreements are no more than contracts, the question becomes one of the responsibility of the state party thereto which modifies or terminates a contract

---

[35] Ms. letters from the Controller of Exchange Operations, National Bank of Egypt, to the Agent Supérieur of the Company, Sept. 18 and Oct. 13, 1947, on file in the headquarters of the Compagnie Universelle (now the Compagnie Financière de Suez) in Paris.

[36] Ms. letter from the Agent Supérieur of the Company to the Controller of Exchange Operations, National Bank of Egypt, Sept. 30, 1947, on file in the headquarters of the Compagnie Universelle (now the Compagnie Financière de Suez) in Paris.

[37] Ms. exchange of notes between the President of the Suez Canal Company and the Egyptian Minister of Finance and Economy, signed at Cairo, May 30, 1956, on file in the headquarters of the Compagnie Universelle (now the Compagnie Financière de Suez) in Paris.

[38] Statement by M. Pineau in the Security Council, Oct. 5, 1956, U.N. SECURITY COUNCIL OFF. REC. 11th year, 735th meeting 18 (S/PV.735) (1956).

[39] Award of July 6, 1864, 2 DE LAPRADELLE AND POLITIS, RECUEIL DES ARBITRAGES INTERNATIONAUX 362, 363 (1924).

with an alien in contravention of the terms of the instrument. But that problem is beyond the scope of the present study.

The Convention of 1866, as has been mentioned above, subjected the Company to Egyptian law but provided that disputes between the Company and its shareholders or regarding its status as a company would be governed by French law. Disputes of the latter two categories were to be judged by arbitrators in France, subject to appeal to the Imperial Court in Paris. The same agreement also provided that disputes in Egypt between the Company on the one hand and individuals or the Egyptian Government on the other would be judged by the local courts "suivant les formes consacrées par les lois et usages du pays et les traités."[40] Litigation between the Suez Canal Company and its shareholders and shipowners has been productive of an inconsistent pattern of judicial decisions and of pronouncements by foreign offices concerning the law applicable to the Company and the tribunals empowered to apply it.

The financial difficulties into which the Company had fallen not long after the completion of the Canal were the occasion for the first important case concerning the Company's status. The Company was sued in a French court by a shipping company, which contended that the tonnage upon which tolls were based was being improperly computed. The Tribunal of Commerce of the Seine held that it had jurisdiction over the matter. De Lesseps, concerned with the immediate problem of protecting the Company from litigation, wrote to Khalil Pasha that "The interpretation of the Act of Concession belonging of right to the Government, the author of such concession, it cannot in any way belong to a foreign Tribunal."[41] The Porte concurred in this view, as did the British Government, which pointed out that if the litigation could be maintained in a French court, the Company "would cease to be Egyptian or Turkish, and become French."[42] An appeal to the

---

[40] Convention du 22 février 1866, art. 16, COMPAGNIE UNIVERSELLE, RECUEIL DES ACTES CONSTITUTIFS 41, 46.

[41] Letter, Nov. 9, 1872, Correspondence relative to the question of the Suez Canal dues together with the Procès-verbaux of the meetings held by the International Commission at Constantinople (1874), GREAT BRITAIN, COMMERCIAL No. 19 (1874) at 19 (C. 1075).

[42] Earl Granville to Sir H. Elliot, Aug. 31, 1872. In the same despatch, Earl Granville wrote that the Government "are of opinion that the Egyptian Government should be supported in maintaining that the Company is for the purposes of jurisdiction Egyptian, and not French." *Id.* at 15 and 16. The concurrence of the Ottoman Porte in the views of De Lesseps was set forth in a letter from Khalil Pasha to De Lesseps, Nov. 15, 1872, *id.* at 21.

Court of Appeal of Paris resulted in a holding that the Tribunal of Commerce of the Seine had jurisdiction of the matter but a dismissal on the merits.[43]

Although the Company was, by virtue of article 16 of the Convention of 1866, considered to be Egyptian and subject to the laws of that country, the Mixed Court of Appeals of Alexandria nevertheless found the Company in the *Magripli* case[44] to be within the jurisdiction of the Mixed Courts of Egypt by reason of the international complexion of its ownership and its international function, which kept it from being assimilated to a native of Egypt. The significance of the case as bearing upon the distinctive character of the Company is severely limited by the fact that "Egyptian" companies of mixed foreign and Egyptian ownership, such as the municipal water, traction, and other public service companies and almost all of the banking system of the country, were likewise considered to be subject to the jurisdiction of the Mixed Courts.[45] The case demonstrates no more than that the Suez Canal Company was treated like any other enterprise of mixed ownership.

In later years, litigation about the proper currency of payment of tolls and shares continued to be a fruitful source of jurisprudence concerning the legal status of the Compagnie Universelle du Canal Maritime de Suez. Currencies depreciated in value after the First World War, and the Company, not unnaturally, demanded the payment of tolls in terms of the gold franc rather than in French francs or their equivalent. Both in Egypt and in France, a shipping company which challenged the Canal Company's right to collect tolls in gold francs or their equivalent lost its case on the ground that the international and "universal" character of the Company's operations required payment of tolls computed in terms of a common medium of exchange.[46] A bondholder thereupon brought an action in the Mixed Courts of Egypt to secure payment on the same basis, that is to say in terms

[43] Compagnie des Messageries maritimes v. Compagnie Universelle du Canal Maritime de Suez, Cour de Cassation (Ch. req.), Feb. 23, 1874, [1874] SIREY, RECUEIL GÉNÉRAL I. 145.

[44] Magripli v. La Compagnie du Canal de Suez, Cour d'Appel Mixte d'Alexandrie, May 20, 1880, 5 JURISPRUDENCE DES TRIBUNAUX DE LA RÉFORME EN EGYPTE 264 (1879-80).

[45] BRINTON, THE MIXED COURTS OF EGYPT 107 (1930).

[46] Compagnie Havraise Péninsulaire de navigation à vapeur v. Compagnie Universelle du Canal Maritime de Suez, Cour d'Appel Mixte, Feb. 9, 1922, 34 BULLETIN DE LÉGISLATION ET DE JURISPRUDENCE ÉGYPTIENNES 160 (1921-22); Compagnie Havraise Péninsulaire v. Compagnie Universelle du Canal Maritime de Suez, Cour d'Appel de Paris, Feb. 15, 1924, [1924] DALLOZ, JURISPRUDENCE I. 189.

of the gold franc, of interest on the bonds of the Company which it held. If tolls were to be paid in that manner, it could not be seriously disputed that interest on the Company's obligations should be set in the same form, and the Mixed Court of Appeals in Egypt directed that the interest on the bonds be computed in terms of gold.[47] The Company was, in the view of the Court, "à la fois égyptienne et universelle."[48] Its universal character, which dictated its utilization of a universal currency in the computation of its tolls and its dividends, was indicated by the concessions themselves, by the financing of the Company by persons of various nationalities, by the use of several languages in its securities, by the composition of its *conseil d'administration,* and by the words of De Lesseps. But it is significant that the Court stated firmly that "juridiquement, elle est une société égyptienne soumise aux lois du pays," even though the Company had been permitted to borrow a form of corporate organization created by French law.[49] The type of currency which should govern the computation of payments to bondholders in the Company once more came in issue in 1935, when an Egyptian decree law of May 2 of that year annulled the gold clause in international contracts and required payments on such contracts in depreciated Egyptian currency. This the Company did, and the bondholders responded by an action against the Company to secure payments in terms of the universal gold franc which had hitherto been prescribed by the Mixed Courts as the proper standard currency for the transactions of the organization. Their contention was upheld by the Court of Appeals,[50] primarily on the ground

[47] A. Debbah et Cts. v. Compagnie Universelle du Canal Maritime de Suez, Cour d'Appel Mixte, June 4, 1925, 37 BULLETIN DE LÉGISLATION ET DE JURISPRUDENCE ÉGYPTIENNES 466 (1924-25). Similarly, the statutory interest and dividends on shares were required to be paid on the basis of the gold franc. Compagnie Universelle du Canal Maritime de Suez v. Joseph Shallam & Sons et autres, Cour d'Appel Mixte, June 18, 1931, 43 *id.* II. 455 (1930-31), ANNUAL DIGEST OF PUBLIC INTERNATIONAL LAW CASES, 1931-32, Case No. 137.

[48] A. Debbah et Cts. v. Compagnie Universelle du Canal Maritime de Suez, *supra* note 47, at 471.

[49] *Id.* at 470.

[50] Crédit Alexandrin v. Compagnie Universelle du Canal Maritime de Suez, Cour d'Appel Mixte, Feb. 26, 1940, 52 BULLETIN DE LÉGISLATION ET DE JURISPRUDENCE ÉGYPTIENNES II. 185 (1939-40). It was apparently in these proceedings, to which the British Government was a party, that that Government asserted, according to a subsequent Egyptian account, that, "It [the Suez Canal Company] is Egyptian because it is granted a concession which has for its object Egyptian public assets and because its legal principal centre is in Egypt," and that, "It would be a legal anomaly to consider the Company at one and the same time Egyptian and non-Egyptian, i.e., universal." U.N. SECURITY COUNCIL OFF. REC. 11th year, 736th meeting 6 (S/PV.736) (1956).

that the decree law was directed to the conservation of the foreign exchange assets of Egypt in the ordinary transaction in which an Egyptian debtor would have to draw upon such resources in order to pay his obligations to a foreign creditor. The Court of Appeals found that the Canal Company represented a special case, not within the purpose of the decree law, in that its revenues, which were in terms of the gold franc, were the source of its payments to bondholders. Adherence to the terms of the gold clause between the bondholders of the Company and that firm would therefore not result in any loss of monetary assets by Egypt, which, moreover, benefited from the activity of the Company as a source of foreign exchange.[51] While the Company was exempted from the application of an Egyptian law on a special basis, the exemption was a consequence of the transactions' not falling within the scope of the decree rather than of any special juridical standing attributed to the Company. The Court of Appeals on the contrary affirmed the Egyptian character of the Suez Canal Company and its subjection to the laws of Egypt.[52] This line of cases can hardly be characterized as establishing that the Compagnie Universelle du Canal Maritime de Suez was accorded a distinctive position under international law or the municipal law of Egypt when its affairs were scrutinized by the Mixed Courts. On each occasion, the Court of Appeals for the Mixed Courts stated firmly that the Company was an Egyptian one, subject to the laws of that country. It was only the type of business it conducted and the manner in which it carried on its activities that on occasion required that its rights and duties as a creditor and debtor under Egyptian law be dealt with in a manner different from that in which other Egyptian corporations might be regulated. Little legal significance can be attached to the fact that the tribunals found the word "universal" in the title of the Company or invoked the shade of De Lesseps.[53]

See also Compagnie Universelle du Canal Maritime de Suez v. Campos, Cour d'Appel Mixte, May 17, 1947, 59 BULLETIN DE LÉGISLATION ET DE JURISPRUDENCE ÉGYPTIENNES II. 219 (1946-47), dealing with the gold franc as a "monnaie de compte internationale" after that coin had ceased to circulate. The Court prescribed the conversion of the gold franc into Egyptian currency on the basis of the price in dollars of the amount of gold in the gold franc on the New York market, the resulting dollar figure being then converted to Egyptian currency at the rate prescribed by the National Bank of Egypt. See p. 281 *infra*.

[51] Crédit Alexandrin v. Compagnie Universelle du Canal Maritime de Suez, *supra* note 50, at 191.

[52] "... à la fois une société égyptienne et universelle ..." *Id.* at 186.

[53] See A. Debbah et Cts. v. Compagnie Universelle du Canal Maritime de Suez, *supra* note 47, at 471.

So far as the administration and control of the Canal itself and of the contiguous area were concerned, the Suez Canal Company and Egypt had carved out separate competences through usage and agreement. Egypt, of course, was at all times responsible for the police of the Canal and of the surrounding area, for the control of immigration, and for the administration of its customs laws.[54] It was Egypt and, during the term of the British protectorate and occupation, Great Britain which were responsible for the defense of the waterway and the security of the adjacent territory.[55] The Company, on the other hand, promulgated the rules of navigation which governed transit through the waterway.[56] It also considered that it had an independent obligation of its own, derived from the terms of the concessions accorded it and from the Convention of Constantinople of 1888, to maintain the neutrality of the Company and freedom of navigation through the Canal. Because the Suez Canal was cut through a desert, the Company was obliged for many years to concern itself with the provision of adequate water supplies, with the creation and administration of municipalities, with the provision of municipal services,[57] and with the administration, jointly with the Egyptian Government, of the *domaine commun,* consisting of those lands which were not needed for the actual exploitation of the Canal.[58] The conduct of these affairs could easily give rise to the impression that the Company was carrying on governmental functions on Egyptian territory and was arrogating to itself some of the perquisites of sovereignty.[59] It is only fair to add that many of these responsibilities had, over the course of years, been gradually turned over to the Egyptian authorities. Subject to the special exemptions from Egyptian law which were provided by the concessions which Egypt had granted and the other agreements which had been arrived

[54] Douanes égyptiennes v. Compagnie Universelle du Canal de Suez, Cour d'Appel d'Alexandrie, Feb. 9, 1887, 2 REVUE INTERNATIONALE DU DROIT MARITIME 728 (1886-87), affirming Tribunal Civil d'Alexandrie, arrêt of May 18, 1886, 2 *id.* 204 (1886-87).

[55] See p. 329 *infra.*

[56] COMPAGNIE UNIVERSELLE, RÈGLEMENT DE NAVIGATION (1956 ed.); and RÈGLEMENT DE NAVIGATION: ANNEXE POUR LES NAVIRES À CARGAISONS DANGEREUSES (1956 ed.).

[57] See p. 288 *infra.*

[58] See Deuxième Convention du 18 décembre 1884, COMPAGNIE UNIVERSELLE, RECUEIL DES ACTES CONSTITUTIFS 62; and *Note relative à l'administration des terrains de la concession, id.* at 352.

[59] Views of this nature were expressed by Dr. El-Hefnaoui in 4 QANAT AL-SUWAIS (The Suez Canal and Its Contemporary Problems) 248-96 (1954). The theme of the colonialism of the Company was joyfully taken up by Mr. Shepilov, the Foreign Minister of the Soviet Union, in the debates in the Security Council. U.N. SECURITY COUNCIL OFF. REC. 11th year, 736th meeting 17 (S/PV.736) (1956).

at with that Government, the Company remained subject to Egyptian law, even in such respects as the taxation of its profits[60] and the fixing of the amount of the tolls it would charge.[61]

The principal argument directed against the nationalization of the Suez Canal Company by Egypt in 1956[62] was that the Company was not merely governed by Egyptian and by French law, as described in the preceding paragraphs, but also by international law.[63] This contention was often supported by assertions which, far from constituting reasons why the Company was outside the control of the Egyptian Government, merely represented conclusions that this was the case. The Canal was spoken of as an "international public service."[64] References were made to the Company as "universal"[65] and "international."[66] Quite aside from the accuracy of these descriptions, they do not establish that the nationalization of the Company,

---

[60] See p. 280 *infra*.

[61] E.g., Décret portant fixation du taux maximum des droits de navigation dans le Canal de Suez, April 28, 1936, GOUVERNEMENT ÉGYPTIEN, MINISTÈRE DE LA JUSTICE, [1936] TABLE DES LOIS, DÉCRETS ET RESCRITS ROYAUX 173. The decree was, however, promulgated only after negotiations with the Company.

[62] Decree Law No. 285 of 1956 respecting the Nationalisation of the Universal Suez Maritime Canal Company, July 26, 1956, in REPUBLIC OF EGYPT, MINISTRY FOR FOREIGN AFFAIRS, WHITE PAPER ON THE NATIONALISATION OF THE SUEZ MARITIME CANAL COMPANY 3 (1956).

[63] Statement by M. Pineau, the French Foreign Minister, at the London Conference, Aug. 17, 1956, THE SUEZ CANAL PROBLEM, JULY 26-SEPTEMBER 22, 1956, at 87 (Dep't of State Pub. 6392) (1956); and see the statements cited in note 66 *infra*.

[64] Statement by M. Pineau cited *supra* note 63; Guillien, *Un Cas de dédoublement fonctionnel et de législation de fait internationale: Le Statut du Canal de Suez*, in 2 LA TECHNIQUE ET LES PRINCIPES DU DROIT PUBLIC: ETUDES EN L'HONNEUR DE GEORGES SCELLE 735, 751 (1950).

[65] Statement by M. Pineau cited *supra* note 63. Cf. Jules Ferry in 1885 speaking of the Canal as having a "caractère essentiellement universel, européen, humanitaire." FRANCE, DOCUMENTS DIPLOMATIQUES, COMMISSION INTERNATIONALE POUR LE LIBRE USAGE DU CANAL DE SUEZ, AVRIL-NOVEMBRE 1885, at 3 (1885).

In 1885, Lawrence spoke of the Company as having "a quasi-diplomatic *status.*" *The Suez Canal in International Law,* in ESSAYS ON SOME DISPUTED QUESTIONS IN MODERN INTERNATIONAL LAW 41, 55 (2d ed. 1885). Cf. the view of Sir Richard Vaux, formerly President of the Mixed Court of Appeal of Egypt: "The Suez Canal Co. is in reality Egyptian, the creature of Egyptian authority, subject in the last resort to Egyptian law, and working under the Egyptian flag." He concluded, ". . . it may fairly be said that the Suez Canal Company ranks with another Egyptian curiosity—the Sphinx." *A Legal Curiosity,* 60 L.Q. REV. 227, 229 (1944).

[66] Three-Power London Talks: Tripartite Statement, Aug. 2, 1956, in THE SUEZ CANAL PROBLEM, JULY 26-SEPTEMBER 22, 1956, at 34 (Dep't of State Pub. 6392) (1956); Radio-Television Report by Secretary Dulles, August 3, *id.* at 38.

whether it be regarded as having the effective national character of France or of Egypt or an effective international character, was beyond the competence of the Egyptian Government. In order to ascertain what international standing, if any, the Company had, it is necessary to look to the question whether states had by their action endowed it with that status. For that purpose, only treaties would possess major relevance,[67] since it is quite clear that the members of the international community had not by their practice exhibited any intention to endow the Company with international personality or to make of it an international organization.[68]

The Convention of Constantinople of 1888[69] contains three references to the concessions granted by Egypt to the Suez Canal Company. The first of these occurs in the preamble in the following words, which follow the recital of the parties to the Convention: ". . . wishing to establish, by a Conventional Act, a definite system destined to guarantee at all times, and for all the Powers, the free use of the Suez Maritime Canal, and thus to complete the system under which the navigation of this Canal has been placed by the Firman of his Imperial Majesty the Sultan, dated the 22nd February, 1866 (2 Zilkadé, 1282), and confirming the Concessions of His Highness the Khedive . . . " Article II of the Convention places the Fresh Water Canal under special protection and refers to the Convention of 1863: "The High Contracting Parties, recognizing that the Fresh-Water Canal is indispensable to the Maritime Canal, take note of the engagements of His Highness the Khedive towards the Universal Suez Canal Company as regards the Fresh-Water Canal; which engagements are stipulated in a Convention bearing date the 18th March, 1863, containing an *exposé* and four Articles." In the second paragraph of the article, it was agreed that the parties would not interfere with the Fresh Water Canal. Finally, it was agreed that "the engagements resulting from the present Treaty shall not be limited by the duration of the Acts of Concession of the Universal Suez Canal Company."[70]

It was contended by a number of the spokesmen for the user nations

---

[67] The concessions themselves are clearly not treaties. Anglo-Iranian Oil Co. Case (United Kingdom v. Iran), [1952] I.C.J. Rep. 92.

[68] See p. 33 *supra,* and the statement of the President of the Board of Trade in 1933, when the question of Suez tolls was under discussion: "The question of the Suez Canal charges is a matter for the Suez Canal Company, which is a private company holding a concession from the Egyptian Government." 276 H.C. DEB. (5th ser.) 492 (1932-33).

[69] Signed Oct. 29, 1888, 15 MARTENS, N.R.G. 2d ser. 557 (1890).

[70] Article XIV.

that the language of the preamble meant that the Convention was intended to add certain international guarantees of freedom of navigation to those already provided by the concession, that the powers would not have concluded the treaty in this form had they been aware that the concession might be unilaterally terminated by Egypt, and that the Convention therefore constituted an international recognition by Turkey and Egypt, as its successor,[71] of the special position of the Company in the general régime of the Canal.[72] In terms of a verbal analysis of the language of the preamble,[73] the words "complete the system" do not readily yield to the interpretation that the system which is being completed was to remain unchanged until 1968, when the concessions would have terminated by their own terms. Unfortunately, there is no indication in the *travaux préparatoires* for the Convention that it was the intention of the draftsmen to place the Company under an international guarantee. Had that purpose been in the contemplation of the powers that signed the Convention, the oblique reference to the completion of the system is a most peculiar way of achieving that purpose. If it were to be assumed that the words "to complete the system" did make the régime established by the concessions immutable, at least until 1968, the consequences of this interpretation would have been to place unreasonable restraints upon Egypt and upon the Suez Canal Company. Presumably, a surrender of the concession to Egypt by the Company would

[71] It appears that the United Arab Republic assumed the obligations of Egypt under the terms of the union between the latter nation and Syria. See 582 H.C. DEB. (5th ser.) 392 (1958).

[72] In the words of Mr. Selwyn Lloyd, speaking in the Security Council, "[T]he Convention [of Constantinople], together with the Company's concessions, constituted a balanced scheme." U.N. SECURITY COUNCIL OFF. REC. 11th year, 735th meeting 5-6 (S/PV.735) (1956). At the London Conference, Mr. Dulles spoke of the Company's concession as "incorporated into and made part of what is called the definite system set up by the 1888 treaty." THE SUEZ CANAL PROBLEM, JULY 26-SEPTEMBER 22, 1956, at 73 (Dep't of State Pub. 6392) (1956). See also, as exemplifying the doctrinal writings on this subject, Hostie, *Notes on the International Statute of the Suez Canal,* 31 TUL. L. REV. 397, 419-24 (1957).

[73] The preamble of a treaty may at least be relied upon in aid of the interpretation of the agreement. Case of the S.S. Lotus, P.C.I.J., ser. A, No. 10, at 17 (1927); Case of the Free Zones of Upper Savoy and the District of Gex, Order of Aug. 19, 1929, P.C.I.J., ser. A, No. 22, at 15 (1929); Case concerning Rights of Nationals of the United States of America in Morocco (France v. United States), [1952] I.C.J. Rep. 175, 196, 198; Fitzmaurice, *The Law and Procedure of the International Court of Justice 1951-4: Treaty Interpretation and Other Treaty Points,* 33 BRIT. Y.B. INT'L L. 203, 227-29 (1957). It is not equally clear that the preamble may serve as the source of legal obligations.

thus be a violation of the Convention, as would the transfer of the concession to another company or the vesting of the administration of the waterway in an international organization pledged to maintain freedom of communications through it. It would not be unreasonable to conclude that if the Convention froze the relations between the Company and Egypt, then all of the agreements between the Suez Canal Company and the Government concluded thereafter constituted unlawful variations of the régime of the Canal, because they had not received the approbation of the signatories of the Convention of 1888. Furthermore, the view that the Company, and only the Company, was at that time regarded as competent to assist the powers in keeping the Canal open to free navigation ill comports with the guerrilla warfare Great Britain had for years been waging with De Lesseps and his company.

The matter of international protection of the right of freedom of navigation is undoubtedly separable from the form of administration under which the territorial sovereign puts the waterway. It was not the Suez Canal Company which might close the Canal to the passage of certain ships, but Great Britain or Egypt or Turkey or another of the powers. The separability of the two matters is quite clear in the cases of the Panama and Kiel canals, for in neither instance is there any requirement that the canal be operated by a governmental agency or by a private company, although both waterways have been opened to free navigation by the ships of all nations. And the fact that freedom of navigation has been maintained in the Suez Canal since the nationalization of the waterway is persuasive evidence that the existence of the Suez Canal Company was not an essential condition of the freedom of the waterway. The Convention of 1888 itself contains internal evidence that the question of freedom of navigation was separable from that of the administration of the waterway. The system of control established by the treaty gives no place to the Suez Canal Company. Under article VIII of the Convention, it is the "Agents in Egypt of the Signatory Powers of the present Treaty" who are "charged to watch over its execution." It is irrelevant for these purposes that this diplomatic body was never constituted; what is important is that the parties drafted this provision of the Treaty as if the Suez Canal Company did not exist. The independence of the international guarantee of free passage through the Canal from the administration of the waterway is further illustrated by article XIV of the Treaty, which has the effect of keeping the agreement in effect after the concession terminated of its own force. From that date, then eighty years distant, freedom of navigation would be revealed as a matter separate and apart from the nature of the operating agency.

The argument in favor of the guarantee of the concessions and of the Company through the Convention of Constantinople must stand or fall on the language of the preamble, for the two other articles referring to the concessions do not sustain the thesis of those who link the Convention and the Company. The effect of article II, dealing with the Fresh Water Canal, is to include this essential adjunct to the Maritime Canal within the international protection offered by the Treaty. The provision is no more than a recognition that Egypt could impede, or even prevent, navigation through the Maritime Canal by drying up this source of fresh water and that the parties to the Convention would see to it that neither Egypt nor one of them should employ this indirect method of obstructing the use of the Canal.[74] At most, article XIV, continuing the Treaty in force after the termination of the concession, may be said to cut two ways. It may mean that the parties were of the opinion that until 1968 freedom of navigation could be secured only with the aid of the Company but that thereafter the Company would be unnecessary for that purpose. This argument, it must be noted, is not an independent one and must rely as well upon the language of the preamble, "to complete the system." On the other hand, article XIV may mean that freedom of navigation is one thing, that the administration of the Canal is another, and that the two do not march together. It may not be out of place to mention that it is attributing remarkable prescience to the signatories of the Convention to think that they supposed the existence of the Company to be essential to freedom of navigation until 1968 but that its essentiality would abruptly cease on a stipulated date eight decades after the signing of the Treaty.

The only other international instrument which might be considered to have placed the Company under international protection is the Turkish Declaration of December 1, 1873. This statement, made at the conclusion of the Tonnage Conference held at Constantinople, was to the following effect: "That no modification, for the future, of the conditions for the passage through the Canal shall be permitted, whether in regard to the navigation toll or the dues for towage, anchorage, pilotage, etc., except with the consent of the Sublime Porte, which will not take any decision on this subject without previously coming to an understanding with the principal Powers

[74] See p. 285 *infra.*

interested therein." [75] The Declaration was cast in the form of an assumption of a legal obligation, and having been accepted by the powers must be considered to have placed the Porte under a legal duty not to make any such changes without the consent of the other parties. Its effect, however, is limited to "conditions for the passage through the Canal," which are thereafter defined in terms of the charges to be imposed. Moreover, following the consultations immediately after the Declaration and its acceptance, there seems to have been no further use made of this method of consultation,[76] and important changes in the "conditions for the passage through the Canal" have been made without "an understanding with the principal Powers interested therein." The Declaration must therefore be considered on two scores to offer no impediment to the nationalization by Egypt of the assets in that country of the Suez Canal Company. The Declaration had ceased to be operative, and it was inapplicable by its terms to alterations in the operating agency for the Canal.

The foregoing discussion of the legal régime to which the Suez Canal Company was subject is not intended to pass judgment on the lawfulness of the nationalization of the Canal Company by the Egyptian Government. Such questions as those of the national character of the Company for the purposes of jurisdiction and protection, the standing of the Company or of its shareholders to secure compensation from Egypt, the extraterritorial effect of the nationalization law, the basis of responsibility as to the taking of property or the termination of the concession, and the quantum of damages would call for a consideration of legal issues transcending the confines of the present study.[77] The controversy between the Company and the

[75] Despatch from the British Delegates on Tonnage at Constantinople, together with the Report and Recommendations of the Commission as to International Tonnage and the Suez Canal Dues, GREAT BRITAIN, PARLIAMENTARY PAPERS, COMMERCIAL No. 7 (1874) at 7 (CMD. No. 943), *The Suez Canal: A Selection of Documents relating to the International Status of the Suez Canal and the Position of the Suez Canal Company*, 5 INT'L & COMP. L.Q., Sp. Supp. 45 (1956).

[76] See p. 261 *infra*.

[77] Among the voluminous literature on this subject may be mentioned Domke, *American Protection Against Foreign Expropriation in the Light of the Suez Canal Crisis*, 105 U. PA. L. REV. 1033 (1957); Finch, *Navigation and Use of the Suez Canal*, [1957] PROCEEDINGS OF THE AMERICAN SOCIETY OF INTERNATIONAL LAW 42; Hostie, *supra* note 72, at 397; Huang, *Some International and Legal Aspects of the Suez Canal Question*, 51 AM. J. INT'L L. 277 (1957); De la Pradelle, *L'Egypte, a-t-elle violé le Droit International en nationalisant la Compagnie Universelle du Canal Maritime de Suez?*, [1958] INTERNATIONALES RECHT UND DIPLOMATIE 20; Olmstead, *Nationalization of Foreign Property Interests, Particularly Those Subject to Agreements with the State*,

Government of the United Arab Republic has since been happily resolved by the payment of compensation to the Company.[78] The old Compagnie Universelle du Canal Maritime de Suez is no more, and the Compagnie Financière de Suez has taken its place.[79]

SECTION B

OPERATION OF AN INTEROCEANIC CANAL BY A FOREIGN SOVEREIGN (THE PANAMA CANAL)

The Panama Canal flows through the territory of the Republic of Panama but is operated by the United States of America. A strip of land ten miles in width has been furnished the United States for the exploitation of the waterway. The rights of the United States derive from two important provisions of the Hay–Bunau-Varilla Treaty of 1903,[1] signed in haste only a few days after Panama had achieved its independence. The first of these is that portion of article II of the Treaty which provides: "The Republic of Panama grants to the United States in perpetuity the use, occupation and control of a zone of land and land under water for the construction, maintenance, operation, sanitation and protection of said Canal [referred to in the preamble of the Treaty] of the width of ten miles . . ." The second is article III of the Treaty: "The Republic of Panama grants to the United States all

---

32 N.Y.U. L. Rev. 1122, 1130-35 (1957); Pinto, *L'Affaire de Suez: Problèmes juridiques,* 2 Annuaire français de droit international 20 (1956); Scelle, *La Nationalisation du Canal de Suez et le Droit International,* 2 *id.* at 3 (1956); Paul de Visscher, *Les Aspects juridiques fondamentaux de la question de Suez,* 62 Revue générale de droit international public 400 (1958); Note, 70 Harv. L. Rev. 480 (1957).

[78] Agreement between the Government of the United Arab Republic and the Compagnie Financière de Suez, signed at Geneva, July 13, 1958 (U.N. Doc. No. A/3898, S/4089) (1958), Lauterpacht, *The Suez Canal Settlement: A Selection of Documents relating to the Settlement of the Suez Canal Dispute, the Clearance of the Suez Canal and the Settlement of Disputes between the United Kingdom, France and the United Arab Republic,* 8 Int'l & Comp. L.Q., Sp. Supp. 6 (1959).

[79] Resolution adopted by the shareholders, July 4, 1958, Compagnie Financière de Suez, Bulletin, No. 1, at 11 (1958).

[1] Convention between the United States of America and the Republic of Panama for the Construction of a Ship Canal to Connect the Waters of the Atlantic and Pacific Oceans, signed at Washington, Nov. 18, 1903, 33 Stat. 2234, T.S. No. 431.

the rights, power and authority within the zone mentioned and described in Article II of this agreement and within the limits of all auxiliary lands and waters mentioned and described in said Article II which the United States would possess and exercise if it were the sovereign of the territory within which said lands and waters are located to the entire exclusion of the exercise by the Republic of Panama of any such sovereign rights, power or authority." The effect of article III is to make a distinction between what was conceded to be the "titular sovereignty" of Panama and the power of the United States to exercise those rights which it would have "if it were the sovereign of the territory." The Treaty was drafted by Bunau-Varilla, the Minister of Panama to the United States, in the course of one day,[2] but it would be naive to suppose that the formula employed was a hasty improvisation.

The establishment of a status for territory in which one state would retain sovereignty but another nation would exercise sovereign rights over the area was not without precedent. The leases of naval bases by China to foreign powers only five years before the Hay–Bunau-Varilla Treaty had made exactly such arrangements. In the Convention respecting the Lease of Kiaochow, for example, it was stipulated that "in order to avoid the possibility of conflicts, the Imperial Chinese Government will abstain from exercising rights of sovereignty in the ceded territory during the term of the lease, and leaves the exercise of the same to Germany . . ."[3] Similar provisions are to be found in the leases to France and to Russia.[4] It may not be irrelevant to observe with respect to the juridical status of the Panama Canal Zone that the view of many learned writers on international law during this general period was that these leases constituted disguised ces-

---

[2] DuVal, CADIZ TO CATHAY: THE STORY OF THE LONG DIPLOMATIC STRUGGLE FOR THE PANAMA CANAL 380 (2d ed. 1947). The abortive Hay-Herran Treaty, which was concluded with Colombia on January 22, 1903, but was not ratified by that nation, contained a similar grant of a canal zone to be subject to the "use and control" of the United States and provided that the "rights and privileges" accorded the United States would not "affect the sovereignty of the Republic of Colombia." *Diplomatic History of the Panama Canal,* S. Doc. No. 474, 63d Cong., 2d Sess. 277, 279 (1914).

[3] Convention between Germany and China respecting the Lease of Kiaochow, signed at Peking, March 6, 1898, sec. I, art. III, 1 MacMurray, TREATIES AND AGREEMENTS WITH AND CONCERNING CHINA, 1894-1919, at 112 (1921).

[4] Convention between France and China for the Lease of Kuang-chou Wan, signed at Peking, May 27, 1898, arts. I and III, 1 MacMurray, *op. cit. supra* note 3, at 128; Convention between Russia and China for the Lease of the Liaotung Peninsula, signed at Peking, March 27, 1898, arts. I and II, *id.* at 119.

sions[5] or, at least, in light of the fact that the leases were for stipulated terms of years, cessions for a limited period of time.[6]

Although certain of the key words of the dispositive provisions of the Treaty of Peace with Japan relating to the Ryukyu Islands are derived from the language of trusteeship agreements, the distinction between sovereignty and sovereign powers is maintained in the establishment of the provisional status of those Islands. Japan agreed to concur in any proposal to place the Ryukyus under trusteeship but was forced to concede: "Pending the making of such a proposal and affirmative action thereon, the United States will have the right to exercise all and any powers of administration, legislation and jurisdiction over the territory and inhabitants of these islands, including their territorial waters."[7] The term "residual sovereignty," which has been applied to the remaining rights of Japan,[8] connotes that same "bare" sovereignty which is evoked by the expression "titular sovereignty," first applied by John Hay to the rights of Panama in the Canal Zone.[9]

Considerable satisfaction was given to the Republic of Panama by the statement in the Treaty of 1936 with the United States that the Canal Zone constituted "territory of the Republic of Panama under the jurisdiction of the United States of America."[10] Further recognition of the "titular sovereignty residing in the Republic of Panama with respect to the Canal Zone" was given symbolic form in the undertaking of the United States in 1960

---

[5] E.g., De Pouvourville, *Les Fictions internationales en Extrême-Orient,* 6 REVUE GÉNÉRALE DE DROIT INTERNATIONAL PUBLIC 113, 118 (1899); 2 MÉRIGNHAC, TRAITÉ DE DROIT PUBLIC INTERNATIONAL 488 (1907).

[6] Memorandum of the Solicitor of the Department of State, inclosure to Secretary of State Hay to Mr. Conger, Feb. 3, 1900, [1900] FOREIGN REL. U.S. 387, 389 (1902); the views of the authorities are extensively discussed in NOREM, KIAOCHOW LEASED TERRITORY 55-86 (1936).

[7] Treaty of Peace with Japan, signed at San Francisco, Sept. 8, 1951, art. 3, 3 U.S.T. 3169, 3172, T.I.A.S. No. 2490.

[8] Statement by Secretary of State Dulles, Sept. 5, 1951, in CONFERENCE FOR THE CONCLUSION AND SIGNATURE OF THE TREATY OF PEACE WITH JAPAN: RECORD OF PROCEEDINGS 78 (Dep't of State Pub. 4392) (1951); and see, for a fuller discussion of this concept, United States v. Ushi Shiroma, 123 F. Supp. 145 (D. Hawaii 1954).

[9] Secretary of State Hay to Mr. de Obaldía, the Panamanian Minister, Oct. 24, 1904, [1904] FOREIGN REL. U.S. 613, 615 (1905).

[10] General Treaty of Friendship and Cooperation between the United States of America and Panama, signed at Washington, March 2, 1936, art. III, para. 6, 53 Stat. 1807, T.S. No. 945. The provision related to the landing of passengers and cargo in the area so defined.

to fly the Panamanian flag together with the United States flag at one point in the Zone.[11]

However, the basis upon which Panama has challenged the exercise of various rights in the Canal Zone by the United States has not been the language of article III, which indicates that the United States enjoys the full range of governmental powers within the Zone, but article II, which grants "the use, occupation and control of a zone of land and land under water for the construction, maintenance, operation, sanitation and protection of said Canal." In these words, according to the consistent Panamanian position over the last half century, are to be found the limits of the authority of the United States in the Zone.[12] Each exercise of jurisdiction by the United States must be measured against this standard and, if not necessary to the construction, maintenance, operation, sanitation, or protection of the Canal, is an infringement of the sovereign rights of Panama. At the time of the Suez crisis of 1956-57, the Panamanian Academy of International Law declared that the cases of Suez and Panama were analogous in that both canals were constructed in territory not belonging to the entity which constructed them and that the Hay–Bunau-Varilla Treaty, which is the foundation of the rights of the United States, is "a contract of concession for international public service."[13]

Panamanian objections to the exercise by the United States of the "rights, power and authority" it would have "if it were the sovereign of the territory" were heard within a few months after the exchange of ratifications of the Hay–Bunau-Varilla Treaty in 1904. The first occasion for these protests was the extension of United States customs laws to the territory of the Canal

---

[11] Note from the United States Ambassador to the Acting Minister of Foreign Relations of Panama, Sept. 17, 1960, 43 DEP'T STATE BULL. 558 (1960). There had been considerable opposition in the Congress to the flying of the Panamanian flag, and the expenditure of funds for the erection of a pole to fly that flag had been forbidden by section 201 of the Department of Commerce Appropriation Act for 1961, 74 Stat. 93. See also *Hearings on United States Relations with Panama Before the Subcommittee on Inter-American Affairs of the House Committee on Foreign Affairs*, 86th Cong., 2d Sess. (1960), which are very largely devoted to a consideration of this issue.

[12] Mr. de Obaldía to Secretary of State Hay, Aug. 11, 1904, [1904] FOREIGN REL. U.S. 598, 599 (1905). The reference in this note to the respective positions of the Republic of Panama and of the United States as those of "lessor and lessee" (at 600) further link the Treaty of 1903 with the international leases of that period.

[13] Declaration of the Panamanian Academy of International Law on the Suez and Panama Canals, Jan. 31, 1957; the same contention is made in a letter from Sr. Arias, the Ambassador of Panama, to The Times (London), May 1, 1957, p. 11, col. 5.

Zone and the creation of ports of entry within the Canal Zone.[14] The operation of the United States postal system within the Zone was shortly added to the list of grievances.[15] Panama maintained that the Zone remained subject to the "fiscal sovereignty"[16] of that country and that articles X and XIII of the Treaty of 1903, which gave the United States immunity from Panamanian taxes and import duties, were to be construed by an argument *a contrario* as recognition of Panamanian authority to levy taxes and customs duties in other respects.[17] The reply of the United States was necessarily that this construction of the Treaty was impossible of reconciliation with the power of the United States to exercise sovereign rights under the language of article III. The restrictions upon the power of the Republic of Panama to tax and to levy customs duties could be explained as providing immunities to the United States within those areas which were subject to the actual exercise of Panamanian sovereignty, that is to say, the territory lying outside the Canal Zone. If the convention were not given that interpretation, the grant to Panama by article IX of the right to establish customs houses in the Zone would be meaningless, for under the Panamanian view, this right would be one already enjoyed by that nation under article II of the Treaty.[18] The words of article II to the effect that the land was granted for certain purposes listed in that article were not terms of limitation but a description of the inducement for the grant.

In this instance, the differences between the two governments were resolved by the so-called Taft Agreement of 1904,[19] whereby the United States agreed not to allow importations through the terminal ports of the

[14] Order of the Secretary of War, June 24, 1904, in EXECUTIVE ORDERS RELATING TO THE PANAMA CANAL 26 (1922), and [1904] FOREIGN REL. U.S. 586 (1905). The first protest by the Panamanian Government was reported to the Department of State on July 25. [1904] FOREIGN REL. U.S. 586 (1905).

[15] Order of the Secretary of War, June 24, 1904, *supra* note 14, at 27; see, as to the Panamanian protest, Mr. de Obaldía to Secretary of State Hay, Aug. 11, 1904, [1904] FOREIGN REL. U.S. 598, 605 (1905).

[16] Mr. Arias, the Panamanian Secretary of Government and Foreign Affairs, to Mr. Barrett, the United States Minister, July 27, 1904, [1904] FOREIGN REL. U.S. 591 (1905).

[17] Note cited *supra* note 12, at 604-05.

[18] Secretary of State Hay to Mr. de Obaldía, Oct. 24, 1904, [1904] FOREIGN REL. U. S. 613, 615, 618-22 (1905).

[19] The terms of the Agreement are embodied in an Executive Order of December 3, 1904. It is referred to as an agreement because its terms were the results of conferences between Mr. Taft, the Secretary of War, and the President of the Republic of Panama and because the text was submitted to the President of Panama for approval prior to promulgation. *Id.* at 640-43.

Canal, except those goods covered by the exemption of article XIII, as being for the use of the United States; goods in transit; and fuel supplies for transiting vessels. The application of the United States tariff to goods entering the Zone from the Republic of Panama was brought to an end, on the condition that Panama would maintain a reciprocal exemption for the United States. The agreement thus gave practical application to the statement of President Roosevelt, made several months prior to the conclusion of the Taft Agreement, that it was not the intention of the United States "to establish an independent colony in the center of the State of Panama." [20]

Although temporarily calmed, the fears of the Republic of Panama concerning the existence of a commercial enterprise established in its territory and in competition with its economy remained a cause of dissension between the two governments. The question of the sovereignty of Panama over the Canal Zone again became a live issue when negotiations were in progress from 1923 to 1926 for a new treaty to replace the Taft Agreement. Twenty years had not changed the legal position taken by the Panamanian Government. Sr. Alfaro wrote to the Secretary of State, "Panama claims as a right derived from the Canal treaty, and confirmed by the Taft Agreement, jurisdiction over the foreign trade of the Canal Zone." [21] In concrete terms, this meant control over the customs of the Canal Zone, the suppression of commissaries and foreign companies in the Canal Zone, restrictions on the operation of the Volstead Act so as not to hamper traffic from, to, and between Panamanian ports, and imposition of Panamanian export taxes on products of the soil exported by the United States from the Zone, over and above the other demands made by Panama in the negotiations between the two countries.[22] Both nations, before and after the abrogation of the Taft Agreement in 1924,[23] expressed a willingness to treat and to compromise

[20] Letter from President Roosevelt to Mr. Taft, Oct. 19, 1904, quoted in [1923] 2 FOREIGN REL. U.S. 640 (1938).

[21] The Panamanian Minister (Alfaro) to the Secretary of State, Jan. 3, 1923, *id.* at 645.

[22] *Id.* at 643-46.

[23] The Secretary of State to the Minister in Panama (South), Oct. 18, 1923, *id.* at 676, directed that the Panamanian Foreign Office be informed that the Taft Agreement would be abrogated on May 1, 1924. It was understood that the Agreement was only a modus vivendi subject to termination by the United States. The Secretary of State in a dispatch of December 14, 1923, informed the United States Minister in Panama of the statements which had been made by the United States negotiators of the Agreement in 1904. The most important of these, made to the President of Panama, was that the Agreement would provide only for a "nonexercise" of powers which

on many of the issues outstanding between them.[24] Each, however, started from a different assumption—the United States from the premise that it enjoyed full powers in the Canal Zone, from which any concessions to Panama would be in derogation, and Panama from the position that the United States had been accorded only limited rights in the Zone and had in fact been exceeding its powers under the Hay–Bunau-Varilla Treaty.

Although the attitude taken by the Republic of Panama toward many of the activities of the United States in the Canal Zone may be justifiable in political or economic terms—questions with which it is not proposed to deal here—its legal case lacks substance. If the language of article II of the Hay–Bunau-Varilla Treaty might seem to suggest the grant of a canal zone for limited purposes, any possible limitations on the authority of the United States are removed by article III, with its sweeping grant of "all the rights, power and authority" which the United States would have if it were the sovereign of the territory "to the entire exclusion of the exercise by the Republic of Panama of any such sovereign rights, power or authority." The Panamanian position is, moreover, inconsistent with the contemporary understanding of grants of the same general character made during the same period. In suggesting that some of the treaties which separated technical sovereignty from the exercise of sovereign rights were actually concealed cessions, some of the authorities who wrote at the turn of the century[25] went far beyond any position ever taken by the United States. The specific exemptions from the law of Panama accorded in various articles of the Treaty, such as the immunity from taxation of the property, officials, and employees of the United States provided by article X, are explainable as protection for these things and persons while they are within the territory of the Republic of Panama, as a defense against a possible extraterritorial application of Panamanian fiscal laws, or simply as restrictions on Panamanian jurisdiction inserted *ex abundantia cautelae*. The express enumeration of the rights enjoyed by Panama within the Canal Zone cannot be recon-

---

would "continue indefinitely until the construction of the canal shall so affect the relations and conditions existing as to require a new adjustment of the relations between the two Governments." The text of the executive order itself stated that it was not a definition of or restriction on the rights of the two nations under the Treaty of 1903. *Id.* at 679, 681-82.

[24] Memorandum by the Secretary of State of a Conversation with the Panamanian Minister (Alfaro), Dec. 15, 1923, *id.* at 682.

[25] See p. 72 *supra*.

ciled with the theory held by that country. It may be observed, however, that the language of article II, that the grant is "for the construction, maintenance, operation, sanitation and protection" of the Canal, upon which such heavy reliance is placed by Panama, does place one important restraint upon the plenitude of powers otherwise enjoyed by the United States. That qualification is temporal. If the zone should cease to be used for the support of the Canal, the rights of the United States in the zone would come to an end. But that is the extent of the limitation imposed.

To look simply to the theoretical extent of the powers of the United States within the Canal Zone under articles II and III of the Hay–Bunau-Varilla Treaty would give an altogether misleading impression of the actual relations of the Republic of Panama and the United States. A complex network of rights and duties of consensual origin controls the exercise of sovereign powers by the two countries. The exercise by the United States of those powers which it would have were it sovereign within the Zone is limited by restraints imposed by treaty, while at the same time the United States has been granted the right to carry on certain of its activities within territory subject to the jurisdiction of the Republic of Panama.

A number of the powers which the United States has in the past exercised outside the Zone and of those which it continues to enjoy today find their justification in being necessary to the operation and the defense of the waterway. Most intimately related to the management of the Canal is the right accorded by article IV of the Treaty of 1903 [26] to use the rivers, lakes, and other bodies of water in Panama for navigation and for the supply of water, without which the operation of the locks would become impossible. The same treaty granted the United States the "use, occupation and control" of any other lands outside the Canal Zone itself which might be necessary for the construction, operation, and maintenance of the Canal and its ancillary works.[27] Within the cities of Panamá and Colón, the adjacent harbors, and territory adjacent to the cities, the United States at one time had the right to acquire needed property by the exercise of the right of eminent domain.[28] Panamanian protests that several decades after the construction of the Canal the United States should have no further need to exercise such powers as these[29] met their response in the Treaty of

[26] Cited *supra* note 1.

[27] Article II.

[28] Article VII.

[29] The Panamanian Secretary of Government and Justice on Special Mission (Alfaro) to the Secretary of State, April 2, 1921, [1922] 2 FOREIGN REL. U.S. 751, 753-

1936, whereby it was agreed that the United States would surrender its power of eminent domain and that the acquisition of needed lands outside the Zone would be the subject of agreement between the two governments.[30] One of the consequences of the surrender of these powers was that the United States was henceforth under the necessity of persuading Panama to make lands available for manoeuver areas and for defense sites by whatever inducements, economic and otherwise, it might be able to hold out. Despite a stipulation in the Treaty of 1936 that the two countries would "in the event of some now unforeseen contingency" reach agreement on additional lands needed for the protection of the Canal,[31] it was only after a year and a half of difficult negotiations and the entry of the United States into the Second World War that land areas needed for defensive purposes were formally granted by Panama.[32] The United States was obliged to withdraw from these after the termination of hostilities, pursuant to the terms upon which they had been made available. The Rio Hato manoeuver area, probably the most important of the lands sought to be retained by the United States during the unsuccessful postwar negotiations, was once more secured in the economic bargaining of the Treaty of 1955.[33] Other land areas secured to the use of the United States include the Boyd-Roosevelt Highway across the Isthmus, for the maintenance of which the United States is now responsible. In return for this undertaking by the United States, that country has been granted the free and unimpeded use of all public roads within the Republic of Panama.[34]

---

54 (1938); The Panamanian Minister (Alfaro) to the Secretary of State, Jan. 3, 1923, *id.* at 638. One of the sore points was that, pursuant to article VI of the Treaty of 1903, the United States was continuing to pay for property secured through eminent domain proceedings according to the value of the property prior to the conclusion of the Treaty. Panamanian note of April 2, 1921, *supra* at 755-56. The position taken by the United States is reflected in 31 Ops. Att'y Gen. 44 (1920), insisting that the plain words of the Treaty could be given but one interpretation.

[30] General Treaty of Friendship and Cooperation, *supra* note 10, arts. II and VI.

[31] Article II.

[32] Agreement for the Lease of Defense Sites in the Republic of Panama, signed at Panamá, May 18, 1942, 57 Stat. 1232, E.A.S. No. 359. The negotiations had begun in October 1940. See Wright, *Defense Sites Negotiations Between the United States and Panama, 1936-1948*, 27 Dep't State Bull. 212, 214-17 (1952).

[33] Treaty of Mutual Understanding and Cooperation between the United States of America and the Republic of Panama, signed at Panamá, Jan. 25, 1955, art. VIII, 6 U.S.T. 2273, 2286, T.I.A.S. No. 3297.

[34] Highway Convention between the United States of America and the Republic of Panama, signed at Panamá, Sept. 14, 1950, arts. I and VII, 6 U.S.T. 480, T.I.A.S. No.

In addition to the lands which the United States has at one time or another utilized, the proprietor of the Canal has exercised certain rights of jurisdiction, now largely of historical interest, outside the Zone. From 1903 until 1936, the United States enjoyed a conventional right to intervene for the maintenance of order in Colón and Panamá, the two terminal cities of the Canal, both of which lay within Panamanian territory.[35] Until 1955, the control of sanitation within these two cities likewise rested with the United States.[36] Still surviving is the right of the United States to operate the Panama Railroad across the Isthmus,[37] a doubtful advantage since the Railroad is being operated at a loss and is of questionable strategic value.

The United States is not alone in enjoying the use of lands and the exercise of rights outside the territory subject to its plenary jurisdiction. Panama has been granted a corridor through the Zone to connect the city of Colón with the rest of its territory.[38] Portions of the Boyd-Roosevelt Highway which intersect the Canal Zone have been placed within narrow corridors subject to Panamanian jurisdiction in order to avoid having vehicles enter the territory of the Canal Zone for limited distances, only to return to Panamanian territory.[39] Within the Zone, the United States has accepted certain limitations on its powers with respect, for example, to the sale of goods to transiting vessels and sales from commissaries to nationals of the Republic of Panama[40] and to the categories of persons who may live in the Canal Zone.[41]

3181. Article VI of the Convention grants the United States a right of way 100 feet in width on each side of the center line of the highway. The highway was constructed at the expense of the United States, and Panama was thereafter to be responsible for the upkeep of the highway. Convention between the United States and Panama concerning the Trans-Isthmian Highway, signed at Washington, March 2, 1936, 53 Stat. 1869, T.S. No. 946.

[35] Treaty cited *supra* note 1, art. VII, para. 3. Concerning the diplomatic history of the right of intervention secured by the Treaty, see PADELFORD, THE PANAMA CANAL IN PEACE AND WAR 60-63 (1943). The right was renounced by the United States in the Treaty of 1936, *supra* note 10, art. VI.

[36] Treaty cited *supra* note 1, art. VII, para. 2. The right was renounced by the United States in the Treaty of 1955, *supra* note 33, art. IV.

[37] Treaty cited *supra* note 1, arts. V and VIII.

[38] Treaty of 1936, *supra* note 10, art. VIII; and see Convention between the United States of America and the Republic of Panama regarding the Colón Corridor and Certain Other Corridors Through the Canal Zone, signed at Panamá, May 24, 1950, 6 U.S.T. 461, T.I.A.S. No. 3180.

[39] Treaty cited *supra* note 38, art. IV.

[40] Treaty of 1955, *supra* note 33, art. XII.

[41] Treaty of 1936, *supra* note 10, art. III, para. 2, which in general excludes from

Others of the arrangements between Panama and the United States are solely a consequence of the fact that the two states are living in close propinquity and that persons subject to their jurisdiction move freely from the territory of one to that of the other. These understandings have their counterparts in similar agreements relating to the administration of boundary areas in other quarters of the world. The United States facilitates the administration of Panamanian customs and immigration controls on goods and persons entering Panama by way of the Zone through the provision of sites for customs houses[42] and by affording immigration officers access to vessels arriving at the piers of Balboa and Cristobal.[43] An extradition treaty[44] between the two governments implements the undertaking of article XVI of the Treaty of 1903[45] to provide for the detention and delivery by the one state of persons charged with crimes in the other. American military police and shore patrols are permitted to maintain order among the members of the United States Armed Forces who are within the territory subject to the jurisdiction of the Republic of Panama.[46] Panamanian nautical inspectors have been authorized by the United States to board and inspect vessels of Panamanian registry within the Canal Zone for the purpose of ascertaining whether there has been compliance with the laws of that country concerning maritime and labor matters.[47]

---

residence in the Canal Zone persons other than employees of the United States Government; members of the United States Armed Forces; contractors' employees; employees of companies doing business in the Zone; persons engaged in religious, educational, and similar work; their families; and their servants.

[42] Treaty of 1936, *supra* note 10, art. V.

[43] *Ibid.*

[44] Treaty between the United States and Panama for the Mutual Extradition of Criminals, signed at Panamá, May 25, 1904, 34 Stat. 2851, T.S. No. 445.

[45] Cited *supra* note 1.

[46] PADELFORD, THE PANAMA CANAL IN PEACE AND WAR 75 (1943); United States military personnel on duty in territory subject to the jurisdiction of the Republic of Panama have been held to be immune from the jurisdiction of the courts of the latter country. Republic of Panama v. Schwartzfiger, Supreme Court of Justice, Aug. 11, 1925, ANNUAL DIGEST OF PUBLIC INTERNATIONAL LAW CASES, 1927-28, Case No. 114; and see, concerning jurisdiction over off-duty personnel, Banks et al. (United States) v. Panama, General Claims Commission, United States and Panama, June 29, 1933, 6 UNITED NATIONS, REPORTS OF INTERNATIONAL ARBITRAL AWARDS 349.

[47] Agreement Authorizing Panamanian Nautical Inspectors to Board Vessels of Panamanian Registry in the Canal Zone for the Purpose of Ascertaining Compliance with Panamanian Maritime and Labor Laws, signed at Panamá, Aug. 5, 1957, 8 U.S.T. 1413, T.I.A.S. No. 3893.

During the half century or more that the United States has exercised jurisdiction over the Canal Zone, the powers of the two countries with respect to the Canal Zone have increasingly been governed by detailed treaty arrangements rather than dictated by abstract considerations of sovereignty or of sovereign rights over this or that area. The jurisdictional competences of the two governments are not separated by the geographical boundary forming the periphery of the Canal Zone. Nevertheless, the basic instrument, the Hay–Bunau-Varilla Treaty of 1903, as modified, constitutes a starting point of legal right for any negotiations for further concessions upon the part of the one country or the other. Once a matter such as the authority of persons to purchase at commissaries makes the transition from the unfettered control of the United States to a right regulated by treaty, the question becomes one which for the rest of time must be governed by agreement between the two countries.

A comparison of the legal powers once enjoyed by the United States with respect to the Canal and the Canal Zone with those it currently exercises would show a steady diminution in the degree of autonomy enjoyed by the operator of the Canal. Thus far, however, Panamanian pressures, against which the territory of the Zone serves as something of a buffer, have not had the purpose or the effect of impeding freedom of passage through the waterway. The Republic of Panama is subject to the same restraints as the United States with respect to the neutralization of the passage and the transit of vessels through the Canal.[48] It is specifically precluded by treaty from imposing any charges or taxes on vessels passing through the Canal, the only exceptions being in favor of port charges and duties on goods entering Panamanian territory.[49] In December 1958 the National Assembly of Panama adopted a law which extends the territorial sea of Panama from three to twelve miles but requires that the Executive Organ of the Government administer the law consistently with international treaties in force.[50] It remains to be seen whether this law, the adoption of which was immediately protested by the United States,[51] will have the effect of establishing a strip of Panamanian territorial sea nine miles in width outside

[48] Treaty cited *supra* note 1, art. XVIII.

[49] Article IX; superseded, but without change in the basic principle set forth in the text, by Treaty of 1936, *supra* note 10, art. V.

[50] Law No. 58, Dec. 18, 1958, Republic of Panama, Gaceta Oficial, No. 13,720 at 1 (1958). Law No. 9, Jan. 30, 1956, *id*. No. 12,939 at 1 (1956), had declared the Gulf of Panama to be a historic bay, subject to Panamanian sovereignty.

[51] United States note of Jan. 9, 1959, 40 Dep't State Bull. 127 (1959).

the three-mile zone of territorial sea granted to the United States at either end of the Canal and whether Panamanian jurisdiction will be exercised in this marginal sea. Were that area to become one in which Panama claimed the right to exercise control over vessels exiting from or about to enter the Canal, the law would be tantamount to a declaration by Panama that it intended to have a hand in determining the conditions under which vessels might be allowed to pass through the Canal. The reference to treaties in force which is made in the decree affords ground for confidence that the law will be interpreted in a manner consistent with the obligations of Panama to maintain freedom of transit through the Canal in accordance with the terms of the Hay-Pauncefote Treaty of 1901.

The operation and navigation of the Panama Canal and those adjacent waters subject to the jurisdiction of the United States are currently regulated exclusively by the United States. The President has exercised the authority delegated to him by the Congress[52] by promulgating regulations dealing with such matters as the arrival and departure of vessels, the control of ships while in transit, navigation within the Canal, the transportation of dangerous cargoes, radio communications, the protection of the Canal and the maintenance of its neutrality,[53] measurement of vessels and the collection of tolls,[54] and sanitation.[55] Because of the military significance of the Canal, the entire Canal Zone has been set apart as a "military airspace reservation." Aircraft entering this area are required to comply with instructions issued by the Governor and other competent authorities of the United States.[56] The control of the airspace over the Canal Zone by the United States has been regarded by the Panamanian Government, grounding its objections on the contentions previously described, as inconsistent with the terms upon which the United States occupies the Zone.[57] The exclusion and control of aircraft is considered not to be necessary to the "construction, maintenance,

[52] Act of Aug. 24, 1912, § 5, 37 Stat. 562, now 2 C.Z.C. § 1331 (1962).

[53] 35 C.F.R., ch. I, pt. 4.

[54] 35 C.F.R., ch. I, pt. 27.

[55] 35 C.F.R., ch. I, pt. 24.

[56] 35 C.F.R. §§ 5.2, 5.11, 5.21, 5.107, 5.201; see Hughes, *Airspace Sovereignty over Certain International Waterways*, 19 J. Air L. & Com. 144, 147-49 (1952).

[57] Wright, *supra* note 32, at 217. In Re Cia. de Transportes de Gelabert, Feb. 22, 1939, Annual Digest and Reports of Public International Law Cases, 1938-1940, Case No. 45, the Supreme Court of Panama held that the courts of Panama had jurisdiction over litigation concerning the crash of a Panamanian aircraft in the Canal Zone. The acts giving rise to liability were considered to have taken place in the airspace over the Canal, an area not subject to the jurisdiction of the United States.

operation, sanitation and protection" of the Canal, as those words are used in the Hay–Bunau-Varilla Treaty.

The restraints which Panama has been successful in imposing upon the authority of the United States within the Zone relate in the main to economic and fiscal matters. In respect to the civil government of the area, the United States has retained that great freedom of action which was contemplated by article III of the Hay–Bunau-Varilla Treaty of 1903. Congress has enacted for the Zone a separate body of law which has been codified as the Canal Zone Code.[58] In addition to those general provisions of the Code governing the operation, maintenance, and government of the Zone, the Code comprehends civil laws, rules of civil procedure and evidence, a code of criminal law, and those other laws having application to the Canal Zone. The law which has been enacted exclusively for the Canal Zone excludes the operation of corresponding provisions of the general law of the United States, such as the Criminal Code embodied in title 18 of the United States Code.[59] In some instances, special legislation adopted for the Zone makes applicable to that area the existing law of the United States,[60] or a statute indicates that the territorial application of the act includes the Zone.[61] Other laws dealing with various aspects of governmental activity are considered applicable in the Zone by reason of the fact that the Canal Zone Government, for example, is an arm of the United States Government.[62] But as to a great many other statutes, the situation is unclear, and it becomes a nice question of construction whether the act should be considered to carry any effect in the Zone.[63] So exceptional is the status of the area that the ports of the Canal Zone have been held to be "foreign ports" for the purpose of at least one statute of the United States.[64]

[58] A new Canal Zone Code was enacted by the Congress and approved by the President on October 18, 1962. 76A Stat. 1.

[59] 18 U.S.C. § 5 expressly excludes the Canal Zone from the territorial application of title 18.

[60] E.g., Act of Feb. 16, 1933, § 1, 47 Stat. 812, now 2 C.Z.C. § 1131 (1962), making the laws and regulations governing the postal service of the United States applicable to the postal service of the Canal Zone.

[61] E.g., Act of July 18, 1956, § 201, 70 Stat. 572, as amended, 18 U.S.C. § 1401, relating to narcotics.

[62] E.g., statutes relating to employment by the Federal Government.

[63] Huasteca Petroleum Co. v. United States, 14 F.2d 495 (E.D.N.Y. 1926) (taking of depositions); Panama Agencies Co. v. Franco, 111 F.2d 263 (5th Cir. 1940) (Merchant Marine Act).

[64] Luckenbach Steamship Company v. United States, 280 U.S. 173 (1930) (transportation of mails between United States and "foreign" ports); but cf. Stafford Allen

The most recent legislation enacted by the Congress for the governance of the Canal Zone and of the Canal itself has separated the functions of operating the Canal and its ancillary facilities from the civil government of the area.[65] The business operation of the Canal and its "supporting operations" rests in the Panama Canal Company, a public corporation of which the United States is the owner.[66] The management of this corporation is vested in a board of directors of between nine and thirteen members, appointed by the Secretary of the Army in his capacity as the stockholder of the corporation.[67] Since laws dealing with the Panama Canal have vested wide authority in the President as regards the administration of the Canal and of the government of the area, he has found it necessary to delegate many of his functions. The individual by whom these duties have been carried out has always been the Secretary of the Army and his predecessor, the Secretary of War.[68] The military character of the operation extends to the management of the Canal, which is entrusted to a managerial staff headed by the President of the Panama Canal Company, by tradition a general officer of the Corps of Engineers of the United States Army.[69] In view of the close relation between the civil government of the Zone and the operation of the Canal, the President of the Canal Company assumes that office by reason of his appointment as the Governor of the Canal Zone.[70]

The Congress has granted to the President the authority to govern the Canal Zone through a Canal Zone Government, of which the Governor is the principal officer.[71] As in the case of the Panama Canal Company, the President has delegated his supervisory functions with respect to the government of the Zone to the Secretary of the Army. Within the Canal Zone, the Government furnishes normal governmental services. Since private land

---

and Sons Ltd. v. Pacific Steam Navigation Co., [1956] 2 All E.R. 716 (C.A.), holding that a port in the Canal Zone is within a "possession" of the United States, as that term is used in the United States Carriage of Goods by Sea Act.

[65] Act of Sept. 26, 1950, 64 Stat. 1038, codified in scattered sections of 2 C.Z.C. (1962).

[66] Act of June 29, 1948, § 2, 62 Stat. 1076, as amended by Act of Sept. 26, 1950, §§ 5-10, 64 Stat. 1042, now 2 C.Z.C. §§ 61-68 (1962).

[67] 2 C.Z.C. § 63 (1962).

[68] Exec. Order No. 9746, July 1, 1946, 11 Fed. Reg. 7329 (1946), as amended by Exec. Order No. 10101, Jan. 31, 1950, 15 Fed. Reg. 595 (1950); see also PANAMA CANAL COMPANY, BYLAWS, art. III (1955 ed.).

[69] PADELFORD, *op. cit. supra* note 46, at 197-98.

[70] 2 C.Z.C. § 64 (1962).

[71] 2 C.Z.C. § 31 (1962).

titles have been extinguished within the Zone, the entire area is in effect one great government reservation.[72] The contents of the annual report of the Governor resemble those of a report by the mayor of an American city—the incidence of disease, the activities of the police and the firefighting service, revenues, education, hospitals—but with an occasional anachronistic section regarding the postal service, customs, or immigration.[73] The President, acting under both his implied powers and those specifically granted him by Congress, supplements the legislation applicable to the Zone and the Canal with regulations dealing with diverse matters ranging from the issuance of licenses for the sale of alcoholic beverages to the control of aliens.[74]

In the event of war or if war is imminent, the President is authorized by statute to vest full control of the Canal and Zone in an army officer.[75] At present, the delimitation of authority as between the Commander in Chief, Caribbean, who commands the troops stationed in the Canal Zone, and the Governor, himself an army officer, is governed by an executive order[76] which stipulates that in the event of a conflict of views between the two as to whether a matter pertains to the military defense of the Zone, the views of the Commander in Chief, Caribbean, are to prevail. The Governor has a right of appeal to the President, who is the ultimate arbiter of the question.

In summation, the Panama Canal is operated by the government of a nation which has been granted full powers of jurisdiction within the Canal Zone—that area which has been provided by Panama as the territorial base from which the waterway might be constructed, managed, and defended. The rights of the United States stem from the broad terms of the irrevocable grant made by the treaty concluded with Panama in 1903. In terms of political jurisdiction over the territory of the Canal Zone and of the provision of free transit through the Canal, the United States has retained, with only insubstantial diminution, the autonomy granted it by that treaty, an autonomy limited by the treaties and customary law governing the Canal. The control of the Canal and of the area through which it flows has been left by the Congress very largely to the discretion of the President, who exercises his

---

[72] Burdick, *The Panama Canal and the Canal Zone: Their Character, Functions, Government and Laws*, 3 FED. B.J. 89, 91 (1937); see 2 C.Z.C. § 2 (1962).

[73] ANNUAL REPORT OF THE PANAMA CANAL COMPANY AND THE CANAL ZONE GOVERNMENT FOR THE FISCAL YEAR 1961, at 105-27 (1961).

[74] See 35 C.F.R., ch. I, pts. 6 and 10. The implied powers of the President in this respect were upheld in McConaughey v. Morrow, 263 U.S. 39 (1923).

[75] Act of Aug. 24, 1912, § 13, 37 Stat. 569, now 2 C.Z.C. § 34 (1962).

[76] Exec. Order No. 10398, Sept. 26, 1952, 17 Fed. Reg. 8647 (1952).

authority through the Secretary of the Army and the Panama Canal Company. Both tradition and the strategic necessities of the waterway[77] have dictated that there should be a strong military flavor to the administration of the Canal and of the Zone. In the present day, however, the plenitude of sovereign powers theoretically enjoyed by the United States under the "as if" formula of the Treaty of 1903 has been qualified by a large number of arrangements with Panama with respect to economic and fiscal matters. It may be expected that those matters which have been regulated by agreements between the two countries will continue to rest upon this conventional basis and that, if the course of events indicated by the treaties of 1903, 1936, and 1955 may be projected into the future, the renegotiation of these agreements will be sought by Panama in the not too distant future.

Debates about the present "internationalization" of the Canal and "sovereignty"[78] over the Zone hold greater interest for the legal metaphysician than for the lawyer.[79] The legal status of the waterway is not to be determined by reference to these abstract concepts but by a consideration of the relations established among Panama, the United States, and user nations by treaty and by the customary law which state practice has established for this and similar waterways. In so far as freedom of passage is concerned, the law governing the Panama Canal is, as is established elsewhere in this study,[80] a general body of canal law, which establishes rights and duties for the proprietor of the waterway and for the users. So long as the present régime of the Panama Canal is maintained, the relation of the operator of the waterway to the adjacent territorial sovereign is, by contrast, not a question of general international law but of particular international law having application to the Republic of Panama and the United States alone. The

[77] It is widely believed, however, that "the Canal today is indefensible in total war and short of total war is less defensible and less strategic than ever before." Travis and Watkins, *Control of the Panama Canal: An Obsolete Shibboleth?*, 37 FOREIGN AFFAIRS 407, 410 (1959).

[78] A term characterized by Lord McNair as "inclined to raise the blood pressure of the person who uses it." *Treaties and Sovereignty*, in SYMBOLAE VERZIJL 222, 226 (1958).

[79] In this category belongs the tiff between the Foreign Minister of Panama and the United States Secretary of State over statements made by Mr. Dulles about sovereignty over the Panama Canal and the internationalization of that waterway and the Suez Canal. News Conference Statement by Secretary Dulles, Aug. 28, 1956, in THE SUEZ CANAL PROBLEM, JULY 26-SEPTEMBER 22, 1956, at 300 (Dep't of State Pub. 6392) (1956); N.Y. Times, Aug. 30, 1956, p. 1, col. 7.

[80] See Chapter III, especially at p. 185.

present legal position would, needless to say, be subject to radical alteration were effect to be given to proposals for the placing of the Canal and of the Canal Zone under the control or administration of the United Nations or a group of American states.[81]

SECTION C

OPERATION OF INTEROCEANIC CANALS BY THE TERRITORIAL SOVEREIGN
(THE KIEL AND SUEZ CANALS)

The administration by the state of an interoceanic canal lying within its territory and subject to its sovereignty and jurisdiction relieves the operator of the waterway of those pressures which may be exerted by a territorial sovereign on a foreign governmental or private operating agency. If, as in the case of the Kiel Canal, vessels of the state concerned are the principal users and the waterway does not shorten trade routes to the same extent as Suez or Panama, conflicts of interest between proprietor and users are largely withdrawn into the domestic domain. If, on the other hand, the waterway is of major importance, there may be a clash of interests, falling within the area of international concern, between the territorial sovereign who administers the canal and the users.

It is perhaps a consequence of this distinction in the roles of territorial sovereign that the operating agency for the Kiel Canal is not prescribed by any international instrument, unless it be article 386 of the Treaty of Versailles.[1] That article required the establishment at Kiel of a local authority to "deal with disputes in the first instance" in order to avoid the reference of such questions to the League of Nations. The Canal, built and since maintained by the German Government,[2] is administered by the Federal Republic of Germany, acting through the Federal Ministry of Transport. Its actual operation falls to the Wasser- und Schiffahrtsdirektion Kiel, which is also responsible for the other waterways of Schleswig-Holstein. Its financing is

[81] See *Report of Senator George D. Aiken on a Study Mission to the Committee on Foreign Relations United States Senate,* 86th Cong., 2d Sess. 15 (1960).

[1] Treaty of Peace between the Allied and Associated Powers and Germany, signed at Versailles, June 28, 1919, 112 BRITISH AND FOREIGN STATE PAPERS 1 (1919).

[2] THE KIEL CANAL AND HELIGOLAND (Foreign Office, Peace Handbook No. 41) at 1-5 (1920).

conducted on the same basis as other government agencies.[3] Subject to the obligations laid upon the Government of the Federal Republic by the Treaty of Versailles, vessels within the Canal are fully subject to German law.[4] Although there has been occasional talk of placing the Canal under international administration, no action has ever been taken to accomplish this.[5]

Since the nationalization of the Suez Canal, that waterway has been administered by the Suez Canal Authority, established by the Government of the United Arab Republic. In this case, the administering authority for the Canal has received international recognition and would appear to be governed by an international instrument which Egypt, the predecessor state, accepted as creating legal obligations. The Egyptian Declaration of April 24, 1957,[6] which, together with the Convention of Constantinople of 1888, forms the basic law for the Canal, provides that "the Canal will be operated and managed by the autonomous Suez Canal Authority established by the Government of Egypt on 26 July 1956."[7] In other paragraphs of the instru-

[3] Lorenzen, *The Administration of the Kiel Canal,* [1953] Nord-Ostsee-Kanal, No. 2, 20 at 21.

[4] Kiel Canal Collision Case, Germany (British Zone), Supreme Court, June 1, 1950, 4 Entscheidungen des Obersten Gerichtshofes für die Britische Zone 194 (1950), [1950] International Law Reports 133 (Case No. 34). This case is significant on two accounts, the first of which is its holding that litigation arising out of a collision of two non-German vessels in the Canal in 1942 was governed by German law. This conclusion rested on two premises, (a) that "internationalization" of the waterway in the sense of guarantee of freedom of passage through it did not affect its quality as German territory, and (b) that article 380 of the Treaty of Versailles, requiring that foreign vessels be treated on a footing of equality with German ships, demanded that a foreign vessel, like a German one, be governed by German law. The other important aspect of the case is its calling in question the continuing force of the provisions of the Treaty of Versailles regarding the Kiel Canal, by reason of the German denunciation in 1936 of the provisions of the Treaty having to do with waterways; see Reichsgesetzblatt, 1936, II, 361, incorporating a German note of November 14, 1936, in English translation in Heald and Wheeler-Bennett, [1936] Documents on International Affairs 283 (1937), and the absence of effective protest by the other parties to that agreement.

[5] President Roosevelt proposed during discussions of the postwar settlement that the Kiel Canal should be governed by the United Nations. Churchill, Closing the Ring 401 (Boston, 1951).

[6] Annexed to letter to the Secretary-General of the United Nations from the Egyptian Minister for Foreign Affairs, April 24, 1957 (U.N. Doc. A/3576, S/3818) (1957), United States Policy in the Middle East, September 1956-June 1957, at 387 (Dep't of State Pub. 6505) (1957). Paragraph 10 of the Declaration states: "This Declaration, with the obligations therein, constitutes an international instrument and will be deposited and registered with the Secretariat of the United Nations."

[7] Paragraph 4.

ment, there are references to the Authority as the agency responsible for the collection of tolls;[8] to nondiscrimination by the Authority; and to the submission to the Authority of complaints of discrimination or violation of the Canal Code, with a right to submit any unresolved dispute to arbitration.[9] It is stipulated that the regulations governing the Canal are embodied in the Canal Code, and there are stipulations for referring to arbitration complaints that a provision of the Code is inconsistent with the Declaration.[10] Since this instrument appears to create international obligations in the same manner as a multilateral treaty to which Egypt might be a party, the United Arab Republic is, it is submitted, precluded from changing the administration of the waterway in a manner inconsistent with the Declaration. Were the operation of the Canal to be transferred from "the autonomous Suez Canal Authority" to a ministry of the Government of the United Arab Republic, for example, that change in régime would be incompatible with the terms of the instrument.

Under its Statutes,[11] the Suez Canal Authority is an independent juridical person governed by a board of directors appointed by the President of the Republic.[12] Its independence finds expression in the stipulations of the Statutes that its property is to be considered as private property[13] and that it is to have an independent budget.[14] The separate character of the Authority was recognized on the international plane in connection with the loan made to it by the International Bank for Reconstruction and Development in 1959; the loan was made to the Authority and guaranteed by the Government of the United Arab Republic.[15] The Authority is managed like a commercial enter-

---

[8] Paragraph 5(a).

[9] Paragraph 7(a) and (b).

[10] Paragraph 6.

[11] Statuts de l'Organisme du Canal de Suez, annexed to Décret-loi du Président de la République No. 146 de 1957, JOURNAL OFFICIEL DU GOUVERNEMENT ÉGYPTIEN, No. 53*bis* "C," July 13, 1957.

[12] Articles 2 and 3.

[13] Article 12.

[14] Article 8; but without prejudice to the control of the Audit Department over its balance sheet.

[15] Loan Agreement (Suez Canal Development Project) between International Bank for Reconstruction and Development and Suez Canal Authority, dated Dec. 22, 1959 (Loan Number 243 UAR); and Guarantee Agreement (Suez Canal Development Project) between United Arab Republic and International Bank for Reconstruction and Development, signed in the District of Columbia, Dec. 22, 1959 (Loan Number 243 UAR).

prise, without being bound by the regulations and methods of operation prescribed for government departments.[16]

The Authority is required to conform its actions to the Convention of Constantinople of 1888 and to the Egyptian Declaration of 1957.[17] These provisions of the Statutes of the Authority are responsive to the obligation undertaken by Egypt in the Declaration of 1957 that the Authority should be "autonomous"[18] and to Number 3 of the "Six Principles" stipulated by the Security Council in its resolution of October 13, 1956—that "the operation of the Canal should be insulated from the politics of any country."[19]

SECTION D

OPERATION OF AN INTERNATIONAL WATERWAY THROUGH INTERNATIONAL COORDINATION

(THE ST. LAWRENCE SEAWAY)

The opening of the St. Lawrence Seaway on June 26, 1959,[1] was the beginning of a new age in the history of a river which had already been used as an avenue of trade and commerce for over 400 years.[2] A waterway navigable by seagoing vessels now linked the Atlantic Ocean and the Great Lakes. To make possible this 27-foot channel between Montreal and Lake Erie, an extensive program of works, including dredging, diversionary canals, and locks, had been carried to completion by the United States and Canada.

The improvement of the St. Lawrence had been of concern on both sides of the Canadian-United States boundary for well over a century.[3] There was a quickening of governmental interest in the provision of a deep waterway

[16] Article 6.
[17] Article 14.
[18] Declaration cited *supra* note 6, para. 4.
[19] U.N. Doc. No. S/3675 (1956).
[1] 41 DEP'T STATE BULL. 75 (1959); Wylie, *The Freshwater Cruise of USS Macon*, 86 UNITED STATES NAVAL INSTITUTE PROCEEDINGS 61 (1960), is a colorful account of the passage of a United States crusier through the Seaway in connection with the opening ceremonies.
[2] MENEFEE, THE ST. LAWRENCE SEAWAY 4-28 (1940); and see generally WILLOUGHBY, THE ST. LAWRENCE WATERWAY (1961); Baxter, *Documents on the St. Lawrence Seaway: A Selection of Documents,* 9 INT'L & COMP. L.Q., Sp. Supp. (1960).
[3] KEYSER, THE ST. LAWRENCE SEAWAY PROJECT 3-8 (Library of Congress, Legislative Reference Service, Public Affairs Bulletin No. 58) (1947).

after the First World War; but the opposition of the railroads,[4] of states which feared loss of business, and of those who feared the competition of cheap power which would be generated as part of the project was long effective in preventing congressional authorization of the project and the conclusion of arrangements between Canada and the United States. The two countries got to the stage of negotiating in 1932 a treaty which looked to the provision of a 27-foot channel by works to be carried on by the two countries. The International Rapids section was to be under the direction of a St. Lawrence International Rapids Section Commission, an international agency to be created by the two governments.[5] But the Treaty, which would have required the United States to pay the full costs of the project, including those of the works in Canadian territory, failed to receive the consent of the Senate.[6] An executive agreement of 1941, the effectiveness of which was conditioned on the necessary authorizing legislation of the Canadian Parliament and the United States Congress, likewise failed of approval.[7]

The impetus to participation by the United States in the construction of the St. Lawrence Seaway was provided by the action of the Canadian Parliament in enacting legislation authorizing the Canadian St. Lawrence Seaway Authority to construct the Seaway "either wholly in Canada or in conjunction with works undertaken by an appropriate authority in the United States."[8] Had the necessary works been wholly on the Canadian side of the international boundary running through the international section of the St. Lawrence, the complete control of that waterway would have rested with Canada,[9] subject to the obligations imposed by the Boundary Waters Convention of 1909.[10] Although that treaty would have required that the waterway be open on a basis of equality to the "inhabitants and to the ships, vessels, and boats of both countries,"[11] the fact that the majority of the

---

[4] See Association of American Railroads, The St. Lawrence Project (1946).

[5] Great Lakes-St. Lawrence Deep Waterway Treaty, signed at Washington, July 18, 1932, arts. I-III (Dep't of State Pub. 347) (1932).

[6] 78 Cong. Rec. 4475 (1934).

[7] Agreement regarding the Great Lakes-St. Lawrence Waterway Project, signed at Ottawa, March 19, 1941, 4 Dep't State Bull. 307 (1941).

[8] The St. Lawrence Seaway Authority Act, 15 & 16 Geo. 6, c. 24, § 10(a).

[9] Senate Committee on Foreign Relations, *St. Lawrence Seaway Manual,* S. Doc. No. 165, 83d Cong., 2d Sess. 75, 78 (1955).

[10] Treaty between the United States and Great Britain relating to Boundary Waters between the United States and Canada, signed at Washington, Jan. 11, 1909, 36 Stat. 2448, T.S. No. 548.

[11] Article I.

shipping would belong to the United States or be destined for ports of the United States would have meant that Canada could, with little reciprocal detriment to its shipping, control the conditions, including tolls, on which vessels could proceed through the Seaway.[12]

The United States responded with legislation establishing the Saint Lawrence Seaway Development Corporation to construct and operate certain of the works needed on the United States side of the river to provide the desired deep waterway.[13] It should be made clear that the generation of power from the river was not made a part of this project in a formal sense and that, although construction both for navigational purposes and for the generation of power proceeded concurrently, the one was dealt with upon the federal level, while the other was a cooperative project of the state and provincial authorities.[14]

The institutions through which the navigational works were constructed and are presently operated are of an exceptional character, which, it is believed, are without counterpart in the administration for navigational purposes of other international waterways. Basically, the Seaway is operated by two national agencies, whose activities have been coordinated by a series of international agreements and less formal arrangements. The works on the Canadian side of the river were constructed by the St. Lawrence Seaway Authority; on the United States side, by the Saint Lawrence Seaway Development Corporation. The United States legislation authorized the construction of the "works solely for navigation" set forth in the joint report of 1941 of the Canadian Temporary Great Lakes-Saint Lawrence Basin Committee and the United States Saint Lawrence Advisory Committee on the condition that assurances were received from Canada that it would construct the works on the Canadian side.[15] That undertaking was secured in an exchange of notes between the two countries in 1954.[16] Since the Boundary Waters Convention of 1909 provides that no submission to the International Joint Commission is required in the case of improvements to navigation solely on one nation's side of the river and not affecting the level or flow of the river,[17] the Seaway

[12] *Op. cit. supra* note 9, at 75, 78.

[13] St. Lawrence Seaway Act, 68 Stat. 92, 33 U.S.C. §§ 981-990.

[14] H.R. Rep. No. 1215, 83d Cong., 2d Sess. 32-33 (1954).

[15] St. Lawrence Seaway Act, § 3(a), 68 Stat. 93, 33 U.S.C. § 983(a).

[16] Agreements between the United States of America and Canada regarding the Saint Lawrence Seaway, signed at Ottawa, Aug. 17, 1954, and at Washington, June 30, 1952, 5 U.S.T. 1784, T.I.A.S. No. 3053.

[17] Treaty of 1909, *supra* note 10, art. III, 2d para.

plans did not have to be approved by the Commission. However, the facilities needed for the generation of hydroelectric power did require such approval, and their relation to the navigational project was outlined in the two national submissions to the Commission[18] and in the order of the Commission approving the power project.[19]

The Saint Lawrence Seaway Development Corporation is a public corporation, managed by an Administrator appointed by the President.[20] During the construction phase of the project, the Corporation operated under the direction and supervision of the Secretary of Defense.[21] Upon the completion of construction, that function passed to the Secretary of Commerce, whose competence includes the general policies of the Corporation, the operation and maintenance of the Seaway, the provision of necessary services and facilities, and the fixing of tolls and rules for the measurement of vessels and cargoes.[22] The Corporation employed the Army Corps of Engineers as its design, contracting, and construction agent[23] and worked out arrangements with other Government departments for the provision of supplies and services.[24] The actual work of construction was carried out by private contractors.

The corresponding Canadian institution, The St. Lawrence Seaway Authority, consisting of three members, has authority to construct and maintain the necessary works, to fix tolls, to expropriate property, and to establish regulations for the navigation of vessels employing the facilities of the Seaway.[25]

Since substantial portions of the work in the International Rapids section of the river were carried on wholly within territory subject to the

---

[18] *Op. cit. supra* note 9, 75 at 78.

[19] Order of Approval of Power Works by the International Joint Commission, Nov. 19, 1952, 27 Dep't State Bull. 1019 (1952); see Kunen, *International Negotiations concerning the St. Lawrence Project,* 33 Detroit L.J. 14 (1955). The Saint Lawrence River Joint Board of Engineers has been established to review and coordinate plans for the power project. Agreement between the United States of America and Canada regarding the Estabishment of Saint Lawrence River Joint Board of Engineers, signed at Washington, Nov. 12, 1953, 5 U.S.T. 2538, T.I.A.S. No. 3116.

[20] St. Lawrence Seaway Act, §§ 1 and 2(a), 68 Stat. 93, 33 U.S.C. §§ 981, 982(a).

[21] Exec. Order No. 10534, June 9, 1954, 19 Fed. Reg. 3413 (1954).

[22] Exec. Order No. 10771, June 20, 1958, 23 Fed. Reg. 4525 (1958).

[23] Saint Lawrence Seaway Development Corporation, Annual Report, 1955, at 3 (1956).

[24] E.g., with the Coast Guard for navigational facilities. Saint Lawrence Seaway Development Corporation, Annual Report, 1956, at 16 (1957).

[25] The St. Lawrence Seaway Authority Act, 15 & 16 Geo. 6, c. 24, §§ 10, 15, 18, 19.

jurisdiction of one of the two cooperating nations, detailed international arrangements were not required as to this construction. In those places, such as the Cornwall Island area, where the works overlapped the international boundary, it was necessary to allocate responsibilities between the two authorities through coordination on the technical level.[26]

Since vessels traveling the length of the Seaway will pass through both Canadian and United States territory, the two countries have agreed upon uniform rules of navigation.[27] A dispute about the sharing of work between Canadian and United States pilots was resolved by the conclusion, two years after the opening of the Seaway, of an agreement on participation in pilotage services and coordination of pilotage pools.[28]

The level of tolls, arrangements for their collection, and their allocation between the two nations have similarly been the subject of international agreement.[29] On the United States side, the St. Lawrence Seaway Act calls for the holding of hearings on proposed tolls before they are established.[30] If it had proved impossible for the two countries to agree on tolls, the United States Corporation would have been authorized to collect separate tolls on the facilities which it had created.[31] Under the arrangements in force

[26] Report cited *supra* n. 24, at 15. This has not, however, precluded the duplication of facilities at some points. See Exchange of Notes between the United States and Canada regarding the Saint Lawrence Seaway: Deep-Water Dredging in Cornwall Island Channels, signed at Ottawa, Nov. 7/Dec. 4, 1956, 7 U.S.T. 3271, T.I.A.S. No. 3708.

[27] The Seaway Regulations and Rules for 1962 are reproduced in 27 Fed. Reg. 2243, 2895 (1962), 33 C.F.R., pt. 401.

[28] Agreement between the United States of America and Canada concerning Pilotage Services on the Great Lakes and the St. Lawrence River, signed at Washington, May 5, 1961, 12 U.S.T. 1033, T.I.A.S. No. 4806; and see p. 296 *infra*.

[29] Exchange of Notes between the United States of America and Canada regarding the Saint Lawrence Seaway Tariff of Tolls, signed at Ottawa, March 9, 1959, 10 U.S.T. 323, T.I.A.S. No. 4192. When Canada proposed to suspend tolls on the Welland Canal, revenues from which accrued solely to Canada, the two countries felt impelled to conclude an agreement on this subject, because tolls on the Canal had been mentioned in the exchange of notes of 1959. Exchange of Notes between the United States of America and Canada regarding Suspension of Tolls on the Welland Canal, signed at Ottawa, July 3/13, 1962, 47 DEP'T STATE BULL. 255 (1962).

Toll committees established within the two national administrations had been responsible for the recommendations. SAINT LAWRENCE SEAWAY DEVELOPMENT CORPORATION, ANNUAL REPORT, 1956, at 24 (1957); Report of United States Tolls Committee, June 12, 1958, annexed to St. Lawrence Seaway Development Corporation Press Release No. SLSDC 141, June 18, 1958.

[30] Section 12(a), 68 Stat. 96, 33 U.S.C. § 988 (a).

[31] *Ibid.*

The St. Lawrence Seaway Authority of Canada acts as collection agent for the tolls, which are then apportioned between the two public authorities.[32] Persons who consider themselves to have been the victims of an improper interpretation of the Tariff of Tolls or of unjust discrimination may appeal to the Joint Tolls Advisory Board, on which the two corporations are equally represented, but the power of the Board is limited to the framing of findings and recommendations to be submitted to the Authority and the Corporation.[33]

The administration of an international waterway, partly artificial in nature, through two national agencies[34] is possible only because of the peculiar geography of the St. Lawrence River and of the international boundary. The development of this organizational framework has been stimulated by the hard-fought battle, on both the domestic and international fronts, which preceded the present arrangements and by the fact that the waterway has never, as a matter of law, been opened as of right to the vessels of all nations.[35] It is unlikely that this institutional framework will lend itself to imitation by other countries desirous of regulating an international waterway of mutual concern.

SECTION E

INTERNATIONAL ADMINISTRATION THROUGH INTERNATIONAL COMMISSIONS

International river commissions and similar international agencies for the administration of other types of waterways offer abundant material for investigation and analysis of the effectiveness of the administration of international waterways by groups of states. The Rhine, Danube, Scheldt, Oder, Elbe, Po, and Pruth have been administered by international bodies composed of littoral and user states and have served international commerce and

---

[32] Memorandum of Agreement, para. 5, annexed to first exchange of notes cited *supra* note 29.

[33] Memorandum of Agreement, para. 6, annexed to first exchange of notes cited *supra* note 29.

[34] See ADAM, LES ETABLISSEMENTS PUBLICS INTERNATIONAUX 41-42 (1957), characterizing these as "établissements nationaux d'intérêt international."

[35] Ports opened to foreign vessels by United States commercial treaties are considered to include those situate along the St. Lawrence Seaway. Piper, *Navigation Provisions in United States Commercial Treaties,* 11 AM. J. COMP. L. 184, 193 (1962).

transport under laws, rules, and regulations supervised or enforced by the operating agency. Through these institutions, it has frequently been possible to adjust conflicts of interest amongst the littoral states and between these nations and the nonriparian users. The international commissions have been particularly successful in the technical, operational, and functional realms.

The magic word "internationalization" has, on the other hand, often been used as a cover by states desiring to extend their control over a waterway for reasons of politics, strategy, or power. The international community was not deluded by the pleas for the internationalization of the Suez Canal which were made by Italy during and after the Ethiopian campaign. The effective domination since the Second World War of much of the navigable length of the Danube by Soviet Russia has meant that its "internationalization" has been very largely a Soviet one.[1] The liberalization of the regulation of that river which has taken place in recent years has come as a consequence of the eagerness of the Soviet Union and the Eastern Bloc riparians to increase the number of vessels available for the transportation of the goods of those countries.[2]

Waterways operated or supervised by international administrations cross or border on the territory of several states. It is only natural that navigation, transportation and transit on such waterways, their maintenance and improvement, and the rules and regulations governing traffic, auxiliary facilities, and obstructions to navigation should become subjects of concerted attention and direct collaboration by those states dependent on or interested in the waterway. This functional interdependence is determined both by the geographic location of the waterway and by the economic interests of the users.

The first international administration, which was established in order to deal with navigation on the River Rhine, had actually been preceded by a number of treaties granting freedom of navigation to the riparians of various rivers. However, the pre-nineteenth century agreements concerning freedom of navigation by the littoral states are of limited importance.[3] The riparian

---

[1] In the period following the Second World War, "of the Danube's total navigable length of 2,380 kilometers, 2,115 or 90 percent were in Soviet-occupied territory." PRIGRADA, INTERNATIONAL AGREEMENTS CONCERNING THE DANUBE 14 (1953); see Gorove, *Internationalization of the Danube: A Lesson in History,* 8 J. PUB. L. 125 (1959); Stolte, *Moscow Regulates Traffic on the Danube,* in 7 INSTITUTE FOR THE STUDY OF THE USSR, BULLETIN, No. 5, at 21 (1960).

[2] Stolte, *supra* note 1, at 28-29; *Danube Blues,* 187 THE ECONOMIST 1199 (1958).

[3] The first agreement concerning mutually guaranteed freedom on navigable water-

countries, despite their formal pronouncements of good will and solidarity with their neighbors, did not intend to give up even a part of the important source of revenue which the waterways represented. An exception was the resolution of the French Convention of 1792 which was formally proclaimed and implemented in practice on two waterways, the Scheldt and the Meuse.[4]

The present-day Central Commission for the Navigation of the Rhine, formally instituted by the Final Act of the Congress of Vienna,[5] was the successor to the first international waterway administration. Established in 1804,[6] L'Administration Général de l'Octroi de Navigation du Rhin had primarily fiscal duties, including the collection of passage tolls on vessels and tolls on cargoes, and responsibilities for police and general control.[7] It was even empowered to make provisional rules and regulations, to be referred promptly to the parent governments for approval.[8] The responsibility for the maintenance and improvement of navigation was divided among the littoral governments or left to the local communities. Only the maintenance of the

ways was contained in article 10 of the Treaty of Peace between Austria and the Ottoman Empire, signed at Vienna, May 1, 1616, which applied to the lower Danube. 1 NORADOUNGHIAN, RECUEIL D'ACTES INTERNATIONAUX DE L'EMPIRE OTTOMAN 113, 117 (1897). Freedom of navigation, prohibition of the holding of vessels and cargoes, and cancellation of all duties, tolls, and customs were formally declared by the signatories to the Treaty of Peace between France and the Holy Roman Empire, signed at Munster, Oct. 24, 1648, so that "pristina securitas, jurisdictio et usus prout ante hos motus bellicos a pluribus retro annis fuit." 6 DUMONT, CORPS UNIVERSEL DIPLOMATIQUE DU DROIT DES GENS, pt. I, 455 (1728).

[4] It guaranteed navigation to all riparians and forbade restrictions on navigation. Le Moniteur universel (Paris), 1792, No. 327, at 1387-88.

[5] Articles concerning the Navigation of the Rhine, annexed to the Final Act of the Congress of Vienna, signed June 9, 1815, 2 MARTENS, N.R. 416, 419 (1818).

[6] Convention between France and Germany on the Octroi of Navigation of the Rhine, signed at Paris, Aug. 15, 1804, 8 MARTENS, RECUEIL DES TRAITÉS 261 (1835). The Netherlands and Switzerland were not signatories to the Convention. The German princes who were parties to the instrument were to some degree subject to the imperial authority.

[7] Convention, *supra* note 6, arts. XLII *et seq.* The director-general was appointed jointly, while two of the four inspectors were appointed by each state; the toll officers on the French bank were assigned by the French Government, and those on the German bank by the German Government. Appeals from the decisions of the toll officials, under the Convention, were brought before a board consisting of the director-general and two inspectors and decided by a majority. An appeal from this body could be made to a commission convened annually at Mayence. This was a purely judicial organ composed of the local French prefect, a commissioner appointed by the Elector Arch-Chancellor, and a legal adviser chosen by the other two members.

[8] Article CXXX.

towpath was financed through the revenue of the Octroi.[9] It is true that the agreement applied only to one section of the Rhine, that if the German principalities were to be lumped together, only two riparian countries were involved, and that a number of matters were not dealt with at all in the treaty. However, the new organization did place restrictions on unilateral governmental control and was in fact a permanent agency.

The Treaty of Paris,[10] while establishing certain basic principles, contained no provisions concerning international administration of the Rhine which would have the effect of preventing future conflicts of interest.[11] It became, however, the frame of reference for the discussion of European international waterways by a select international committee at the Congress of Vienna in 1815.[12] The Committee of Navigation and the Congress itself decided to retain a modified version of the Rhine administration of 1804 but changed its organization and lessened its powers considerably in creating the new Central Commission of the Rhine.[13] All riparian countries gained representation on the Commission with equal voting power. In theory, the decisions of the Commission were to be reached by an absolute majority.[14] In practice, unanimity was the rule,[15] and states which were members of the Commission were not bound by the Commission's vote if they had not consented to the decision taken.[16] The Commission was to be an advisory, cooperative organiza-

---

[9] Article XXXIV. In the Treaty between France and the Rhine Princes, signed at Paris, April 29, 1813, a mixed commission of engineers was established to study and approve all plans for improvements on the Rhine. French control of the commission was assured by its appointment of seven out of the twelve members. The duties of the commission were advisory only. 1 RHEINURKUNDEN 35 (1918).

[10] Treaty of Peace between France and Austria and its Allies, signed at Paris, May 30, 1814, 2 MARTENS, N.R. 1 (1818).

[11] Article 5 of the Treaty of Paris contained provisions bearing on freedom of navigation on the Rhine, equitable regulation of tolls on that river, and application of these principles to other navigable rivers which separated or traversed different states.

[12] The *procès verbaux* of the Committee of Navigation are reported in 3 KLÜBER, ACTEN DES WIENER CONGRESSES 11-275 (1815); see also VAN EYSINGA, LA COMMISSION CENTRALE POUR LA NAVIGATION DU RHIN 11-22 (1935); KAECKENBEECK, INTERNATIONAL RIVERS 40-71 (1918); CHAMBERLAIN, THE REGIME OF THE INTERNATIONAL RIVERS: DANUBE AND RHINE 175-200 (1923).

[13] Articles concerning the Navigation of the Rhine, *supra* note 5, arts. X *et seq.*

[14] Article XVII.

[15] However, the Commission did act by majority vote in its capacity as a court of appeals and probably in connection with minor administrative matters, such as allowances and pensions. VAN EYSINGA, *op. cit. supra* note 12, at 50.

[16] Article XVII expressly stipulated that "ses décisions ne seront obligatoires pour les états riverains que lorsqu'ils auront consenti par leur Commissaire."

tion of the littoral states with general administrative powers, final judicial authority in deciding appeals, and general supervision of the river. Between sessions of the Central Commission, authority on behalf of the Commission was to be exercised by the river inspectors, whose duty it was to see to the execution of the regulations, and to organize the police of navigation.[17] Thus, the fluvial community of interest of the old Octroi was to be kept alive, based less on autonomous decisions of an administrative organ than on the actual necessity of real cooperation among the littoral states. Decentralization, brought about by the Congress of Vienna, meant the end of the centralized administration of the Octroi. The Commission, meeting annually,[18] was not, in the language of the twentieth century, an international or regional organization but a diplomatic conference of the riparians,[19] finding the basis for its operations in reciprocity of interests.[20]

Not until 1831 did the Central Commission take up its new functions when the required *règlement de navigation*[21] was adopted in the Treaty of Mayence.[22] According to this agreement, and in keeping with the Articles concerning the Navigation of the Rhine adopted at Vienna, freedom of navigation, the right to ship goods on the river, was the only recognized common ground for cooperation among the riparian states. The Treaty of

[17] Article XV. In addition, articles XII, XIII, XIV, XV, and XVIII governed the rank, appointment, pay, pensions, and functions of the chief inspector and the inspectors.

[18] Article XI.

[19] As a matter of fact, it was impossible at the time for observers to regard the Central Commission as other than a small diplomatic corps. VAN EYSINGA, *op. cit. supra* note 12, at 46.

[20] Biays, *La Commission centrale du Rhin,* 56 REVUE GÉNÉRALE DE DROIT INTERNATIONAL PUBLIC 223, 227-28 (1952).

[21] Articles concerning the Navigation of the Rhine, *supra* note 5, art. XXXII. This article stipulated that the Commission would first draw up a definitive and detailed *règlement de navigation* according to the principles adopted by the Congress, to be approved by the riparian states. The long delay was caused by the controversy between the Netherlands and Prussia over the interpretation of the Articles adopted at Vienna. In this dispute, which related to the words "jusqu'à la mer" employed in describing the principle of freedom of navigation established at Vienna (art. I), Holland maintained that the rivers connecting the Rhine with the sea, other than the Old Rhine, were not subject to the Articles and that the crucial words meant free navigation as far as the point to which the sea tides flowed, but not into the sea. This latter contention allowed the Dutch a free hand in taxing goods passing through the maritime ports of Rotterdam and Amsterdam. For details of these difficulties, see KAECKENBEECK, *op. cit. supra* note 12, at 62-63; CHAMBERLAIN, *op. cit. supra* note 12, at 190-98; VAN EYSINGA, *op. cit. supra* note 12, at 25-36.

[22] Convention of Mayence, signed March 31, 1831, 9 MARTENS, N.R. 252 (1833), I RHEINURKUNDEN 212 (1918).

Mayence dealt only with an international administration of fluvial law; it included provisions for a judicial authority with a general jurisdiction over the river[23] and for a chief inspector appointed by the Commission to watch over the execution of the Treaty.[24] But beyond this there was no common administrative authority, no common organ for the collection of tolls, no common revenue source, no common maintenance and improvement service, no common agency watching out for such impediments to navigation as mills, bridges, and other constructions interfering with traffic on the river. The sovereign riparians were left in charge of all these matters. Freedom of navigation and equality of treatment were secured to the riparians; the rest was left to the discretion of the littoral states.

Chronologically, the second river commission to be established by the riparian states was that on the River Elbe,[25] created by the Elbe Act of 1821.[26] Article XXX of this Act established the Revision Commission, composed of delegates of all the riparians, meeting "from time to time" and reaching decisions by a majority vote. The purpose and responsibility of the Commission were, like those of the Rhine Commission, to safeguard the observance of the Act, to become a center of cooperation in matters concerning navigation, to remove obstacles to traffic, to propose new measures, and eventually to revise tariffs.[27] The Revision Commission met only in 1824, 1842-1844, 1850-1854, 1858, 1861-1863, and 1870.[28] Without any delegated administrative authority, the Commission remained no more than a conference of member states, and its responsibilities did not increase in the

[23] Title VIII, arts. LXXXI-LXXXVIII.

[24] Title IX, art. XCVIII. The chief inspector's duties were to safeguard freedom of navigation, to help to settle complaints, and to investigate irregularities and interference with traffic. His job was to persuade, not to order, and he reported to the Central Commission and to the local governments concerned.

[25] The Treaty between Prussia and Saxony, signed at Vienna, May 18, 1815 (and annexed to the Final Act of the Congress of Vienna), provided in article XVII: "Les principes généraux qui ont été adoptés au Congrès de Vienne pour la libre navigation sur les fleuves, serviront de norme à la Commission établie en vertu de l'Article XIV pour régler sans délai tout ce qui est relatif à la navigation, et sont particulièrement appliqués à celle sur l'Elbe . . ." 2 MARTENS, N.R. 258, 267 (1818). The provisions of article XVII were partially incorporated in a convention of 1819 concluded by the representatives of the two governments. Principal Convention between Prussia and Saxony in Execution of the Treaty of Peace Concluded at Vienna, May 18, 1815, signed at Dresden, Aug. 28, 1819, art. XXXIV, 5 MARTENS, N.R. Supp. 117 (1829).

[26] Act for the Free Navigation of the Elbe, signed at Dresden, June 23, 1821, 5 MARTENS, N.R. 714 (1824).

[27] Article XXX of the Elbe Act.

[28] DÜERKOP, DIE INTERNATIONALISIERUNG DER ELBE 21 (1931).

period preceding the First World War. The Revision Commission remained, needless to say, considerably less influential than the Central Commission for the Navigation of the Rhine. The Commission merely drafted rules of fluvial law, for the adoption of which the consent of all of the fluvial nations was necessary.[29]

Similarly, a mixed commission for navigation on the Scheldt was provided for in the Treaty of London of 1839.[30] It was to be composed of delegates of both governments and to be charged with supervision of navigation and of pilotage and with the drawing up of rules and regulations for traffic on the river and on its tributaries.[31] When the regulations concerning navigation were signed by both governments four years later,[32] the organization for the "common supervision" of the river[33] was elaborated in detail. The members of the commission were to meet at least once every three months, alternately in Antwerp and Flushing. Their duty in these two chief pilot stations was to inspect the state of navigation facilities (buoys, beacons, pilot service, channels, and the like).[34] The commissioners were also to inspect the river beyond the channels of the Scheldt and its mouth whenever they deemed it proper and to observe any changes which might have occurred in the river bed and in the position of buoys and beacons.[35] However, the commissioners were not permitted to change, replace, or correct the position of any facilities directly, except in cases of emergency.[36] Under normal con-

---

[29] The other provisions of the Elbe Act did not reflect any significant departures from the Articles drafted at the Congress of Vienna.

[30] Treaty between Belgium and Holland respecting the Separation of their Respective Territories, signed at London, April 19, 1839, 16 Martens, N.R. 773 (1841). This treaty is sometimes referred to as the "Charter of the Scheldt."

[31] Article 9. Article 9, paragraph 6, authorized the Commission also to regulate fishing rights and commercial fisheries for subjects of both governments "on the basis of absolute reciprocity."

[32] Only after many details left unresolved by the Treaty of London were settled by special conventions, the most significant of which was the Treaty between Belgium and The Netherlands concerning Limits and Navigation of Interior Waters, signed at The Hague, Nov. 5, 1842, 3 Martens, N.R.G. 613 (1845), 31 British and Foreign State Papers 815 (1842-43).

[33] Articles 67-73 of the Rules established between Belgium and The Netherlands for the Execution of the Dispositions of Articles 9 and 10 of the Treaty of April 19, 1839, and of Chapter II, Sections 1, 2, 3, and 4 of the Treaty of Nov. 5, 1842, signed at Antwerp, May 20, 1843, 5 Martens, N.R.G. 294, 326 (1847).

[34] Article 68.

[35] Article 69.

[36] That is, when the commissioners of both states could agree on the urgency of the situation (art. 69).

ditions, only the governments dealt with these matters. The "common super-vision" of the Scheldt, although entirely technical and expert in nature and with no administrative functions, did not work well in practice. There was a series of disagreements between the two governments with regard to main-tenance and improvement of the river, the effect of the outbreak of war, distribution of the financial burden, and the navigation dues,[37] with the result that the work of the mixed commission was carried on under great difficulties.

The River Po Commission was in existence for less than twenty years and therefore is only of historical interest.[38] Two out of its four members, including the chairman, were appointed by Austria; the Commission made its decisions by a "majority of votes." Its duty was to supervise the execution of the convention as it pertained to navigation, to direct the necessary works, and to serve as the means of communication among the signatory states.[39] The Commission met twice a year to inspect the river, to determine the necessity of improvement, to deal with the removal of obstacles, and to supervise the officials collecting the navigation dues. It then reported to the respective governments.[40] Tolls were paid into the treasury of the Com-mission directly; the proceeds were used to cover all expenses connected with navigation on the Po.[41] In 1866, the Po passed entirely within Italian juris-diction, and the Commission was dissolved.[42]

The precedents of the Treaty of Paris of 1814 for the Rhine and the Treaty of London for the Scheldt were followed in the Treaty of Paris of 1856,[43] proclaiming the lower Danube open to the vessels of all nations. The

[37] See KAECKENBEECK, *op. cit. supra* note 12, at 81-82; VOŠTA, MEZINÁRODNÍ ŘEKY 160-67 (1938).

[38] Between 1849 and 1866. See Convention between Austria and the Duchies of Modena and Parma for Free Navigation on the Po, signed at Milan, July 3, 1849, 14 MARTENS, N.R.G. 525 (1856). In 1850, the Holy See acceded to the Convention. Act of Accession, signed at Portici, Feb. 12, 1850, 14 MARTENS, N.R.G. 532 (1856), 38 BRITISH AND FOREIGN STATE PAPERS 136 (1862).

[39] Article V.

[40] Article VI.

[41] Article X.

[42] The Po, however, remained open to foreign vessels. KAECKENBEECK, *op. cit. supra* note 12, at 200.

[43] General Treaty of Peace between Austria, France, Great Britain, Prussia, Russia, Sardinia, and the Ottoman Porte, signed at Paris, March 30, 1856, 15 MARTENS, N.R.G. 770 (1857), 46 BRITISH AND FOREIGN STATE PAPERS 8 (1865).

For the preliminary discussions see Protocols of the Conferences Held at Vienna between the Plenipotentiaries of Austria, France, Great Britain, Russia, and the Otto-

Treaty provided for the establishment of two river commissions, the European Commission of the Danube and the Riparian Commission. The European Commission, composed of representatives of Austria, France, Great Britain, Prussia, Russia, Sardinia, and Turkey—thus including both riparian and nonriparian states—was charged with putting the Danube "in the best possible conditions of navigability" between Isatcha and the mouths[44] in the Black Sea. To that end, the Commission was empowered to fix the tolls.[45] After having accomplished its duty, the Commission, intended as a temporary technical agency, was to be dissolved.[46] The Riparian Commission, composed of delegates of Austria, Bavaria, Turkey, and Wurttemberg, and of commissioners from the three Danubian principalities, was intended as a permanent body to implement the treaty provisions. Its functions were to include preparation of rules of navigation and of fluvial police, the removal of impediments to navigation, the carrying out of improvements throughout the whole course of the river, and the assumption of the duties of the European Commission after it had been dissolved.[47] However, the regulations elaborated in the Act of Navigation of the Danube by the Riparian Commission[48] proved highly controversial, primarily because of clear cases of discrimination in favor of littoral states,[49] and they were not ratified. Consequently,

---

man Porte, Protocols Nos. 4 (March 21, 1855) and 5 (March 23, 1855), 15 MARTENS, N.R.G. 633, 646, 651 (1857).

[44] Article XVI of the Treaty of Paris, *supra* note 43.

[45] By majority vote. Article XVI.

[46] Article XVIII.

[47] Article XVII.

[48] Act of Navigation of the Danube, concluded between Austria, Bavaria, the Ottoman Porte, and Wurttemberg, signed at Vienna, Nov. 7, 1857, 16 MARTENS, N.R.G. pt. 2, at 75 (1858).

[49] Compare with article I of the Elbe Act, cited *supra* note 26, article 8 of the Danube Act, *supra* note 48, reserving the right of navigation between Danube ports to the littoral states to the entire exclusion of nonriparian users. This provision was further fortified by provisions of the Danube Act concerning licensing and monopolies (arts. 11-18). The tributaries of the Danube were not covered by the Act. And finally, article 34 reserved to the riparians the right to adopt more detailed rules of navigation and of fluvial police. Shortly thereafter, the riparian states re-examined the Act and modified some of its discriminatory provisions. All vessels coming from or going to the high seas were to be permitted to transport passengers and merchandise from any port of the Danube to any other port in the direction of their voyage (art. 1). The system established by the Final Act of the Congress of Vienna was to apply to the navigable tributaries of the Danube (art. 6). Additional Articles to the Act of Navigation for the Danube of Nov. 7, 1857, signed at Vienna, March 1, 1859, STURDZA, RECUEIL DE DOCUMENTS RELATIFS À LA LIBERTÉ DE NAVIGATION DU DANUBE 78 (1904).

each riparian state drew up its own rules of navigation; in addition, the unfortunate Act brought an end to the Riparian Commission for all practical purposes.[50]

The European Commission, on the other hand, soon ceased to be considered temporary and merely technical. The life of the Commission was extended;[51] it was endowed by the signatory states with a statute setting forth its rights and duties;[52] and it was empowered to issue navigation and police regulations.[53] The Public Act of 1865 provided: "The exercise of the navigation on the Lower Danube is placed under the authority and the superintendence of the Inspector-General of the Lower Danube, and of the Captain of the port of Sulina," under the general "superintendence" of the Commission.[54] Warships stationed by the signatory states at the river mouth were placed at the disposal of the Commission to enforce the authority of its agents.[55] In addition to the engineering works which the Commission carried on in the river, it possessed responsibility for lighthouses and other navigational facilities at the mouth of the river. Through its licensing power, it controlled tugs, pilotage, and lighterage. It administered the port of Sulina in so far as shipping and the handling of cargo were concerned.[56] From a temporary and purely technical agency which was anticipated to be of limited importance had developed an effective and vigorous river administration. Endowed as it was with its own budget,[57] ships, flag, agents, and personnel, the

[50] After drawing up the proposed Act of Navigation, *supra* note 48, the Commission did not meet for a prolonged period of time, and at the Conference of London in 1871, its convocation was indefinitely postponed.

[51] The Second World War brought its downfall.

[52] Public Act of the European Commission of the Danube, relative to the Navigation of the Mouths of the Danube, signed at Galatz, Nov. 2, 1865, 18 MARTENS, N.R.G. 144 (1873), 55 BRITISH AND FOREIGN STATE PAPERS 93 (1864-65).

[53] Which were ". . . binding as law, not only in what concerns the River Police, but also for the judgment of cases of civil procedure arising from the exercise of the navigation" (art. VII). The Rules of Navigation and of Police applicable to the Lower Danube were annexed to the Public Act. The European Commission had already, in the previous year, supplanted the various national regulations with Rules of Navigation and of Police applicable to the Lower Danube, signed at Galatz, Nov. 21, 1864, 18 MARTENS, N.R.G. 118 (1873).

[54] Article VIII. Both of these officers were appointed and paid by the Sublime Porte and pronounced sentences in the name of the Sultan.

[55] Article XI. This police function had no counterpart in the case of the Rhine.

[56] See CHAMBERLAIN, THE REGIME OF THE INTERNATIONAL RIVERS: DANUBE AND RHINE 91, 95-96 (1923).

[57] By Articles XIII and XIV of the Public Act of 1865, *supra* note 52, the Commission was authorized to impose navigation dues on Danube traffic to provide the neces-

Commission enjoyed a considerable degree of independence from the sovereigns within the territory of which it carried on its activities. Although it did not control the whole navigable course of the Danube, the longest waterway in Europe, the Commission did administer with efficiency and independence the more important part of that river.[58]

Two further international commissions, both of them non-European, must also be mentioned. A highly ambitious blueprint for an international administration was the portion of the General Act of the Conference of Berlin dealing with the River Congo,[59] many of the provisions of which were copied from the successful example of the European Commission of the Danube. The International Commission for the Navigation of the Congo was charged in particular with undertaking all measures for ensuring the navigability of the river, the determination of the fees for pilotage and navigation dues, administration of the revenues derived from tolls, supervision of the quarantine establishment, and appointment of its officials and employees.[60] As in the case of the European Commission of the Danube, the International Commission was authorized to use the war vessels of the signatory powers for enforcement of its decisions on the river,[61] and it was empowered to negotiate loans, to be "exclusively guaranteed by the revenues

---

sary funds for its activities, including the payment of interest on and the redemption of loans the Commission had secured. In 1856, Turkey had advanced to the European Commission sums of money for the execution of its work. In 1868, all the signatory states except Russia guaranteed the interest and sinking fund of the loan. Convention for the Guaranty of a Loan Contracted by the European Commission of the Danube, signed at Galatz, April 30, 1868, 18 MARTENS, N.R.G. 153 (1873).

[58] In this connection, it may be mentioned that the Pruth, a tributary of the Danube, was also administered by an international commission, composed of representatives of the littoral states (Austria-Hungary, Rumania, and Russia). The "international authority" on the navigable Pruth decided what improvements of the river should be undertaken, fixed the tolls, drew up police regulations, and supervised the maintenance and fluvial works. The inspector and the toll collector, appointed by and responsible to the Commission, had judicial authority and international status. Stipulations concerning the Navigation of the Pruth, signed at Bucharest, Dec. 15, 1866, 20 MARTENS, N.R.G. 296 (1875). This agreement was modified by the Convention relative to the Navigation of the Pruth, signed at Bucharest, Feb. 18, 1895, STURDZA, *op. cit. supra* note 49, at 760. See also Rules of Navigation of the Pruth, April 1/13, 1896, *id.* at 768.

[59] General Act of the Conference at Berlin, signed Feb. 26, 1885, Act of Navigation of the Congo, 10 MARTENS, N.R.G. 2d ser. 414 (1885-86), 76 BRITISH AND FOREIGN STATE PAPERS 4, 12 (1884-85).

[60] Article 20 of the General Act.

[61] Article 21.

raised by the said Commission."[62] However, unlike the European Commission, this highly independent, quasi-sovereign organization, representing the most liberal interpretation of the principles of the Congress of Vienna, "never had an effective life,"[63] and its value as a guide to the organization of river commissions in the future was therefore slight.

In 1905, Great Britain and the United States established an International Waterways Commission[64] to "investigate and report upon the conditions and uses" of the boundary waters between the United States and Canada. The Commission was solely an advisory and investigative body which merely reported to the two governments; it had no administrative, legislative, or judicial functions. Four years later, Canada and the United States agreed to replace the International Waterways Commission with a permanent agency endowed with more extensive powers over the boundary waters— the International Joint Commission.[65] The duties of the Commission were "to prevent disputes . . . to settle all questions which are now pending . . . and to make provision for the adjustment and settlement of all such questions as may hereafter arise."[66] In particular, the Commission was to pass upon all cases involving the use, obstruction, or diversion of boundary waters, by either private persons or governmental organs, according to certain rules set forth in the Treaty.[67] In this respect, the International Joint Commission acts in a quasi-judicial capacity,[68] resembling that of an administrative agency of the United States Government. Its decisions are binding upon the two

[62] Article 23.

[63] Chamberlain, Book Review of KAECKENBEECK, *op. cit. supra* note 12, 28 YALE L.J. 519, 522 (1918-19). The riparians were left to supervise for themselves the application of the Act of Navigation. See Convention Revising the General Act of Berlin of Feb. 26, 1885, and the General Act and the Declaration of Brussels of July 2, 1890, signed at St. Germain-en-Laye, Sept. 10, 1919, art. 8, 8 L.N.T.S. 25.

The Central Commission for Navigation on the Main (1829), although planned in detail, also never met.

[64] The United States proposed the establishment of the Commission upon the request of the Congress. Rivers and Harbors Appropriations Act, 1902, § 4, 32 Stat. 331, 373; see CHACKO, THE INTERNATIONAL JOINT COMMISSION BETWEEN THE UNITED STATES OF AMERICA AND THE DOMINION OF CANADA 74-75 (1932).

[65] Treaty between the United States and Great Britain relating to Boundary Waters between the United States and Canada, signed at Washington, Jan. 11, 1909, 36 Stat. 2448, T.S. No. 548; see BLOOMFIELD AND FITZGERALD, BOUNDARY WATER PROBLEMS OF CANADA AND THE UNITED STATES 11-14 (1958).

[66] Preamble.

[67] Articles III and IV; the applicable rules are set forth in article VIII.

[68] Bloomfield and FitzGerald characterize the International Joint Commission in this role as a "judicial body." *Op. cit. supra* note 65, at 17-37.

governments, but it has no power to hear civil actions for damages and it lacks criminal jurisdiction.[69] The Commission is also charged with the responsibility of examining and reporting upon "any other questions or matters of difference" between the two countries relating to the frontier between the United States and Canada, by which was meant essentially the boundary waters defined by the Treaty. Its function in this latter regard is purely advisory, the Treaty providing that the reports of the Commission are not decisions and are not to have "the character of an arbitral award."[70]

The arrangements which have been described constituted the initial endeavors to create international agencies, both permanent and temporary, to deal with international rivers.[71] Some of them, like the Central Commission for Navigation on the Main and the International Commission for the Navigation of the Congo, never left the blueprint stage. Others, like the Mixed Commission for the River Douro, which was set up only to bring about unification of regulations, tolls, and charges between Spain and Portugal,[72] were discharged after accomplishing the tasks for which they were created. Some international river commissions, like the one for the River Po, were in existence for only a limited number of years; others, like the European Commission of the Danube, the Central Commission for the Navigation of the Rhine, the Scheldt Commission, and the International

[69] The Commission may, however, require an indemnity by way of compensation. BLOOMFIELD AND FITZGERALD, *op. cit. supra* note 65, at 29-31; CHACKO, *op. cit. supra* note 64, at 372. H. A. Smith remarks that the functions of the Commission have been extended under article VIII "to a point at which they are practically equivalent to those of an arbitral tribunal." THE ECONOMIC USES OF INTERNATIONAL RIVERS 126 (1931).

[70] Article IX.

[71] It may be observed that the one international commission which was contemplated in the Far East during this period, the Whangpoo River Conservancy Board, was never constituted, and other arrangements were shortly made for the improvement and maintenance of the river. The Rules for the Amelioration of the Course of the Whangpoo, signed at Peking, Sept. 7, 1901, Annex No. 17 to the Final Protocol between the Foreign Powers and China for the Resumption of Friendly Relations, signed at Peking, Sept. 7, 1901, 94 BRITISH AND FOREIGN STATE PAPERS 686, 709 (1900-01), were superseded by the Agreement regarding the Establishment of a Board of Conservancy for the Whangpoo River at Shanghai, signed at Peking, Sept. 27, 1905, 1 CHINA, INSPECTORATE GENERAL OF CUSTOMS, TREATIES, CONVENTIONS, ETC., BETWEEN CHINA AND FOREIGN STATES 342 (2d ed. 1917).

[72] The Mixed Commission consisting of two Portuguese and two Spanish delegates was established to provide uniform regulations for the river. Convention between Portugal and Spain for the Free Navigation of the Douro, signed at Lisbon, Aug. 31, 1835, 14 MARTENS, N.R. 97 (1839). These regulations were published a year later and revised in 1840.

Joint Commission, have had long and often distinguished records. And still others, like the Octroi Administration on the Rhine, the Riparian Commission on the Danube, and the Revision Commission of the Elbe, prepared the ground for permanent administrations which followed.

The historical origins of the various international river commissions and their development from infrequent conferences of delegates to strong regional organizations with judicial and executive powers were rooted in the common concern of the riparian states with navigation on the rivers bordering on or crossing their territories. In creating these specialized international organizations, the fluvial community was merely institutionalizing the shared interest of its members, which had outgrown the bonds imposed by rigid notions of state sovereignty. It might be argued with some basis that the legal institutions created for navigable waterways, and for rivers in particular, should offer a rich source of precedent concerning organs of control for uses of international waters other than navigation. However, it must be borne in mind that in those early days navigation itself was by far the most vital concern of the riparian states in navigable rivers and that for the most part other uses of the rivers, as for the generation of power, for sanitary purposes, or for irrigation, were not taken into account in the establishment of these international agencies.[73]

The establishment of the first international river commissions came as a response to two important demands. The first of these was the need to reconcile a riparian state's inherent control and undisturbed jurisdiction over its own section of the river with the needs it had as a user for rights on other stretches of the river. It is in these terms, rather than in those of legal abstractions, that riparian states approached the problem of "free navigation." The second concern of littoral states, thus conceived as users and as territorial sovereigns of these waterways, was with the rights they would have to surrender and the obligations they would have to assume in order to give effect to the principle of free navigation.

Navigable international rivers were the natural channels of intercourse for the riparian states and often the only available routes for commerce between these nations and other countries across the seas. States which were dependent

[73] Two exceptions are the mixed commission established pursuant to article 9, paragraph 6, of the Treaty between Belgium and Holland respecting the Separation of their Respective Territories, signed at London, April 19, 1839, which authorized the commission also to regulate commercial fishing and fishing rights (16 Martens, N.R. 773, 781 (1841)) and the International Joint Commission, Canada-United States (see note 65 *supra*).

on these waterways were acutely aware of the danger presented to their commercial shipping by notions of the unabridged sovereignty of riparians over sections of the river which they controlled. Solemn exhortations concerning freedom of navigation for the riparians proved important in principle but were far from complete solutions. Maintenance of the towpath, maintenance and improvement of the river and its mouth, operation and inspection of navigational facilities, rules and regulations of navigation, pilotage and police, judicial settlement of disputes relative to navigation—all these matters ancillary to freedom of navigation demanded the collective attention of the riparians. The first international river commissions were instituted precisely for these reasons; without institutionalized controls and additional facilities, freedom of navigation could lose its meaning.

Free navigation for the commercial shipping of subjects of the riparian states was necessarily the first premise of the early treaties establishing international river commissions.[74] However, freedom of navigation was conceived as a relative term. Severe limitations were imposed upon it and restricted its scope considerably. Thus, the Octroi Administration guaranteed free navigation on the Rhine only to the two signatories, France and Germany, and only on that part of the river which they shared.[75] Despite lip service to the principle of free navigation into the high seas, the Final Act of the Congress

[74] Convention between France and Germany on the Octroi of Navigation of the Rhine, signed at Paris, Aug. 15, 1804, preamble and art. II, 8 MARTENS, RECUEIL 261 (1835), 1 RHEINURKUNDEN 6 (1918) (applying only to that part of the river common to France and Germany); Act for the Free Navigation of the Elbe, signed at Dresden, June 23, 1821, art. 1, 5 MARTENS, N.R. 714 (1824); Treaty between Belgium and Holland respecting the Separation of their Respective Territories, signed at London, April 19, 1839, art. 9, para. 1, 16 MARTENS, N.R. 773 (1841) (Scheldt); Convention between Austria and the Duchies of Modena and Parma for Free Navigation on the Po, signed at Milan, July 3, 1849, art. 1, 14 MARTENS, N.R.G. 525 (1856); General Treaty of Peace between Austria, Great Britain, Prussia, Russia, Sardinia, and the Ottoman Porte, signed at Paris, March 30, 1856, art. XV, 15 MARTENS, N.R.G. 770 (1857), 46 BRITISH AND FOREIGN STATE PAPERS 8 (1865) (Danube); Stipulations concerning the Navigation of the Pruth, signed at Bucharest, Dec. 15, 1866, art. 1, 20 MARTENS, N.R.G. 296 (1875); General Act of the Conference of Berlin, signed at Berlin, Feb. 26, 1885, Act of Navigation of the Congo, art. 13, 10 MARTENS, N.R.G. 2d ser. 414 (1885-86), 76 BRITISH AND FOREIGN STATE PAPERS 4, 12 (1884-85); Treaty between the United States of America and Great Britain relating to Boundary Waters between the United States and Canada, signed at Washington, Jan. 11, 1909, art. I, 36 Stat. 2448, T.S. No. 548.

[75] Convention between France and Germany on the Octroi of Navigation of the Rhine, signed at Paris, Aug. 15, 1804, 8 MARTENS, RECUEIL 261 (1835), 1 RHEINURKUNDEN 6 (1918); CHAMBERLAIN, *op. cit. supra* note 56, at 167.

of Vienna was drafted in such a form as to extend the right only to the riparian nations.[76] Discriminatory and burdensome Dutch tolls on vessels passing between the Rhine and the high seas over the Leck and the Waal were not consistent with freedom of navigation but were not eliminated definitively until the Convention of Mayence of 1831.[77] Even at that stage, navigation was effectively reserved to the riparians through a licensing requirement for boats which confined the issuance of licenses to nationals of the riparian states.[78] Finally, article 1 of the Act of Mannheim of 1868 proclaimed: "The navigation of the Rhine and its mouths, from Basle to the high seas . . . shall be free to the ships of all nations for the transport of merchandise and persons."[79] This liberal principle, seemingly establishing complete freedom of navigation on the Rhine, was restricted by the clause immediately following, ". . . on the condition of conforming to the stipulations contained in the present convention and the measures prescribed for the maintenance of the general security."[80] In consequence, so-called "riparian national treatment"[81] was differentiated from treatment of non-riparian shipping by a number of restrictions. Boatmen on the Rhine had to "have sailed on the Rhine for a definite period";[82] they had to receive a license from that riparian state in which they were domiciled;[83] and all boats navigating the Rhine had to be inspected for riverworthiness.[84]

According to the 1821 Elbe Act, navigation on the Elbe was free only for vessels and boatmen licensed by the governments of the riparian nations, and

[76] This result was achieved through the interpretation in this sense of the expression "sous le rapport de commerce" in article 1 of the Articles concerning the Navigation of the Rhine, annexed to the Final Act of the Congress of Vienna, signed June 9, 1815, 2 MARTENS, N.R. 416 (1818); see the minutes of the seventh meeting of the Committee on the Freedom of Fluvial Navigation, March 3, 1815, 1 RHEINURKUNDEN 124 (1918), and KAECKENBEECK, INTERNATIONAL RIVERS 46-47 (1918).

[77] Convention of Mayence, signed March 31, 1831, 9 MARTENS, N.R. 252 (1833), 1 RHEINURKUNDEN 212 (1918).

[78] Convention of Mayence, arts. 3 and 42.

[79] Revised Convention for the Navigation of the Rhine, signed at Mannheim, Oct. 17, 1868, 20 MARTENS, N.R.G. 355 (1875), 2 RHEINURKUNDEN 80 (1918).

[80] Article I.

[81] Provided for "tout bateau ayant le droit de porter le pavillion d'un des Etats riverains et pouvant justifier ce droit au moyen d'un document délivré par l'autorité compétente" (art. II).

[82] Article XV. The Protocole de Clotûre declared that the period was at least four years.

[83] Article XV.

[84] Article XXII.

cabotage was reserved to the riparians.[85] In 1844, free navigation was extended to all flags, but the right of cabotage remained restricted, all transport between Elbe ports being reserved to Elbe vessels.[86] The Riparian Commission on the Danube attempted to follow the example of the early treaties relating to international rivers by reserving the right of navigation on the river to the riparian states, with an exception made in the case of navigation by seagoing vessels between the high seas and river ports. When a treaty incorporating these principles[87] was laid before the signatories of the Treaty of Paris, it was opposed by the nonriparian states. These nations maintained that their approval of the proposed *règlement de navigation* was necessary to its entry into force.[88] Denied a basic law to administer, the Riparian Commission never entered upon its administrative functions.[89] It was symbolic of the conflict engendered by the transition to a wider right of free navigation that the Commission should have foundered upon this shoal.

Even after the River Po passed within exclusive Italian jurisdiction and its International Commission was abolished, the river remained open to all flags.[90] Similarly liberal to nonriparian traffic was the Treaty of London of 1839 for the Scheldt and the Treaty of Paris of 1856 for the Danube. In both these cases, the mouth and the lower part of the respective rivers were free to all vessels.[91] And from 1866, the River Pruth was opened to vessels of all nations as well; restrictions introduced in the Stipulations and in the Rules of Navigation[92] were of minor significance.

[85] Act for the Free Navigation of the Elbe, signed at Dresden, June 23, 1821, art. 1, 5 MARTENS, N.R. 714 (1824).

[86] Additional Act of the Treaty of June 23, 1821, signed at Dresden, April 13, 1844, 6 MARTENS, N.R.G. 386 (1849).

[87] Act of Navigation of the Danube, concluded between Austria, Bavaria, the Ottoman Porte, and Wurttemberg, signed at Vienna, Nov. 7, 1857, 16 MARTENS, N.R.G. pt. 2, at 75 (1858).

[88] Protocols of the Meetings held at Paris, from May 22 to August 19, 1858, for the Organization of Moldavia and Walachia, Protocol No. 18, Meeting of Aug. 16, 1858, 16 MARTENS, N.R.G. pt. 2, 40 at 49 (1858).

[89] HAJNAL, THE DANUBE: ITS HISTORICAL, POLITICAL AND ECONOMIC IMPORTANCE 80-87 (1920); CHAMBERLAIN, *op. cit. supra* note 56, at 52-54. See note 49 *supra* concerning the later revision of the Act of Navigation.

[90] Convention between Austria and the Duchies of Modena and Parma for Free Navigation on the Po, signed at Milan, July 3, 1849, *supra* note 74.

[91] Treaty between Belgium and Holland respecting the Separation of their Respective Territories, signed at London, April 19, 1839, *supra* note 74, art. 9; General Treaty of Peace, signed at Paris, March 30, 1856, *supra* note 74, art. XV.

[92] Rules of Navigation of the Pruth, April 1/13, 1896, STURDZA, RECUEIL DE DOCUMENTS RELATIFS À LA LIBERTÉ DE NAVIGATION DU DANUBE 768 (1904).

By the Treaty of Washington of 1871, the St. Lawrence was proclaimed permanently "free and open for the purposes of commerce to the citizens of the United States."[93] The Boundary Waters Treaty of 1909, which established the International Joint Commission, confirmed this freedom and extended the principle of free navigation to "all navigable boundary waters . . . for the purposes of commerce to the inhabitants and to the ships, vessels, and boats of both countries equally."[94]

The initial treaties and agreements establishing the first international river commissions legalized free navigation for the signatory riparians. A community of interest became the basis of legal rights of the members. The riparians located on the upper part of these navigable waterways fought successfully for access to the sea. Their membership on the river commission destroyed the preferential position with which geography had endowed the other riparians and established their equality in law with respect to the primary use of the waterway—that of navigation. Here any general principle of free navigation stopped. The hope for absolute freedom of navigation on international rivers, expressed for the first time in the Treaty of Paris[95] and

[93] Treaty between the United States of America and Great Britain for the Amicable Settlement of All Causes of Difference between the Two Countries, signed at Washington, May 8, 1871, art. XXVI, 17 Stat. 863, T.S. No. 133. Navigation on the rivers Yukon, Porcupine, and Stikine was secured to nationals of both countries, and arrangements were to be made for "the use of the Welland, St. Lawrence, and other canals in the Dominion . . . [and] the St. Clair Flats Canal on terms of equality . . ." (arts. XXVI and XXVII).

Previously free navigation had been secured on the St. Lawrence by article 4 of the Treaty between the United States and Great Britain relative to Fisheries, Commerce, and Navigation in North America, signed at Washington, June 5, 1854, 10 Stat. 1089, T.S. No. 124. This treaty was terminated on March 17, 1866, following appropriate notice given by the United States. The privilege of navigation for American vessels was consequently extinguished. 1 MOORE, DIGEST OF INTERNATIONAL LAW 634 (1906). See generally as to navigation on the St. Lawrence, Lawford, *Treaties and Rights of Transit on the St. Lawrence*, 39 CAN. B. REV. 577 (1961).

[94] Treaty between the United States and Great Britain relating to Boundary Waters between the United States and Canada, signed at Washington, Jan. 11, 1909, art. I, 36 Stat. 2448, T.S. No. 548. The preliminary article states: "For the purposes of this treaty boundary waters are defined as the waters from main shore to main shore of the lakes and rivers and connecting waterways, or the portions thereof, along which the international boundary between the United States and the Dominion of Canada passes, including all bays, arms, and inlets thereof, but not including tributary waters which in their natural channels would flow into such lakes, rivers, and waterways, or waters flowing from such lakes, rivers, and waterways, or the waters of rivers flowing across the boundary."

[95] Treaty of Peace between France and Austria and its Allies, signed at Paris, May 30, 1814, 2 MARTENS, N.R. 1 (1818).

again articulated in the Final Act of the Congress of Vienna,[96] was implemented on only five rivers, the Scheldt, Po, Danube, Pruth, and Rhine. Only on the Po, however, were the vessels of riparians and nonriparians treated unconditionally alike. On the Scheldt and Danube, solely the mouths and the lower streams were free to all vessels. The régime of the Rhine, although professedly making the river "free to the ships of all nations," favored the vessels of riparians. Navigation on the Pruth was also restricted, but in lesser degree. Needless to say, the importance of the Po and Pruth could not compare with that of the Danube, Rhine, and Scheldt. The first international river administrations, while introducing or confirming the principle of freedom of navigation of littoral states as a legal right, treated the equality of all flags in practice as a privilege or a concession. In most cases, it was granted to seagoing vessels of nonriparians only to enable them to reach ports in the river estuary.

A river must be kept navigable and in good repair. Maintenance and improvement cannot be dispensed with if free navigation is to be preserved. Any revenue for such services should come in the form of dues, tolls, charges, or tariffs from those who derive direct benefit from the waterway. Necessary arrangements of this nature, together with the establishment of reasonable and uniform compensation for special facilities and services, were the second important set of problems with which the original international river commissions had to deal. In the presence of an established international administration, should necessary maintenance and improvement of the river be carried out by the respective states, by local authorities, or by the commission itself? Who should determine the rates of tolls and tariffs, and who should collect these?

All of the first international river commissions had some degree of compentence to deal with the necessary maintenance and improvement of the waterway for which they were created. This competence ranged from the simple receipt of reports from local authorities and from the littoral states by the Central Commission for the Navigation of the Rhine to the complete planning, designing, and carrying out of all works of maintenance and improvement by the European Commission of the Danube. In the latter instance, international execution of the necessary improvements was at the outset a consequence of disturbed conditions in that area and of the inability or unwillingness of the local authorities to deal with the engineering prob-

[96] Final Act of the Congress of Vienna, signed June 9, 1815, art. CIX, 2 MARTENS, N.R. 361, 408 (1818).

lem.[97] Under the more stable conditions prevailing along the Rhine, the Central Commission needed only to inspect and to coordinate. A body of engineers, representing the riparian nations, examined the river from time to time and informed the Central Commission on its navigable condition.[98] While technical supervision of the waterway was thus one of the primary duties of the international river administrations, the functions of each were determined by the circumstances under which that particular commission was created.

The determination, unification, and collection of passage tolls were the corollary of this technical duty and one of the direct causes for instituting international river administrations.[99] Again, a considerable range of tech-

[97] CHAMBERLAIN, *op. cit. supra* note 56, at 41, 54-57; H. A. Smith, *The Danube*, [1950] YEAR BOOK OF WORLD AFFAIRS 191, 206.

[98] Revised Convention for the Navigation of the Rhine, signed at Mannheim, Oct. 17, 1868, art. XXXI, 20 MARTENS, N.R.G. 355 (1875), 2 RHEINURKUNDEN 80 (1918). On the Elbe, improvements of the river were left to the littoral states, who had a right of complaint to the Elbe Commission about the acts and omissions of the other riparians. Act for the Free Navigation of the Elbe, signed at Dresden, June 23, 1821, *supra* note 74, art. 28. The Scheldt Commission supervised facilities on the river and the state of its navigability and reported to the two governments. Treaty between Belgium and Holland respecting the Separation of their Respective Territories, signed at London, April 19, 1839, *supra* note 74, art. 9, para. 2. The Po Commission was entrusted with periodic inspection of the river and determination of what improvements were necessary. It then submitted its report to the riparian governments. Convention between Austria and the Duchies of Modena and Parma for Free Navigation on the Po, signed at Milan, July 3, 1849, *supra* note 74, art. VI. Like the European Danube Commission, the Pruth Commission planned and executed works of improvement on the river and then supervised the maintenance of these works. The general plan, as well as the special plans and specifications, had to be approved by the riparians. Stipulations concerning the Navigation of the Pruth, signed at Bucharest, Dec. 15, 1866, *supra* note 74, arts. 7 and 11. Reference has already been made (see p. 107 *supra*) to the power of the International Joint Commission, with respect to the St. Lawrence, to pass upon "all cases involving the use or obstruction or diversion of the waters." Treaty between the United States of America and Great Britain relating to Boundary Waters between the United States and Canada, signed at Washington, Jan. 11, 1909, *supra* note 94, art. VIII.

[99] As in the case of the Octroi Administration on the Rhine, the first of the international river commissions. Convention between France and Germany on the Octroi of Navigation of the Rhine, signed at Paris, Aug. 15, 1804, 8 MARTENS, RECUEIL 261 (1835), 1 RHEINURKUNDEN 6 (1918). Maintenance of the fiscal régime previously existing had proved impossible after the extension of French sovereignty to the left bank of the Rhine. A unified and common organization for the collection of tolls seemed the easiest and most convenient solution. See Treaty between Belgium and Holland respecting the Separation of their Respective Territories, signed at London, April 19, 1839, *supra* note 74, art. 9, para. 3, concerning the right of the Netherlands to levy tolls on ships on the Scheldt.

niques developed in practice. In some cases, as on the rivers Po, Danube, and Pruth, the commissions determined the rates of passage tolls, collected them, and used them for the maintenance and improvement of the river.[100] On the Elbe, with the creation of the Commission in 1821, the existing tolls were abolished and a uniform navigation tariff substituted.[101] Tolls were subsequently reduced several times, and by 1870 they were entirely abolished by a treaty between Austria and the North German Confederation.[102] A similar development occurred on the Rhine. By the Treaty of Mannheim, tolls were abolished in 1868.[103] On the maritime Scheldt, a fixed toll was paid to Holland from 1839 onward. The right to collect tolls was bought and abolished by Belgium with the help of contributions from twenty-one states and free cities in 1863.[104] Tolls remained authorized on the southern Scheldt and on the waters between the Scheldt and the Rhine.[105] With the important exception of the St. Lawrence River,[106] the fiscal problems of international rivers normally constituted an important stimulus to the establishment of river commissions. With the development of the international river ad-

[100] Convention between Austria and the Duchies of Modena and Parma for Free Navigation on the Po, signed at Milan, July 3, 1849, *supra* note 74, art. X; General Treaty of Peace, signed at Paris, March 30, 1856, *supra* note 74, arts. XV and XVI; Stipulations concerning the Navigation of the Pruth, signed at Bucharest, Dec. 15, 1866, *supra* note 74, arts. 2 and 18.

[101] Act for the Free Navigation of the Elbe, signed at Dresden, June 23, 1821, *supra* note 74, art. 7.

[102] Charges remained only for the use of certain services for the facilitation of traffic. Treaty between Austria and the North German Confederation for the Abolition of Elbe Dues, signed at Vienna, June 22, 1870, art. 1, 20 MARTENS, N.R.G. 345 (1875), 63 BRITISH AND FOREIGN STATE PAPERS 594 (1879).

[103] "Aucun droit basé uniquement sur le fait de la navigation ne pourra être prélevé sur les bateaux ou leurs chargements . . ." Revised Convention for the Navigation of the Rhine, signed at Mannheim, Oct. 17, 1868, *supra* note 98, art. III.

[104] The right to levy tolls was redeemed in the Treaty between Belgium and the Netherlands for the Redemption of the Scheldt Toll, signed at The Hague, May 12, 1863, 53 BRITISH AND FOREIGN STATE PAPERS 15 (1868). The indemnification of Belgium was effected by the General Treaty for the Redemption of the Scheldt Toll, signed at Brussels, July 16, 1863, 17 MARTENS, N.R.G. pt. 2, at 223 (1869). See, on the history of Scheldt tolls, GRANDGAIGNAGE, HISTOIRE DU PÉAGE DE L'ESCAUT (1868).

[105] Treaty between Belgium and Holland respecting the Separation of their Respective Territories, signed at London, April 19, 1839, *supra* note 74, art. 9, paras. 4 and 5.

[106] There were no navigation tolls in the past on the St. Lawrence prior to the construction of the St. Lawrence Seaway. The Treaty between the United States and Great Britain relating to Boundary Waters between the United States and Canada, signed at Washington, Jan. 11, 1909, 36 Stat. 2448, T.S. No. 548, provided in article 1: "Either of the High Contracting Parties may adopt rules and regulations governing the

ministrations, toll rates tended to decrease to the level of cost needed for the maintenance and improvement of the waterway. Special dues were charged for ancillary navigational facilities, and then only on a cost basis. On rivers where they were not strongly needed, tolls were in time abolished entirely.

Equally important as technical supervision of the navigability of an international river is the supervision of river traffic and the elaboration and enforcement of rules and regulations governing navigation. Most of the first river commissions were entrusted with general supervision of the traffic on their respective waterways. This called for the drafting and enforcement of rules of the road on the river and of regulations dealing with such matters as the licensing of vessels and boatmen, the control of pilots, and towage and safety measures for the vessels and the waterway. Even under the Octroi Administration on the Rhine, "regulations concerning administration of the traffic and police rules" were to be prepared by the director-general and inspectors for the approval of the two governments and were to be applied and enforced by the Octroi inspectors.[107] The Articles concerning the Navigation of the Rhine of the Congress of Vienna empowered the Central Commission to draw up regulations for the navigation of the Rhine, consistent with the Articles,[108] which were to be adopted only with the consent of all of the members. These basic regulations, originally incorporated in the Convention of Mayence,[109] have been replaced by the Act of Mannheim of 1868[110] which, as amended, is still today the basic instrument

---

use of such canals within its own territory and may charge tolls for the use thereof, but all such rules and regulations and all tolls charged shall apply alike to the subjects or citizens of the High Contracting Parties and the ships, vessels, and boats of both of the High Contracting Parties, and they shall be placed on terms of equality in the use thereof."

[107] Convention between France and Germany on the Octroi of Navigation of the Rhine, signed at Paris, Aug. 15, 1804, arts. CXXX, XLIII, and XLIV, 8 Martens, Recueil 261 (1835), 1 Rheinurkunden 6 (1918).

[108] Articles concerning the Navigation of the Rhine, annexed to the Final Act of the Congress of Vienna, signed June 9, 1815, art. XXXII, 2 Martens, N.R. 416 (1818). The Central Commission of the Rhine was to enter upon its ordinary functions only when the *règlement de navigation* had been adopted by the nations represented in the Commission. Difficulties created by Prussia and the Netherlands postponed this event for fifteen years until the Act of Mayence of 1831. Van Eysinga, La Commission centrale pour la navigation du Rhin 29-35 (1935).

[109] Convention of Mayence, signed March 31, 1831, 9 Martens, N.R. 252 (1833), 1 Rheinurkunden 212 (1918).

[110] Revised Convention for the Navigation of the Rhine, signed at Mannheim, Oct. 17, 1868, 20 Martens, N.R.G. 355 (1875), 2 Rheinurkunden 80 (1918).

governing navigation on the river. The Rhine inspectors were directed "to see to the execution of the regulations, and to organize everything relating to the police of navigation."[111] The river administrations on the Elbe, Scheldt, and Po were instructed to supervise the execution of navigation treaties and agreements and to prevent violations.[112] The most autonomous position in this respect, however, was attained by the European Commission of the Danube. After the failure of the Riparian Commission, the European Commission was empowered to draw up, "d'un commun accord,"[113] its own rules of navigation and of river police.[114] These rules were executed and supervised by two Turkish officials under the control of the Commission.[115] The Mixed Commission on the River Pruth drew up and administered its navigation and police regulations, patterned on the Danubian regulations and almost as far-reaching.[116]

The International Joint Commission, dealing with the boundary waters between the United States and Canada, has exercised a compulsory jurisdiction over the boundary waters, for without its approval no change in the use

---

[111] Articles concerning the Navigation of the Rhine, annexed to the Final Act of the Congress of Vienna, signed June 9, 1815, *supra* note 108, art. XV.

[112] See especially the detailed provisions of the Convention on the Publication of Uniform Ordinances for the Police of Navigation on the Elbe, signed at Dresden, April 13, 1844, 6 MARTENS, N.R.G. 463 (1849).

[113] General Treaty of Peace, signed at Paris, March 30, 1856, *supra* note 74, arts. XVII and XIX. The requirement of an *accord commun* meant that the Commission's competence was not actually a legislative one, particularly when considered in light of the fact that the representatives sitting on the Commission presumably acted under the instructions of their governments.

[114] Rules of Navigation and of Police applicable to the Lower Danube, signed at Galatz, Nov. 21, 1864, 18 MARTENS, N.R.G. 118 (1873); see Public Act of the European Commission of the Danube, relative to the Navigation of the Mouths of the Danube, signed at Galatz, Nov. 2, 1865, art. 7, *id.* at 144, 55 BRITISH AND FOREIGN STATE PAPERS 93 (1864-65). See also Rules of Navigation and of Police applicable to the Part of the Danube between Galatz and its Mouths, promulgated by the European Commission of the Danube, May 19, 1881, STURDZA, *op. cit. supra* note 92, at 136. The jurisdiction of the Commission was extended as far as Galatz by the Additional Act to the Public Act of the European Commission of the Danube of Nov. 2, 1865, relative to the Navigation of the Mouths of the Danube, signed at Galatz, May 28, 1881, 8 MARTENS, N.R.G. 2d ser. 207 (1883).

[115] Rules of Navigation and of Police applicable to the Lower Danube, signed at Galatz, Nov. 21, 1864, *supra* note 114, arts. 1-3.

[116] Rules of Navigation of the Pruth, April 1/13, 1896, STURDZA, *op. cit. supra* note 92, at 768, promulgated pursuant to the Stipulations concerning the Navigation of the Pruth, signed at Bucharest, Dec. 15, 1866, 20 MARTENS, N.R.G. 296 (1875).

of the waters may be effected.[117] However, it has not exercised a general legislative function. Navigation on the boundary waters is subject to the laws, regulations, and rules of "either country, within its own territory, not inconsistent with . . . [the] privilege of free navigation and applying equally and without discrimination to the inhabitants, ships, vessels, and boats of both countries."[118] The International Joint Commission was thus not called upon to elaborate any navigation and police regulations.[119]

No common pattern can be discerned in the creation and application of rules of navigation on international rivers. The river agencies on occasion administered the waterways under the laws, rules, and regulations of the littoral states. In other instances laws and rules specifically for river navigation were enacted by all of the riparian states in common. The European Commission of the Danube had broad power not only to supervise and administer, but also to enact by common accord, the rules of navigation and of police for the waterway. In all of these instances, however, supervision of the laws and regulations governing traffic proved at least as important as the supervision of the navigability of the waterway in a technical sense.

Fluvial laws, rules, and regulations have variously been enforced by regular tribunals of the riparian nations, by special river courts established by the riparians, by local customs officials, by toll collectors, by permanent authorities established by the river commission to supervise the waterway, and by the commission itself. Appeals were permitted to the local courts of appeal or to the commission or to both. An early precedent for the assumption of judicial functions by an international river administration had been established by the Octroi Administration on the Rhine, under the authority of which toll collectors were authorized to deal with contraventions of the treaty.[120] Appeals could be taken by the aggrieved parties to the Commission itself.[121] In the Act of Mayence, the judicial authority was to be vested in a

[117] Treaty between the United States and Great Britain relating to Boundary Waters between the United States and Canada, signed at Washington, Jan. 11, 1909, arts. III, IV, and VIII, 36 Stat. 2448, T.S. No. 548.

[118] Article I.

[119] Since the signals and steering rules of the United States and Canada are practically identical (The New York, 175 U. S. 187, 199 (1899)), there has been no difficulty on that score. Concerning the matter of pilotage, see p. 291 *infra*.

[120] Convention between France and Germany on the Octroi of Navigation of the Rhine, signed at Paris, Aug. 15, 1804, art. CXXII, 8 MARTENS, RECUEIL 261 (1835), 1 RHEINURKUNDEN 6 (1918).

[121] Article CXXIV granted a right of appeal to the Commission with regard to the levying of tolls and the police of navigation.

"fonctionnaire de l'ordre judiciaire" appointed by the governments of the states which were members of the Central Commission. These officials were competent to deal with violations of the *règlement de navigation,* with disputes about tolls and other charges, and with litigation arising out of accidents.[122] This civil and criminal jurisdiction was maintained by the Act of Mannheim, again to be exercised by special national river courts acting according to a simple and expeditious procedure.[123] Appeals in matters involving more than a given amount might then, and still may, be made either to the Central Commission for the Navigation of the Rhine or to municipal courts of appeal.[124] Although the number of such special national courts has diminished over the course of years and the Commission hears only a comparatively small proportion of the appeals,[125] the Commission, acting in its judicial capacity, has built up a considerable body of jurisprudence over the years.

On the Elbe, navigational matters were decided at first by judicial officers of each toll station[126] and later by regular joint river courts. Appeals lay to the national courts of appeal.[127] On the Po, navigational disputes were settled by local chief customs officials, with an appeal to the Commission.[128] The Pruth toll collectors enforced toll regulations, while rules of navigation were enforced by the inspector of navigation with an appeal to the Mixed Commission.[129] On the Danube, fluvial rules and regulations were at first enforced by the local authorities, and only later by the captain of the port and the inspector general in the first instance, with a right of appeal to the

---

[122] Convention of Mayence, signed March 31, 1831, art. LXXXI, 9 MARTENS, N.R. 252 (1833), 1 RHEINURKUNDEN 212 (1918).

[123] Revised Convention for the Navigation of the Rhine, signed at Mannheim, Oct. 17, 1868, arts. 33 and 34, 20 MARTENS, N.R.G. 355 (1875), 2 RHEINURKUNDEN 80 (1918).

[124] Article 37.

[125] WALTHER, LA JURISPRUDENCE DE LA COMMISSION CENTRALE POUR LA NAVIGATION DU RHIN, 1832-1939, at 15 (1948).

[126] Act for the Free Navigation of the Elbe, signed at Dresden, June 23, 1821, art. XXVI, 5 MARTENS, N.R. 714 (1824).

[127] Additional Act to the Treaty of June 23, 1821, signed at Dresden, April 13, 1844, §§ 46-51, 6 MARTENS, N.R.G. 386 (1849).

[128] Convention between Austria and the Duchies of Modena and Parma for Free Navigation on the Po, signed at Milan, July 3, 1849, art. XX, 14 MARTENS, N.R.G. 525 (1856).

[129] Rules of Navigation of the Pruth, April 1/13, 1896, arts. 84-86, STURDZA, *op. cit. supra* note 92, at 768 (1904).

European Commission.[130] However, the scope of the judicial power on the Danube, comprising *only* offenses against the rules and regulations for the Danube, was narrower than that of the Rhine courts, which had jurisdiction over civil and criminal matters and the power not only to fine but also to imprison.[131] The discernible pattern in these instances of judicial authority over civil and criminal cases relating to river navigation appears to be that either permanent officials or the local courts enforced the police and navigation regulations for the river but that the regular local courts or special national navigation courts dealt with civil litigation concerning navigation. The commission was not infrequently charged with judicial responsibilities as an appellate tribunal, a function which it might perform to the exclusion of, or in conjunction with, national courts of appeal.

A number of other functions were assumed by various river commmissions, according to the particular circumstances under which they functioned. The prevalence of disease in the area of the mouth of the Danube made it natural that the European Commission should have some interest in the establishment of quarantines and in reconciling sanitary measures with the free movement of river transport.[132] The large number of jurisdictions through which vessels navigating the Rhine might find it necessary to pass required simplification of customs clearance for boats and their relief from certain customs formalities.[133] Finally, a number of treaties establishing the commissions devoted attention to ancillary facilities, such as free ports,[134] which, while not requisites of free navigation, might be said to promote the development of river commerce.

The original international river commissions were set up to facilitate commercial shipping under normal conditions in time of peace. Several of the treaties establishing or enlarging the functions of these agencies had, however, anticipated the possibility of hostilities and included provisions

[130] Rules of Navigation and of Police applicable to the Lower Danube, signed at Galatz, Nov. 21, 1864, arts. 5, 108, and 110, 18 MARTENS, N.R.G. 118 (1873).

[131] See Affaire Noll, May 10, 1899, cited in WALTHER, *op. cit. supra* note 125, at 97 (1948). Since 1927, the Central Commission of the Rhine has regarded the Act of Mannheim as not giving it such authority. *Ibid.*

[132] Public Act of the European Commission of the Danube, relative to the Navigation of the Mouths of the Danube, signed at Galatz, Nov. 2, 1865, *supra* note 114, arts. 18-20.

[133] Revised Convention for the Navigation of the Rhine, signed at Mannheim, Oct. 17, 1868, *supra* note 123, arts. 9-13.

[134] Article 8 of the Revised Convention.

neutralizing the waterway and immunizing the commission's personnel, works, and establishments in time of war. The agreement establishing the Octroi Administration on the Rhine provided that in time of war, personnel engaged in the toll service and their boats would be neutralized in order that tolls might continue to be collected without molestation. Protection was to be provided "for the offices and the safes of the toll administration."[135] The same provision was adopted in the Convention of Mayence,[136] but was dropped in the Act of Mannheim. The Public Act of the European Commission of the Danube decreed that "all the works and establishments" of the European Commission "shall continue to be devoted exclusively to the use of the navigation of the Danube, and can never be turned aside from this object for any motive whatever; to this end they are placed under the guarantee and protection of international law."[137] These works were neutralized, as were all the personnel of the European Commission.[138] This provision was reaffirmed in 1871, when the operations and jurisdiction of the Commission were extended to Galatz,[139] and again in 1878, when the Danube was neutralized up to the Iron Gates. At that time, all warships were excluded from that part of the Danube, and all fortifications were ordered to be demolished.[140] Other agreements concerning navigation on international rivers under the jurisdiction of international commissions were directed to changed conditions in time of war. On the Scheldt, for example, Holland, while admitting the necessity of agreement concerning alterations and improvements of buoys, beacons, and low-water marks in time of peace, claimed full liberty of action in dealing with these aids to navigation in time of war. Belgium acceded to this demand in 1891, and

---

[135] Convention between France and Germany on the Octroi of Navigation of the Rhine, signed at Paris, Aug. 15, 1804, *supra* note 120, art. CXXXI.

[136] Convention of Mayence, signed March 31, 1831, *supra* note 122, art. CVIII.

[137] Public Act of the European Commission of the Danube, relative to the Navigation of the Mouths of the Danube, signed at Galatz, Nov. 2, 1865, art. I, 18 Martens, N.R.G. 144 (1873), 55 British and Foreign State Papers 93 (1864-65).

[138] Article XXI.

[139] Treaty for the Revision of the Stipulations of the Treaty Concluded at Paris, March 30, 1856, relative to the Navigation of the Black Sea and the Danube, signed at London, March 13, 1871, art. VII, 18 Martens, N.R.G. 303 (1873).

[140] Treaty of Berlin, signed July 13, 1878, art. LII, 3 Martens, N.R.G. 2d ser. 449 (1878-79). A flag and arm band were provided in 1881 as aids to establishing the identity of personnel and establishments of the Commission. Additional Act to the Public Act of the European Commission of the Danube of Nov. 2, 1865, relative to the Navigation of the Mouths of the Danube, signed at Galatz, May 28, 1881, art. 8, 8 *id.* at 207 (1883).

Holland became entitled to remove all these navigational facilities in time of war or when war might be threatened. This compact scarcely promoted free navigation at such times and eventually was productive of a heavy burden on Scheldt navigation.[141]

THE PERIOD BETWEEN THE TWO WORLD WARS

The First World War saw widespread interruption of navigation on the international rivers of Europe and the breakdown of a number of the international river commissions which had been established to supervise them. The belligerency of the riparians proved to be incompatible with the provisions of the treaties neutralizing the waterways, and the immunities of personnel, works, and establishment of the international commissions became illusory under wartime conditions. The River Scheldt and its mouth were closed throughout most of the war.[142] The lower Danube, administered by the European Commission, was first crippled by the loss of personnel of the Commission who had to report for military duty and then by the closing of the Bosporus and Dardanelles by Turkey. The Commission met for the last time in the spring of 1915.[143] The upper Danube became a theater of military operations, with the result that the entire river was closed to navigation.[144] The mouth of the Elbe became one of the battle harbors of the German high-seas fleet.[145] Navigation on the Oder, Niemen, and Vistula was equally crippled by the war.[146] Although the Central Com-

---

[141] BLONDEAU, L'ESCAUT: FLEUVE INTERNATIONAL ET LE CONFLIT HOLLANDO-BELGE 20 (1932). This question was closely related to that of the right of the nations which guaranteed Belgian neutrality to send warships into the Scheldt, a right which was denied by the Netherlands. The problems of neutralization and defense bulked large in the controversy between the Netherlands and Belgium over the régime of the Scheldt following the war. Telders, *La Revision des traités de 1839*, [1935] GROTIUS ANNUAIRE INTERNATIONAL 37.

[142] 7 HACKWORTH, DIGEST OF INTERNATIONAL LAW 473 (1943).

[143] When Rumania entered the war, it interned all the German, Austrian, Hungarian, and Turkish employees of the Commission and deported the German and Austro-Hungarian representatives on the Commission. At the end of the year 1916, the activities of the Commission ceased completely. On October 10, 1917, the headquarters and records of the Commission were destroyed by bombardment. LA COMMISSION EUROPÉENNE DU DANUBE ET SON OEUVRE DE 1856 À 1931, at 37 (1931).

[144] CHAMBERLAIN, THE REGIME OF THE INTERNATIONAL RIVERS: DANUBE AND RHINE 123-26 (1923).

[145] CRUTTWELL, A HISTORY OF THE GREAT WAR, 1914-1918, at 64 (1934).

[146] VOŠTA, MEZINÁRODNÍ ŘEKY 350 (1938).

mission for the Navigation of the Rhine did not meet during the years 1914-1919,[147] navigation on the river remained possible because the waterway remained outside the area of actual military operations. Under pressure from the British Government, the Netherlands was obliged to place restrictions on the shipment between occupied Belgium and Germany of coal, ore, and other materials which were susceptible of military use.[148]

Although the problems of fluvial navigation facing the Paris Peace Conference bore some resemblance to those with which the Congress of Vienna had had to deal, the passage of one hundred years had added immeasurably to the complexities of the situation.[149] The year 1815 had seen the establishment of principles and the deferment of details. What more could be asked of the Paris Peace Conference, faced with a problem grown infinitely more difficult? The Elbe, Vltava (Moldau), Oder, Niemen, and Danube were declared international.[150] On these waterways, freedom of navigation was guaranteed to all commercial shipping on the footing of the absolute equality of all flags. Other provisions of a general character dealt with the limitation of charges to those reasonably necessary for the maintenance and improvement of the river and of navigation, customs formalities, other dues and charges, the maintenance of the waterway in good condition, and disputes about obstacles to navigation.[151] This régime, which was to be essentially of a provisional character, was to be "superseded by one to be laid down in a General Convention drawn up by the Allied and Associated Powers, and approved by the League of Nations, relating to the waterways recognised in such Convention as having an international character," to include those rivers just mentioned.[152] The Elbe, Oder, and Niemen were to be placed under the administration of international commissions[153] consisting of both riparians and nonriparians, a measure designed to protect in particular the landlocked riparians, such as Czechoslovakia and Poland, which might otherwise be outvoted by the downstream states.[154] However, the Niemen

---

[147] VAN EYSINGA, LA COMMISSION CENTRALE POUR LA NAVIGATION DU RHIN 58-59 (1935).

[148] See p. 202 *infra.*

[149] 2 TEMPERLEY, A HISTORY OF THE PEACE CONFERENCE OF PARIS 92-94 (1920).

[150] Treaty of Peace between the Allied and Associated Powers and Germany, signed at Versailles, June 28, 1919, art. 331, 112 BRITISH AND FOREIGN STATE PAPERS 1 (1919).

[151] Articles 332-37.

[152] Article 338.

[153] Articles 340-42.

[154] 2 TEMPERLEY, *op. cit. supra* note 149, at 93.

Commission, the establishment of which was contingent upon a request by one of the riparian nations, was never created.[155] The Oder, which provided communication between Upper Silesia and the Baltic, was to be administered by the International Commission of the Oder.[156] The Commission, which first met in 1920, became embroiled in disputes about the extent of its jurisdiction over the Oder system. The chief source of controversy was the jurisdiction of the Commission over the Warthe and Netze, two tributaries of the Oder, sections of which were in Polish territory. The Advisory and Technical Committee for Communications and Transit of the League of Nations did not succeed in its mission of conciliation, and the matter was brought before the Permanent Court of International Justice. The Court confirmed the jurisdiction of the International Commission over the portions of the Warthe and Netze situated in Polish territory.[157] The Commission had not completed the drawing up of an act of navigation before Germany rejected its authority in 1936.[158] The Statute of Navigation for the Elbe, establishing the permanent régime envisaged by the Treaty of Versailles, was drawn up in 1922.[159] The Commission, acting by a majority of the weighted votes accorded the various states represented on the body,[160] was to supervise freedom of navigation, to hear complaints arising out of the application of the treaty, to decide whether tariffs were in accordance with the treaty, and to hear appeals from national courts of navigation.[161] Rules of navigation were to be drafted by the riparian states but submitted to the Commission for approval.[162] While the Commission was

[155] THE TREATY OF VERSAILLES AND AFTER: ANNOTATIONS OF THE TEXT OF THE TREATY 663 (Dep't of State Pub. 2724) (1947). The Port of Memel, which included a portion of the course of the Niemen, was, however, placed under the jurisdiction of a Harbour Board of international composition. Convention concerning the Territory of Memel, signed at Paris, May 8, 1924, Annex II, arts. 1, 3, and 5, 29 L.N.T.S. 85, 109.

[156] Treaty of Versailles, *supra* note 150, art. 341. The states represented and the number of their representatives were: Poland (1); Prussia (3); Czechoslovakia (1); Great Britain (1); France (1); Denmark (1); and Sweden (1).

[157] Territorial Jurisdiction of the International Commission of the River Oder, P.C.I.J., ser. A, No. 23 (1929).

[158] THE TREATY OF VERSAILLES AND AFTER: ANNOTATIONS OF THE TEXT OF THE TREATY 662 (Dep't of State Pub. 2724) (1947).

[159] Convention Instituting the Statute of Navigation of the Elbe, signed at Dresden, Feb. 22, 1922, 26 L.N.T.S. 221.

[160] The states represented on the Commission and the number of votes accorded each were: German states bordering on the Elbe (4); Czechoslovakia (2); Great Britain (1); France (1); Italy (1); and Belgium (1). Article 2 of the Elbe Convention.

[161] Articles 2 and 46.

[162] Article 37.

not empowered to carry on works of improvement on its own account, it was given a veto power over improvements proposed by the members and a right, subject to a substantial number of limitations, to require the carrying out of certain works.[163]

Immediately after the First World War, commercial navigation on the Danube decreased considerably.[164] Under the authority of the Treaty of Versailles,[165] the reconstituted European Commission of the Danube[166] was soon deep in the work of attempting to bring the river and its installations back to their prewar state.[167] That portion of the river not under the jurisdiction of the European Commission was directed by article 347 of the Treaty of Versailles to be placed under the administration of the International Commission, which was to consist of two representatives of German riparian states, one representative of each other riparian state, and one representative of each nonriparian state represented on the European Commission. Two years after the signing of the peace treaty, the definitive statute for the Danube called for by that agreement was signed at Paris.[168] While the powers of the European Commission remained unchanged,[169] the International Commission was given considerably less authority than the corresponding agency on the maritime section of the river. Both the initiation and execution of works of improvement remained responsibilities of the individual nations bordering on the river. The International Commission merely reviewed the program of works.[170] The Commission was called upon to fix the dues which might be levied by the riparians on navigation and to control the collection and application of such dues.[171] A *règlement de navigation* was

---

[163] Articles 39-43.

[164] HINES, REPORT ON DANUBE NAVIGATION SUBMITTED TO THE ADVISORY AND TECHNICAL COMMITTEE FOR COMMUNICATIONS AND TRANSIT OF THE LEAGUE OF NATIONS 16 (LEAGUE OF NATIONS DOC. No. C. 444 (a) M. 164 (a). 1925. VIII).

[165] Treaty of Versailles, *supra* note 150, art. 346.

[166] As a "provisional measure," the Commission was made up of representatives only of Great Britain, France, Italy, and Rumania.

[167] The Commission cleaned the river bed, replaced or repaired the beacons and buoys, repaired and rebuilt its buildings and works in Sulina and elsewhere, and established an Advisory Committee of Engineers to advise it on the works necessary at the mouth of the Danube. LA COMMISSION EUROPÉENNE DU DANUBE ET SON OEUVRE DE 1856 à 1931, at 50-51 (1931).

[168] Convention Instituting the Definitive Statute of the Danube, signed at Paris, July 23, 1921, 26 L.N.T.S. 173.

[169] Article 5.

[170] Articles 11-17.

[171] Article 18.

to be drawn up by the Commission on the basis of drafts submitted to it by the riparian nations.[172]

Both the European and International commissions had a record of substantial achievement in the technical domain having to do with the maintenance and improvement of their respective parts of the Danube.[173] They were less successful in maintaining free navigation along the waterway and in maintaining their jurisdiction and functions in the face of attacks from the riparian states. The fluvial nations were able to place practical limitations on freedom of navigation through their power of reserving cabotage to their own nationals.[174] Germany first put difficulties in the way of international cooperation in the International Commission and then withdrew from that body in 1936. The withdrawal of the Reich from the Commission was followed by determined German efforts to dissolve the Commission through the exclusion of the nonriparians or the withdrawal of the Danubian states. The eventual domination by the Axis of southeastern Europe permitted Germany to conclude an agreement in September 1940 with Bulgaria, Hungary, Italy, Rumania, Slovakia, and Yugoslavia which purported to bring the Commission to an end.[175] As for the maritime Danube under the jurisdiction of the European Commission, Rumania, which had long harbored strong resentment of the European Commission, carried on a battle with the other members of the Commission, Great Britain, France, and Italy, over the extent of the jurisdiction of the Commission. The Advisory and Technical Committee for Communications and Transit of the League of Nations reached the conclusion in 1924 that the European Commission's competence extended to the section of the Danube between Galatz and

---

[172] Article 24.

[173] DUVERNOY, LE RÉGIME INTERNATIONAL DU DANUBE 213-16, 223-24 (1941).

[174] As authorized by article 22 of the Convention Instituting the Definitive Statute of the Danube, *supra* note 168; see HAJNAL, LE DROIT DU DANUBE INTERNATIONAL 302 (1929). Further difficulties were thrown in the path of the International Commission by the increasing truculence of Germany, the problem of German representation on the Commission (see Bacon, *Representation in the International Commission of the Danube*, 31 AM. J. INT'L L. 414 (1937)), and by the refusal of Germany to comply with the provisions of the Treaty of Versailles regarding fluvial navigation (Note of Dec. 14, 1936, REICHSGESETZBLATT, 1936, II, 361).

[175] Provisional Arrangement regarding the Danube Régime, signed at Vienna, Sept. 12, 1940, 1 DEPARTMENT OF STATE, DOCUMENTS AND STATE PAPERS 273 (1948); see MANCE, INTERNATIONAL RIVER AND CANAL TRANSPORT 54-55 (1945); PRIGRADA, INTERNATIONAL AGREEMENTS CONCERNING THE DANUBE 13 (1953). Under article 3 of the Provisional Arrangement, a Special Committee of the Fluvial Danube was formed to deal with the Iron Gates for the duration of the war.

Braila and that the dividing line between the competence of the Commission and of the Rumanian authorities in ports could not be drawn on a geographical basis but would have to be fixed in functional terms which would take account of the Commission's responsibility for the maintenance of freedom of navigation.[176] The question was then referred to the Permanent Court at The Hague, which reached substantially the same result as the Committee of the League.[177] The members of the Commission thereafter drew up in 1930 a draft treaty regulating the respective competences of the Commission and of Rumania,[178] by the terms of which regulations for the police of the Danube were to be drawn by the international body and regulations for the policing of ports and banks by the territorial sovereigns.[179] Navigation tribunals were to be established by the Rumanian Government, with an appeal provided to a Navigation Court which the European Commission and Rumania would jointly constitute.[180] The convention attached to the declaration made at that time did not, however, come into force. Further negotiations between Rumania on the one hand and the other members of the Commission on the other led to a modus vivendi whereby Rumania agreed "to abstain from disputing the complete jurisdiction of the European Commission of the Danube from the sea to Braila," while the Commission agreed to refrain from exercising its judicial authority between Braila and Galatz and to limit territorially the authority of the Commission's inspector of navigation.[181] This slow attrition in the powers of the Commission came to a logical conclusion before the Second World War in a new agreement, entering into force on May 13, 1939, on the powers

[176] Resolution adopted July 28, 1925, LEAGUE OF NATIONS OFF. J., 6th year 1221 (1925).

[177] Jurisdiction of the European Commission of the Danube between Galatz and Braila, P.C.I.J., ser. B, No. 14 (1927); see Radovanovitch, *Le Danube maritime et le règlement du différend relatif aux compétences de la Commission européenne sur le secteur Galatz-Braila*, 13 REVUE DE DROIT INTERNATIONAL ET DE LÉGISLATION COMPARÉE, 3d ser. 564 (1932).

[178] Declaration of March 12, 1931, LEAGUE OF NATIONS OFF. J., 12th year 736 (1931). The signatories to the Convention instituting the Definitive Statute of the Danube (*supra* note 168) declared that if the provisions agreed upon by the nations represented on the European Commission should be put into force, they would be substituted for the stipulations of previous treaties in so far as they might be inconsistent.

[179] Draft Convention relating to the Maritime Danube, annexed to the Declaration of March 12, 1931, *supra* note 178, art. 1.

[180] Articles 2-5.

[181] Modus vivendi agreed upon by the delegates of France, Great Britain, Italy, and Rumania, May 17, 1933, P.C.I.J., ser. E, No. 9 at 115 (1933).

of the European Commission.[182] The European Commission of the Danube became a ghost. It was shorn of its powers regarding supervision of navigation, the port of Sulina, the enactment of regulations,[183] the furnishing of pilots, the execution and planning of improvements in the river works,[184] sanitary matters, dues, and the exercise of judicial authority. These functions were assumed by the Rumanian Government, and with the conclusion of the agreement came German membership on the Commission.[185] The period between the two World Wars saw only a slow decline in the powers and importance of the Commission. With the coming into force of the new agreement in 1939, the Commission which had been created in the nineteenth century ceased to exist for all practical purposes.

Although the Rhine was not free of its political problems, the Central Commission managed, unlike the commissions on the Danube, to survive the interwar years. The Treaty of Versailles gave the vessels and cargoes of nonriparians the same rights as those of riparians[186] and provided for a reconstituted Central Commission, on which the Allied and Associated Powers controlled a majority of the votes.[187] The Commission, in which voting was conducted on a weighted basis according to the number of representatives authorized each country,[188] included two nonriparians in addition to the littoral states. Their purpose was not only to increase the voting power of the Allied and Associated Powers but also to provide a stabilizing influence in disputes between the riparian states.[189] The Convention of Mannheim

[182] Arrangement between the Governments of Great Britain and Northern Ireland, France, and Rumania relative to the Exercise of the Powers of the European Commission of the Danube, signed at Sinaia, Aug. 18, 1938, 196 L.N.T.S. 113.

[183] Article 3. The Navigation and Police Regulations were to be drawn up by the European Commission "on the basis of proposals presented by the Rumanian Government." Rumania was to put them in force and to be responsible for their application.

[184] Article 6. While the Commission was authorized to modify the plans for works of improvement and upkeep proposed by Rumania, this action could be taken only by a majority vote of the Commission, which had to include the vote of Rumania!

[185] Agreement relative to the Entry of Germany into the European Commission of the Danube, the Accession of the Governments of Germany and Italy to the Arrangement signed at Sinaia on August 18th, 1938, between the Governments of Great Britain and Northern Ireland, France and Rumania, and the Amendment of Articles 4 and 23 of that Arrangement, signed at Bucharest, March 1, 1939, 196 L.N.T.S. 127.

[186] Article 356, Treaty cited *supra* note 150.

[187] Article 355.

[188] The Central Commission was to consist of 19 members: two from the Netherlands, two from Switzerland, four from the German riparian states, four from France, two from Great Britain, two from Italy, and two from Belgium.

[189] VAN EYSINGA, *op. cit. supra* note 147, at 119.

was to be revised by the Central Commission within six months after the coming into force of the Treaty of Versailles.[190] However, this task could not be effectively accomplished without the participation of the Netherlands, and that country looked with suspicion upon what seemed to be the increasing power of the Central Commission.[191] Concessions had to be made. The Netherlands was accorded an additional representative on the Commission and was promised that the Treaty of Versailles was not to "be interpreted as determining the territorial extent of the competence of the Central Commission of the Rhine or the juridical force of its regulations" and that the jurisdiction of the Commission was not to extend to the territory of the Netherlands without the consent of that country. In return the Netherlands agreed to participate in the revision of the Convention of Mannheim and, subject to a number of reservations in matters of detail, to adhere to the articles of the Treaty of Versailles relating to the Rhine.[192]

Closely allied to the question of representation in the Central Commission was that of voting. Prior to the First World War, this matter had been regulated by the provisions of the Act of Mannheim alone. Under that agreement the effect of a stipulation that decisions would be taken by a majority vote was in practice vitiated by the requirement that no such decisions would be obligatory until approved by the government in question.[193] In fact, unanimity was necessary if a decision binding on all was to be taken. Suspicions about a change in existing procedures had been raised by the provisions of the Treaty of Versailles which spoke of regulations "drawn up by the Central Commission" and of the Commission's ensuring that vessels comply with the "general regulations applying to navigation on the Rhine." [194] To satisfy objections on this score, particularly those raised by the Netherlands, a protocol was concluded in 1923, in which it was declared that decisions of the Central Commission would be taken by a majority of votes but that no state would be "obliged to take steps for the

[190] Article 354 of the Treaty of Versailles, *supra* note 150.

[191] VAN EYSINGA, *op. cit. supra* note 147, at 112-13.

[192] Protocol regarding the Adhesion of the Netherlands to the Modifications Introduced by the Treaty of Versailles in the Convention of Mannheim, signed at Paris, Jan. 21, 1921, 20 L.N.T.S. 111.

[193] Revised Convention for the Navigation of the Rhine, signed at Mannheim, Oct. 17, 1868, art. 46, 20 MARTENS, N.R.G. 355 (1875), 2 RHEINURKUNDEN 80 (1918), which provided: "Les résolutions de la Commission centrale seront prises à la pluralité absolue des voix, qui seront émises dans une parfaite égalité. Ces résolutions ne seront toutefois obligatoires qu'après avoir été approvées par les Gouvernements."

[194] Article 356.

execution of any resolution which it may have refused to approve."[195] The statement appears to be no more than declaratory of existing law. The juridical situation cannot, however, be seen in perspective unless it be mentioned that during the interwar years the members of the Central Commission did not exercise their privilege of refusing to give effect to a resolution against which they had voted.[196]

Somewhat paradoxically, the Central Commission of the Rhine assumed a position of greater independence with respect to the governments of member states at the very time that the atmosphere of the body became increasingly political.[197] Complaints were investigated, changes in the *règlement de police* considered, and technical committees appointed without reference of the matters to the member states.[198] In addition to its normal concern with technical and navigational matters on the Rhine, the Commission also gave consideration to various economic matters, such as subsidies[199] and the sharing of traffic,[200] which might be thought to affect freedom of navigation on the river. All such matters were not dealt with exclusively through the medium of the Central Commission of the Rhine. A long-standing dispute between the Netherlands and France concerning French taxes on goods which had entered the country without passing through a French port[201] was brought to an end by an agreement signed in 1939. This protocol, concluded by direct intergovernmental negotiation, gave to the Netherlands the benefits of the same exemption from this tax burdening goods passing up the Rhine to Strasbourg as had previously been accorded Belgium.[202]

It will be recalled that one of the functions assigned to the Central Com-

---

[195] Additional Protocol concerning the Adhesion of the Netherlands to the Modifications Introduced by the Treaty of Versailles in the Convention of Mannheim of 1868, signed at Paris, March 29, 1923, 20 L.N.T.S. 117.

[196] VAN EYSINGA, LA COMMISSION CENTRALE POUR LA NAVIGATION DU RHIN 126 (1935).

[197] *Id.* at 116.

[198] *Id.* at 120-21.

[199] *Compte rendu de l'activité de la Commission centrale pour la navigation du Rhin en 1936,* 15 NAVIGATION DU RHIN 144, 147 (1937).

[200] *Compte rendu de l'activité de la Commission centrale pour la navigation du Rhin en 1934,* 13 NAVIGATION DU RHIN 135, 137 (1935).

[201] The taxes were *surtaxes d'origine et d'entrepôt.* See 17 NAVIGATION DU RHIN 160 (1939).

[202] Agreement between Belgium, France, and the Netherlands regarding Certain Questions Connected with the Régime Applicable to Navigation on the Rhine, signed at Brussels, April 3, 1939, 195 L.N.T.S. 471.

mission had been the revision of the Convention of Mannheim of 1868.[203] It was not until 1936, thanks to German obstructionism, that a modus vivendi was signed which placed in provisional operation portions of a Revised Convention for the Navigation of the Rhine.[204] The Convention to which this limited effect was given reduced the powers of the Commission to the establishment of rules of application for the Convention, the consideration of complaints, the making of recommendations on economic matters and on the interpretation of the Convention and the rules adopted thereunder, and the preparation of an annual report.[205] The Commission, thus reduced to an advisory and consultative function, had jurisdiction over the portion of the river from Basel to Krimpen and Gorinchem, the rest of the river being governed by the Act of Mannheim.[206] In November 1936 Germany announced its refusal to comply further with the provisions of the Treaty of Versailles having application to waterways in German territory and with the modus vivendi.[207] The agreement of 1939 between France, Belgium, and the Netherlands, to which reference has previously been made,[208] provided that certain provisions of the Revised Convention of 1936 would be given effect as to the ports of Rotterdam, Amsterdam, Dordrecht, Antwerp, and Ghent, as well as to Rhine navigation proceeding to or from those ports, the open sea, or Belgium.[209] Despite the same German hostility which had for all practical purposes destroyed the international administration of the Danube, the Central Commission survived the years between the two World Wars, although not without attempts, as have been seen, to reduce the degree of autonomy and jurisdiction which it had enjoyed under the Convention of Mannheim.

The special provisions which were made by the Treaty of Versailles with respect to various international rivers of Europe were expected to yield to a general convention regarding the international régime of waterways to be concluded by the Allied and Associated Powers with the approval

---

[203] Treaty of Versailles, *supra* note 150, art. 354.

[204] Signed at Strasbourg, May 4, 1936, 36 MARTENS, N.R.G. 3d ser. 769 (1939).

[205] Revised Convention for the Navigation of the Rhine, art. 78, 36 MARTENS, N.R.G. 3d ser. 770 (1939).

[206] Article 1 of the Revised Convention.

[207] REICHSGESETZBLATT, 1936, II, 361, incorporating a German note of Nov. 14, 1936. See also *La Dénonciation par l'Allemagne de l'internationalisation des fleuves*, 14 NAVIGATION DU RHIN 417 (1936).

[208] See note 202 *supra*.

[209] Articles 6 and 7 of the Revised Convention, *supra* note 205.

of the League of Nations.[210] This treaty was in particular to apply to the Elbe, Oder, Niemen, and Danube. Representatives from 44 states met at Barcelona in 1921 for the purpose of drawing up such a convention.[211] The Convention on the Régime of Navigable Waterways of International Concern, which was the fruit of the conference's deliberations, imposed no new requirement that navigable waterways of international concern be placed under the supervision of an international administration. Indeed, the Convention took as the general situation the absence of an international administration and then proceeded to make special provisions for those cases in which a waterway was already or might in the future be subjected to such a régime, provided nonriparians were represented on the body.[212] To the general principle of freedom of navigation recognized by the Convention[213] an exception was made in favor of the reservation of cabotage to the vessels of each riparian, but the right of cabotage was still further limited in the case of waterways under the administration of an international commission to the "local transport of passengers or of goods which are of national origin or are nationalized."[214] The international commission was recognized to be ruled by the governing treaty as regards the maintenance and improvement of the waterway, but it was expressly declared that, in the absence of any provision to the contrary, decisions with regard to works would be taken by the commission.[215] Finally, the Convention provided that if a commission was established, the act of navigation for the river was to stipulate that the commission was to be entitled to draw up navigation regulations, to indicate to the riparian states the measures necessary for the upkeep and maintenance of the river, to be furnished information by the riparians on the improvements they intend to make in the waterway, and to approve the levying of dues.[216] The Convention contained the mild injunction that it was "highly desirable" that the riparian states "come to an understanding

---

[210] Treaty of Versailles, *supra* note 150, art. 338. Under article 379 of the Treaty, Germany undertook to adhere to any such general convention that might be concluded within five years after the coming into force of the Treaty.

[211] Convention and Statute on the Régime of Navigable Waterways of International Concern, signed at Barcelona, April 20, 1921, 7 L.N.T.S. 35, 1 HUDSON, INTERNATIONAL LEGISLATION 638 (1931); see Toulmin, *The Barcelona Conference on Communications and Transit and the Danube Statute*, 3 BRIT. Y.B. INT'L L. 167 (1922-23).

[212] Statute, art. 2.

[213] Statute, art. 3.

[214] Statute, art. 5, para. 1.

[215] Statute, art. 10, subpara. 5(*a*).

[216] Statute, art. 14.

with regard to the administration of the navigable waterway." [217] It is quite clear that the signatories of the Convention did not consider an international administration as a *sine qua non* or even as an important adjunct to the administration of an international river.

The Barcelona Convention was at once too conservative and too liberal. On the one hand, it extended the right of freedom of navigation only to signatories, in this respect derogating from a more general right of freedom of navigation established by a number of previous treaties. [218] On the other hand, it called for the assumption of new obligations by the riparian nations of navigable waterways of international concern which had not yet been opened to free use. [219] As a consequence of its vulnerability from these two perspectives, it was ratified by a disappointingly small group of states [220] and is thus of comparatively slight practical importance. The refusal of one riparian to become a party to the agreement could leave the relations of the riparians *in statu quo*. Thus, when Poland did not ratify the Convention, it became impossible for the other members of the International Commission of the Oder, in the view of the Permanent Court of International Justice, to invoke the provisions of the Barcelona Convention against the dissenter. [221] The Convention was, however, given application to the Niemen,

[217] Statute, art. 12.

[218] HAJNAL, LE DROIT DU DANUBE INTERNATIONAL 252 (1929); Van Eysinga, *Les Fleuves et canaux internationaux*, 2 BIBLIOTHECA VISSERIANA 121, 138-39 (1924).

[219] This point caused particular difficulties for the Latin American states. Alvarez, speaking for Chile at the Barcelona Conference, drew a distinction, in respect to freedom of navigation, between the European and American systems, the former of which was, for political, geographic, and economic reasons, more liberal than the latter. Only those rivers mentioned in the Treaty of Versailles he thought could be called "really international." LEAGUE OF NATIONS, BARCELONA CONFERENCE, VERBATIM RECORDS AND TEXTS RELATING TO THE CONVENTION ON THE RÉGIME OF NAVIGABLE WATERWAYS OF INTERNATIONAL CONCERN AND TO THE DECLARATION RECOGNISING THE RIGHT TO A FLAG OF STATES HAVING NO SEA-COAST 20-21 (1921); and see SOSA-RODRIGUEZ, LES FLEUVES DE L'AMÉRIQUE LATINE ET LE DROIT DES GENS 108 (1935); KASAMA, LA NAVIGATION FLUVIALE EN DROIT INTERNATIONAL 229-30 (1928).

[220] Within the ten years following the opening of the Barcelona Convention for signature, only 21 nations had ratified the treaty or acceded to it.

[221] Territorial Jurisdiction of the International Commission of the River Oder, P.C.I.J., ser. A, No. 23 at 22 (1929). It had been contended by the other members of the commission that article 338 of the Treaty of Versailles, *supra* note 150, providing that the provisional régime established by the Treaty "shall be superseded by one to be laid down in a General Convention" and that the Convention "shall apply" to the river systems which included the Oder, meant that Poland automatically became a party to the Barcelona Convention even though it did not ratify the agreement. This submission was rejected by the Court, according to the general rule that conventions are binding only by virtue of their ratification.

and Germany was in 1925 directed by the Conference of Ambassadors to observe the Convention with regard to the Rhine.[222]

THE PERIOD AFTER THE SECOND WORLD WAR

German hostility[223] had dealt harshly with the international river commissions before the Second World War. Part of the task of reconstruction had therefore to be, under the changed pattern of international politics created by the war, the reconstitution of the river commissions. One of these, the Danube, had passed from the German into the Russian sphere of influence.[224] The peace treaties with Bulgaria, Hungary, and Rumania spoke in general terms of a right of free navigation on the Danube, cabotage excepted, but laid no injunction upon the parties with respect to the rebuilding of the International and European commissions.[225] Unable to agree on this point, the Council of Foreign Ministers called a conference,[226] which met at Belgrade in August 1948, to draft a new convention governing the Danube.[227] The dominance of the river by the Soviet Union and its satellites —seven states as opposed to the three nonriparians, the United States, Great Britain, and France, which were invited—left no doubt of the outcome of the conference.[228] The Soviet Union had already made it quite clear that it regarded nonriparians as entitled neither to navigate the river nor to

---

[222] The Treaty of Versailles and After: Annotations of the Text of the Treaty 663, 671 (Dep't of State Pub. 2724) (1947).

[223] The freeing of Germany from the obligations imposed by the Treaty of Versailles with respect to international rivers running through German territory became one of the most important points of the National Socialist foreign policy. Wiese, Die völkerrechtliche Stellung der Elbe 28 (1927); Triepel, Internationale Wasserläufe 32 (1931); Lederle, *Die Zukunft der deutschen internationalisierten Ströme*, 20 Zeitschrift für Völkerrecht 65 (1936).

[224] H. A. Smith, *The Danube*, [1950] Year Book of World Affairs 191, 211. As had happened in the past, the silting of the Danube at the end of the hostilities of the Second World War posed a technical obstacle to navigation. 427 H.C. Deb., 5th ser. 1507 (1945-46).

[225] Treaty of Peace with Bulgaria, signed at Paris, Feb. 10, 1947, art. 34, 61 Stat. 1915, T.I.A.S. No. 1650, and corresponding provisions of the other two peace treaties.

[226] Hadsel, *Freedom of Navigation on the Danube*, 18 Dep't State Bull. 787, 792-93 (1948).

[227] *Id.* at 793; see, concerning the work of the conference, Kunz, *The Danube Régime and the Belgrade Conference*, 43 Am. J. Int'l L. 104 (1949); Gorove, *Internationalization of the Danube: A Lesson in History*, 8 J. Pub. L. 125, 146-53 (1959).

[228] The three nonriparians had proposed that Austria be included among the participants. The Soviet Union refused to allow Austria, which was still under occupation, to have other than a consultative status.

participate in its administration.[229] At the conference, the noncommunist nations which had been parties to the Definitive Statute of the Danube of 1921 maintained that that agreement was still in force and could not be abrogated or amended without the concurrence of all of the signatory states,[230] while the Soviet Union contended that the Sinaia Agreement of 1938 and the decision of the Council of Foreign Ministers in calling the conference had effectively terminated the Definitive Statute.[231] This latter argument was offered despite the fact that the Sinaia Agreement had not been concurred in by a majority of the states parties to the Definitive Statute. By reason of the preponderant position of the Soviet Union in the conference, the convention which was adopted over the protests of the nonriparians paid lip service to the principle of free navigation for the vessels of all states but subjected such navigation to regulations to be framed by the riparians— regulations which could make freedom of transit illusory.[232] A Danube River Commission, which was to include only the seven communist riparians, was charged with the responsibilities of making recommendations, studies, and investigations concerning the navigation, commerce, works, and regulations for the river.[233] The convention was signed by the seven communist riparians alone, and the three nonriparians voting at the conference maintained their position that the new treaty had no valid international effect.[234]

---

[229] Molotov had maintained in the Council of Foreign Ministers that the previous Danube régime was "the expression of imperialism." PARIS PEACE CONFERENCE, 1946, SELECTED DOCUMENTS 818 (Dep't of State Pub. 2868) (1946). The same position was echoed by Yugoslavia, *id.* at 817; and see Hadsel, *supra* note 226, at 792-93, 797.

[230] See the remarks of the representative of the United Kingdom on July 31, 1948 in MINISTÈRE DES AFFAIRES ETRANGÈRES DE LA RÉPUBLIQUE POPULAIRE FÉDÉRATIVE DE YUGOSLAVIE, CONFÉRENCE DANUBIENNE, BEOGRAD 1948, at 61 (1949); concerning the requisite participation in the revision of the régime of the river, see HOYT, THE U-NANIMITY RULE IN THE REVISION OF TREATIES: A RE-EXAMINATION 147-55 (1959).

[231] See the remarks of Mr. Vyshinsky, CONFÉRENCE DANUBIENNE 63-64, cited *supra* note 230.

[232] Convention regarding the Régime of Navigation on the Danube, signed at Belgrade, Aug. 18, 1948, arts. 1, 23, and 24, 33 U.N.T.S. 196.

[233] Articles 5 and 8. Special river administrations were also created for the mouth of the Danube and for the Iron Gates sector, which had historically been the subjects of special régimes.

[234] United States note of Nov. 15, 1949, 21 DEP'T STATE BULL. 832 (1949). Similar notes were delivered by the governments of France and the United Kingdom. The European Commission continued in existence after 1948 for the purpose of maintaining the juridical position of the nonriparians and liquidating its financial affairs. COMMISSION EUROPÉENNE DU DANUBE, UN SIÈCLE DE COOPÉRATION INTERNATIONAL SUR LE DANUBE, 1856-1956, at 50-52 (1956).

A Soviet monopoly of navigation was established over the river through the medium of shipping and other companies jointly constituted by the Soviet Union and other Danube states.[235] Coincident with a rapprochement in 1953 between the Soviet Union and Yugoslavia, which had been at odds over navigation on the Danube and participation in the new commission,[236] there was a gradual freeing of navigation over the portion of the river controlled by the communist states.[237] The conclusion of the treaty of 1957 between Austria and the Soviet Union, providing reciprocal rights of free navigation[238] and the full membership of Austria in the commission,[239] merely confirmed the privileges which had already been accorded to Austrian vessels.[240] As a matter of law, the Danube remains a river from which vessels of nonriparians may be excluded and in the administration of which nonriparians play no part.

Meetings of the Central Commission of the Rhine were suspended between April 1940 and the general termination of hostilities in 1945. The decision was taken to invite the Commission to resume its activities, and the United States was added to the membership of the body in recognition of its interest and role in Europe.[241] The newly reconstituted organization

---

[235] Hadsel, *supra* note 226, at 791.

[236] *Situation de la navigation danubienne*, 24 REVUE DE LA NAVIGATION 12 (1952); PRIGRADA, INTERNATIONAL AGREEMENTS CONCERNING THE DANUBE 23-26 (1953).

[237] 25 REVUE DE LA NAVIGATION 501 (1953); 26 *id.* at 631, 675 (1954); 27 *id.* at 163 (1955); see, concerning navigation on the Danube, Feuerstein, *Le Danube: Son aménagement, son trafic, son rôle dans l'activité économique de l'Europe centrale*, 31 *id.* at 302 (1959); Stolte, *Moscow Regulates Traffic on the Danube*, 7 INSTITUTE FOR THE STUDY OF THE U.S.S.R., BULLETIN, No. 5, 21 at 27-29 (1960).

[238] Agreement between the Union of Soviet Socialist Republics and Austria concerning the Settlement of Technical and Commercial Questions relating to Navigation on the Danube, signed at Moscow, June 14, 1957, 285 U.N.T.S. 169.

[239] N. Y. Times, Jan. 24, 1960, p. 17, col. 1.

[240] SHEPHERD, RUSSIA'S DANUBIAN EMPIRE 240 (1954).

[241] Arrangement providing for Participation by the United States of America in the Central Commission of the Rhine, signed at London, Oct. 4/29, Nov. 5, 1945, 60 Stat. 1932, T.I.A.S. No. 1571.

One of the curiosities of membership in the C.C.R. through the medium of instruments other than the Convention of Mannheim itself was the discovery in 1952 that that treaty, to which Belgium had become a party through ratification of the Treaty of Versailles, had never been published in Belgium and hence that it was not applicable to the private rights of Belgians in a case arising out of a collision between Belgian ships on the Rhine. Boileau v. Mélard et Veuve Wyckmans, Cour d'Appel de Liège, March 23, 1952, PASICRISIE BELGE, 1952, II, 64. The Convention was published in 1954. 26 REVUE DE LA NAVIGATION 746 (1954).

was called upon[242] to cooperate with the European Central Inland Transport Organization, which concerned itself with the allocation, utilization, and return to the owners of the fleet operating on the inland waterways of Europe.[243] During the early days of the Central Commission's new life, there was also in existence the Rhine Interim Working Committee, constituted under British sponsorship, which dealt with certain problems relating to the river fleet and navigation on the waterway.[244]

Some of the most urgent administrative tasks facing the Commission were engendered by the occupation of Germany and the disturbed conditions which prevailed in Europe immediately after the close of hostilities. Prolonged negotiations with the military authorities were necessary before a uniform *laissez-passer* was established for boatmen.[245] The Commission concerned itself with the provisioning and supplying of vessels in Germany[246] and for a time established an international rationing card for the boatmen who plied the Rhine.[247] Interventions were made with the military authorities to minimize the effect of military exercises on navigation of the river.[248] Special provisions had to be made for the obstacles to navigation, such as destroyed bridges, which the war had left in its path.[249] Over and above these exceptional tasks arising out of the occupation and the aftermath of war, the Central Commission was called upon to resume its normal functions. The tribunals of navigation in the various riparian countries were reconstituted—in the case of Germany only after somewhat involved arrangements with the occupation authorities.[250] The Commission resumed

[242] United States note of Oct. 29, 1945, 60 Stat. 1933, T.I.A.S. No. 1571.

[243] Agreement Concerning the Establishment of an European Central Inland Transport Organization, signed at London, Sept. 27, 1945, 59 Stat. 1740, E.A.S. No. 494.

[244] Address by Brigadier Walter at Congrès International de Navigation, Basel, Oct. 11-12, 1946, 18 NAVIGATION DU RHIN 389, 391 (1946).

[245] *L'Activité de la Commission centrale du Rhin en 1951*, 24 REVUE DE LA NAVIGATION 548, 549 (1952).

[246] See, e.g., Commission Centrale pour la Navigation du Rhin, 3d sess., 1949, *Communiqué du Secrétariat*, 21 NAVIGATION DU RHIN 475 (1949).

[247] First sess., 1947, *Communiqué du Secrétariat*, 19 NAVIGATION DU RHIN 41 (1947). The card was found to be no longer needed in 1951 and was discontinued.

[248] Session of April 1953, *Communiqué du Secrétariat*, 25 REVUE DE LA NAVIGATION 289 (1953).

[249] See, e.g., *Compte rendu sur l'activité de la Commission Centrale en 1946*, 19 NAVIGATION DU RHIN 85, 87 (1947), concerning the marking of navigable channels past destroyed bridges.

[250] Third sess., 1947, *Compte rendu du Secrétariat*, 19 NAVIGATION DU RHIN 411, 412 (1947).

its own juridical functions.[251] A new *règlement de police* for the river was framed and came into force in 1955,[252] and new safety regulations were drafted.[253] There were a number of complaints about delays caused by customs controls, particularly on the German-Netherlands border, with which the Commission had to contend.[254]

More significant than these technical and administrative questions, however, were the economic problems which were assuming an increasing importance for the Commission, as cargo was diverted to other means of transport and to other ports.[255] An epidemic of what Professor van Eysinga referred to as "neo-protectionism"[256] was reflected in attempts by various riparian nations to reserve navigation to their own vessels.[257] The Commis-

[251] The Comité des Rapporteurs, the judicial organ of the Commission, met at Basel for the first time after the war on March 23, 1948. Five cases were pending before it. 20 NAVIGATION DU RHIN 155 (1948).

[252] See *Le Nouveau règlement de police pour la navigation du Rhin*, 26 REVUE DE LA NAVIGATION 191 (1954). The new *règlement* replaced that of 1939.

[253] A new set of rules for the inspection of vessels was drawn up in 1947. Fourth sess., 1947, *Compte rendu du Secrétariat*, 19 NAVIGATION DU RHIN 493 (1947).

[254] A special committee of the Commission was set up to deal with customs questions. Second sess., 1949, *Communiqué du Secrétariat*, 21 NAVIGATION DU RHIN 292 (1949). Concerning the amelioration of conditions on the border between Germany and the Netherlands, see, e.g., 3d sess., 1950, *Communiqué du Secrétariat*, 22 *id.* 532, 533 (1950).

[255] *Recent Trends in the Use of Some European Inland Waterways for the Transport of Goods*, 3 TRANSPORT AND COMMUNICATIONS REVIEW, No. 1, at 3, 5 (1950).

[256] Address at Rotterdam, Oct. 16, 1948, 21 NAVIGATION DU RHIN 64-65 (1949).

[257] Second sess., 1951, *Communiqué du Secrétariat*, 23 REVUE DE LA NAVIGATION 460 (1951).

The Netherlands created a system for the sharing of freight within its own shipping industry. Customs clearance was denied if the proprietor of the ship could not prove that he was a member of one of the two organizations charged with administering this system and that the organization had authorized the particular voyage. In 1950, a Dutch tug drawing a Belgian barge was denied customs clearance to leave the Netherlands because it lacked the requisite authorization. The Belgian barge-owner, who had been forced to procure another tug, brought an action against the Netherlands. The Dutch Supreme Court held for the barge-owner on the ground that the requirement of approval by the Stichting Nederlandse Particuliere Rijnvaartcentrale or the Vereniging Centraal Bureau voor de Rijn-en Binnenvaart constituted a restriction on navigation forbidden by article 1 of the Convention of Mannheim. De Staat der Nederlanden v. Joh. Boon en Société Anonyme Chantiers Navals du Rupel, Jan. 25, 1952, Nederlandse Jurisprudentie, 1952, No. 125; see Fortuin, *L'Artisanat néerlandais et la liberté de navigation sur le Rhin*, 24 REVUE DE LA NAVIGATION 214 (1952).

In 1953, the Netherlands complained that interior traffic in Germany had remained closed to Dutch vessels in contravention of the Convention of Mannheim. Note from the Netherlands Embassy to the German Foreign Ministry, July 28, 1953, 25 REVUE DE LA NAVIGATION 497 (1953).

sion called an economic conference to deal with these matters in 1952,[258] but the implementation of the recommendations made by the conference proved difficult of attainment, thanks in part to the close relation between the economic problems of the Rhine and those of Europe as a whole. It accordingly became necessary for the Commission to convene a further such conference in 1959.[259]

The Commission was not alone in its concern with these matters. The High Authority of the European Coal and Steel Community found that its effort to avoid discrimination between producers was obstructed by the disparity which existed between inland and international rates on the inland waterways of Europe, including the Rhine.[260] An agreement concluded in 1957 had as its purpose the establishment of procedures whereby the regulated internal rates and the free international rates for the transport of coal and steel might be harmonized.[261] This question of coordination of rates in turn fell within the purview of the European Conference of Transport Ministers, which found itself burdened as well with problems, similar to those which had existed throughout the past 150 years, concerning the extent to which cabotage should be reserved to individual riparians.[262] When

[258] *Conférence économique de la navigation rhénane,* 24 REVUE DE LA NAVIGATION 154 (1952); see Walther, *Le Statut international de la navigation du Rhin,* 2 EUROPEAN YEARBOOK 3, 26 (1956).

[259] RAPPORT ANNUEL DE LA COMMISSION CENTRALE POUR LA NAVIGATION DU RHIN, 1959, at 5.

[260] HIGH AUTHORITY, EUROPEAN COAL AND STEEL COMMUNITY, MONTHLY REPORT, Jan./Feb. 1955, 3d year, No. 1, at II, 3, I-II, 3, 4.

In 1955 the Common Assembly of the European Coal and Steel Community caused an expert committee to be established to draw up proposals for the coordination and integration of all European transport (COMMUNAUTÉ EUROPÉENNE DU CHARBON ET DE L'ACIER, DÉBATS DE L'ASSEMBLÉE COMMUNE, Aug. 1955, No. 9, at 381). The committee rendered a substantial report two years later: RAPPORT FAIT AU NOM DE LA COMMISSION DES TRANSPORTS SUR LA COORDINATION DES TRANSPORTS EUROPÉENS, Doc. No. 6, 1957-58.

[261] *Accord relatif aux frets et conditions de transport pour le charbon et l'acier sur le Rhin,* 7 JOURNAL OFFICIEL DE LA COMMUNAUTÉ EUROPÉENNE DU CHARBON ET DE L'ACIER 49/58 (1958); see EUROPEAN COAL AND STEEL COMMUNITY, HIGH AUTHORITY, NINTH GENERAL REPORT ON THE ACTIVITIES OF THE COMMUNITY 185-87 (1961).

[262] Commission centrale pour la navigation du Rhin, sess. d'automne 1954, *Communiqué du Secrétariat,* 26 REVUE DE LA NAVIGATION 717 (1954). The Economic Commission for Europe also constituted a special committee to look into the matter of putting on a uniform basis the regulation of navigation, of police, and of the transport of dangerous goods and fluvial law for all European waterways, an activity in which the Central Commission for the Navigation of the Rhine cooperated. 27 REVUE DE LA NAVIGATION 192 (1955); RAPPORT ANNUEL DE LA COMMISSION CENTRALE POUR LA NAVIGATION DU RHIN, 1959, at 3.

the Central Commission of the Rhine took up the matter of social security for boatmen, collaboration with the International Labor Organization was necessary in order to permit the conclusion of satisfactory arrangements.[263] Such is the interdependence of Rhine navigation with the economy of Europe that the integration of European inland transport may be a necessary step in the economic unification of that area. The inability of the Central Commission of the Rhine to deal with these problems in isolation suggests that an international body should be brought into existence to concern itself not only with the fluvial system of Europe but with the entire transport system of that area. With the creation of such a body, it might be expected that the Central Commission might be limited in its competence to the technical function of dealing with navigation and works along the water-course.

The most recently established of the international river commissions has been the Commission of the Moselle, which France, Germany, and Luxembourg agreed in 1956 should be established at least one year before the opening of the river to large-scale navigation as the consequence of the canalization of the watercourse.[264] The Commission was accorded only limited powers. Sitting twice yearly and acting by a rule of unanimity, the Commission is to regulate the collection of the tolls established by the treaty;[265] to verify the consistency of works along the river with freedom of navigation;[266] to look generally after the economic health of river navigation;[267] and to hear appeals from national navigation tribunals.[268]

STRAIT AND CANAL COMMISSIONS

The international commission, which has historically played such an important part in maintaining free passage through international rivers, has only rarely been adopted, and then only with doubtful efficacy, for the

[263] Agreement concerning the Social Security of Rhine Boatmen, done at Paris, July 27, 1950, 166 U.N.T.S. 73; see *La Conférence de Genève sur la sécurité sociale et les conditions de travail des bateliers du Rhin*, 21 NAVIGATION DU RHIN 467 (1949).

[264] Convention between the Federal Republic of Germany, the French Republic, and the Grand Duchy of Luxembourg concerning the Canalization of the Moselle, signed at Luxembourg, Oct. 27, 1956, art. 39, 11 VERTRÄGE DER BUNDESREPUBLIK DEUTSCHLAND 34 (1959); LES ACTES DU RHIN 46 (1957).

[265] Article 40, subpara. 1a.

[266] Article 37, para. 2.

[267] Article 40, subpara. 1c.

[268] Article 34, para. 4.

supervision of passages between two stretches of the high seas. The only one which actually functioned was the Straits Commission established under the Treaty of Lausanne.[269] The agreement itself was a compromise between the demands of the Western powers for free and unimpeded navigation through the Turkish Straits for all vessels of commerce and of war and the desire of the Black Sea nations to preserve their security and their sovereignty.[270] Vessels of commerce were to be allowed free passage in time of peace and in time of war, but limits were placed on the number of war vessels which might be sent through the Straits into the Black Sea, and Turkey was permitted to take defensive measures against enemy warships in the event of war.[271] The Bosporus and Dardanelles were demilitarized.[272] The Straits Commission, consisting of representatives of Turkey, France, Great Britain, Italy, Japan, Bulgaria, Greece, Rumania, Russia, and Yugoslavia,[273] was "to see that the provisions relating to the passage of warships and military aircraft are carried out." [274] Unlike the river commissions, which were in the main operational, the Straits Commission was no more than a supervisor of transit, responsible for assuring that warships would pass through the Straits without hinderance and for reports on the numbers and types of warships using the waterway and present within the Black Sea. It was occasionally called upon to make representations to the Turkish Government about such matters as the consistency of sanitary inspection of vessels of war[275] and of prohibited zones[276] with the right of free passage guaranteed by the Treaty of Lausanne, but for the most part it led an unruffled existence as inspector of the free navigation of warships. The termination of its existence by the Montreux Convention of 1936 was not the result of any dissatisfaction with the way in which it had performed its task.[277]

---

[269] Convention relating to the Régime of the Straits, signed at Lausanne, July 24, 1923, 28 L.N.T.S. 115.

[270] Routh, *The Montreux Convention Regarding the Regime of the Black Sea Straits (20th July, 1936)*, in TOYNBEE, SURVEY OF INTERNATIONAL AFFAIRS, 1936, 584 at 597 (1937); see also BERKOL, LE STATUT JURIDIQUE ACTUEL DES PORTES MARITIMES ORIENTALES DE LA MÉDITERRANÉE (LES DÉTROITS–LE CANAL DE SUEZ) 119-55 (1940).

[271] Annex to art. 2 of the Treaty of Lausanne, *supra* note 269, arts. 1 and 2.

[272] Treaty of Lausanne, art. 4.

[273] Treaty of Lausanne, arts. 10 and 12.

[274] Treaty of Lausanne, art. 14.

[275] RAPPORT DE LA COMMISSION DES DÉTROITS À LA SOCIÉTÉ DES NATIONS, ANNÉE 1926, at 11 (1927).

[276] *Id.*, ANNÉE 1927, at 12 (1928).

[277] When the revision of the Treaty of Lausanne was formally proposed by the Turkish Government, the existence of the Straits Commission was not mentioned in a

On the contrary, the even tenor of the Commission's life suggested that the function of supervising passage through the Straits might safely be handed over to Turkey. More than this, however, the changing pattern of international politics, the understandable desire of Turkey to reassert its sovereignty over a waterway flowing through its territory, and the need of that country to substitute more effective measures of defense for the power vacuum left by the demilitarization of the Straits called for a new régime for the Bosporus and Dardanelles.[278] Restrictions were maintained by the Montreux Convention of 1936[279] on the passage of the Straits and on the number of warships in the Black Sea, but it seemed not only convenient, but proper as well, that Turkey, once more in command of the waterway, should like any other littoral state bear the responsibility for assuring free passage through the Straits. The Straits Commission had hardly been a dramatic experiment; its passing was unmourned.

Proposals have from time to time been made that interoceanic canals be placed under the supervision of an international commission. Such, for example, was the French position with regard to the Kiel Canal at the Paris Peace Conference of 1919.[280] The unwillingness of the Council of Four to provide such a régime for the Kiel Canal may well have been inspired by the suspicion that this form of internationalization might set a bad precedent for Suez and Panama, to which canals allusions were made in the course of the discussions.[281] History repeated itself in a peculiar form

---

lengthy exposition of the reasons why Turkey considered a change in the régime of the Straits desirable. See Telegram, dated April 10th, 1936, from the Turkish Government to the Secretary-General, LEAGUE OF NATIONS OFF. J., 17th year, 505 (1936).

[278] On the diplomatic history of the position of Turkey toward the Treaty of Lausanne, see Routh, *supra* note 270, at 598-612.

[279] Convention concerning the Régime of the Straits, signed at Montreux, July 20, 1936, 173 L.N.T.S. 213, 7 HUDSON, INTERNATIONAL LEGISLATION 386 (1941). Article 24 transferred the functions of the "International Commission" (i.e. the Straits Commission) to the Turkish Government, which was required to furnish certain statistics and information regarding passages of the Straits.

[280] Draft Articles Concerning the Kiel Canal for Insertion in the Preliminary Treaty of Peace with Germany, Appendix I to Notes of a Meeting Held at President Wilson's House, April 24, 1919, [1919] 5 FOREIGN REL. U.S., PARIS PEACE CONFERENCE 206 (1944).

[281] Lloyd George inquired whether it was wise to set up an international commission to control a purely German canal and whether it was "worth while to hurt German pride and add to our own difficulties for so small a matter." Clemenceau thereupon withdrew the French draft. President Wilson had previously spoken of the analogy between the régime of the Panama Canal and the proposed arrangements for

when President Roosevelt suggested at the Teheran Conference[282] that "trustees" be appointed for Kiel.

It should not be forgotten that article VIII of the Convention of Constantinople,[283] which remains the basic law for the Suez Canal, contemplated that the "Agents in Egypt of the Signatory Powers" would be "charged to watch over its [the treaty's] execution." These persons were to meet once a year to "take note of the due execution of the Treaty," were to demand "the suppression of any work or the dispersion of any assemblage on either bank of the Canal" which might interfere with navigation, and were to meet, on the summons of three of their number, "in case of any event threatening the security or the free passage of the Canal." This rudimentary form of international commission never became operative. In later years, Italian fears about the vulnerability of Suez as a portion of its line of communication between the colonial empire and Italy itself led to demands for "a truly international administration" of the Canal and for "placing it under a truly international control" in which Italy would be represented.[284]

One of the responses to the nationalization by Egypt of the Suez Canal Company was a proposal by France, the United Kingdom, and the United States that the Canal be operated by an "International Authority for the Suez Canal," in which the nations chiefly interested in navigation and trade through the Canal would be represented.[285] It was envisaged that such an authority would carry out the necessary works, generally administer and control the waterway, and fix tolls which would assure an "equitable return" to Egypt. This plan was indignantly rejected by Egypt.[286] A watered-down proposal for a "Suez Canal Board," made by eighteen powers at the

---

Kiel. Notes of a Meeting Held at President Wilson's Residence, April 25, 1919, *op. cit. supra* note 280, at 236-37. Cf. the views of Mr. Lansing, who had observed that he was being "very careful not to have the Kiel Canal given any status similar to the Panama Canal, which might perhaps involve a whole discussion of the Panama Canal question by the Conference." Minutes of the Daily Meetings of the Commissioners Plenipotentiary, March 20, 1919, [1919] 11 FOREIGN REL. U.S., PARIS PEACE CONFERENCE 121 (1945).

[282] CHURCHILL, CLOSING THE RING 381 (Boston, 1951).

[283] Signed Oct. 29, 1888, 15 MARTENS, N.R.G. 2d ser. 557, 562 (1891).

[284] ITALIAN LIBRARY OF INFORMATION, ITALY AND THE SUEZ CANAL 55 (1941).

[285] Tripartite Proposal for the Establishment of an International Authority for the Suez Canal, Aug. 5, 1956, THE SUEZ CANAL PROBLEM, JULY 26-SEPTEMBER 22, 1956, at 44 (Dep't of State Pub. 6392) (1956).

[286] Statement by President Nasser Rejecting Invitation to the London Conference, Aug. 12, 1956, *id.* at 50-51.

conclusion of the First London Conference,[287] proved to be equally unacceptable.[288] The Board, on which Egypt and other interested States would serve, would have been responsible for "operating, maintaining and developing the Canal and enlarging it." The Suez Canal Users Association, which was the product of the Second London Conference,[289] had not the physical means of administering the Canal and must be regarded as nothing more than a negotiating weapon in the diplomatic war being waged with Egypt.

THE LESSONS OF THE PAST

This lengthy description of the organization, functions, and history of the international administrations set over international waterways has been necessitated by the undoubted importance of these bodies to the development of free fluvial and maritime communication between nations. These institutions reached their high point in the nineteenth and early twentieth centuries, for the history of the period following the First World War was a succession of withdrawals, defeats, and disappointments for the international river commissions, unaccompanied by the emergence of any new and vigorous organizations. If one speaks of the effectiveness of international river commissions, one's mind is drawn to an earlier era; no one can speak with enthusiasm of what the last forty years have brought.

Despite this generally discouraging pattern, "internationalization" or the establishment of an "international commission" is not infrequently suggested as a panacea for the political difficulties to which a particular waterway has given rise.[290] It may be well, therefore, to attempt to set down a number of conclusions suggested by the foregoing discussion of the role of international commissions in the administration of international waterways.

1. The international commissions have been most successful in their

---

[287] Five-Power Proposal, Aug. 21, 1956, *id.* at 291; and Statement Submitted by New Zealand, Aug. 23, 1956, *id.* at 293.

[288] Letter from President Nasser to Prime Minister Menzies, Sept. 9, 1956, *id.* at 319.

[289] Declaration Providing for the Establishment of a Suez Canal Users Association, Sept. 21, 1956, *id.* at 365.

[290] See generally Chapter VII *infra*. One example, drawn from many, is the proposal of Tchirkovitch that the Turkish Straits should be placed under an International Straits Commission (*La Question de la révision de la Convention de Montreux concernant le régime des Détroits Turcs: Bosphore et Dardanelles*, 56 REVUE GÉNÉRALE DE DROIT INTERNATIONAL PUBLIC 189, 219 (1952)).

functions as organs of consultation and coordination with respect to matters of a strictly technical order—rules of navigation, safety standards, the suppression of obstructions to navigation, and investigation of the state of the waterway. While free navigation, in both a legal and a technical sense, can be maintained on an international river without an international commission, such a body is a convenient means of concerting the activities of the riparians.[291] The achievements of the commissions in this sphere reflect an undoubted reality of international relations—that states can most easily and effectively work through common organs in matters of common interest when conflicting vital interests are not at stake.

2. The maintenance of free navigation, as the recent history of the Rhine has shown, is today increasingly bound up with economic problems going beyond the single river or river system. The conventional international river commissions, by reason of their preoccupation with the problems of individual rivers, may not be qualified to deal with these economic questions. The view that fluvial transportation must ultimately be organized on a regional basis[292] has much to commend it, provided it be understood that the economics of fluvial navigation cannot be separated from the larger question of the establishment of a viable transportation system and of a sound economy within the region served by the river.

3. The international commissions are not, and never have been, legislative bodies in the sense that a dissenting state would be bound by the decision of the commission. Even in the European Commission of the Danube, which is often held up as a model of the autonomous river commission, votes on important matters, save only the fixing of tolls, were taken by unanimous consent, and regulations for the river were likewise adopted by common agreement.[293] In essence, a similar requirement of unanimity

---

[291] Eagleton expressed the view that an administrative body of some sort is "a necessary part of the solution of the problems concerning the use of waters of an international river." An Inquiry Concerning the Legal Principles Governing the Use of the Waters of International Rivers, A Talk at the Canadian Branch of the International Law Association, Montreal, March 3, 1955, at 9 (mimeographed).

[292] Wehle, *International Administration of European Inland Waterways,* 40 Am. J. Int'l L. 100, 116-20 (1946).

[293] General Treaty of Peace between Austria, France, Great Britain, Prussia, Russia, Sardinia, and the Ottoman Porte, signed at Paris, March 30, 1856, art. XVI, 15 Martens, N.R.G. 770 (1857), 46 British and Foreign State Papers 8 (1865); Internal Rules of the European Commission of the Danube, signed at Galatz, Nov. 10, 1879, art. 12b, 9 Martens, N.R.G. 2d ser. 712 (1884). Chamberlain points out that the autonomy of the Commission was still further diminished by the fact that the members of the

has existed in the case of the Central Commission for the Navigation of the Rhine, whose decisions are not binding upon those states which did not approve them.[294]

4. The presence of nonriparians on the commission is a useful but not necessarily completely efficacious manner of maintaining a right of free navigation in favor of the ships of states other than the riparians. It is inevitable that such nonriparians should be looked upon, to greater or less degree, as intruders within the riparian community.[295]

5. The placing of a river under the supervision or control of an international commission does not necessarily insulate the waterway from politics.[296] It may do no more than to institutionalize the conflict and provide a further arena in which it may be fought. An international commission will not prevent internal strife. It will not protect the other riparians from the attacks of a Russia or a Rumania or a Germany determined to exploit its geographical or political position to gain control of the waterway or to exact concessions from the other riparians. A genuine community of interest has, on the other hand, had much to do with the success of the Central Commission of the Rhine over the century and a quarter of its existence.

6. Neither an international river nor the international commission set over it can be effectively neutralized in such a way as to protect it from violence and from the consequences of war. The activities of every river commission have been either suspended or permanently terminated by the outbreak of a war which involves the particular river.

7. A state which desires to dominate a waterway may give a coloring of legitimacy to its claims by seeking membership on the commission[297]

Commission acted under the instructions of their governments. THE REGIME OF THE INTERNATIONAL RIVERS: DANUBE AND RHINE 97 (1923).

[294] Revised Convention for the Navigation of the Rhine, signed at Mannheim, Oct. 17, 1868, art. 46, 20 MARTENS, N.R.G. 355, 369 (1875), 2 RHEINURKUNDEN 80, 98 (1918).

[295] This proved to be true in the experience of the European Commission of the Danube and of the [Turkish] Straits Commission.

[296] "C'est que les affaires du Danube étaient, plus que celles d'aucune autre voie navigable, dominées par les considérations de la politique: elles faisaient partie intégrante de la question d'Orient et elles touchaient, au point de vue économique aussi bien qu'au point de vue national, aux intérêts contradictoires d'un grand nombre d'Etats. Si l'on voulait atteindre un résultat, il fallait donc s'efforcer de concilier par des transactions plus ou moins chaotiques les exigences egoïstes des différents intéressés." 1 FAUCHILLE, TRAITÉ DE DROIT INTERNATIONAL PUBLIC, pt. II, 553 (1925); and see H. A. Smith, *The Danube*, [1950] YEAR BOOK OF WORLD AFFAIRS 191, 207.

[297] As in the case of German membership in the European Commission of the Danube, toward the end of the existence of that body. See p. 129 *supra*.

or by demanding "internationalization" of the waterway. The most delicate balancing of the various interests which should be represented in a commission—those of the territorial sovereigns, and of the users—is necessary in order to avoid domination of the river by a single power.

These conclusions suggest that proposals for the creation of international commissions should be viewed with an appropriate mixture of skepticism and realism. Facile references to ancient successes in one sphere do not necessarily mean that the same techniques that have worked successfully in the past may be transplanted to other waterways or other types of waterways. Above all, the creation of an international commission is not a cure-all for a troublesome political situation which cannot be solved by other means. Nevertheless, the not inconsiderable body of wisdom which has grown up about the administration of international waterways should have the positive effect of suggesting workable techniques of international ordering and the negative one of protecting the creators of new institutions from repeating some of the worst blunders of the past.

CHAPTER III

PASSAGE OF SHIPS THROUGH INTERNATIONAL WATER‑WAYS IN TIME OF PEACE

FREEDOM of passage through international waterways in time of peace must be understood in two senses. The first of these is the legal obligation to permit free passage by the vessels of other nations in the sense of not placing legal prohibitions or unreasonable restrictions on the use of the waterway. The second aspect of free passage, which may be referred to as the "technical" aspect of transit, is concerned with the prevention of physical impediments to transit and with the maintenance and improvement of the waterway to make it suitable for navigation. It is only the first of these two means of maintaining freedom of transit which concerns us at this point.[1]

Although the principle of freedom of transit has frequently been asserted to have application to rivers, canals, and straits, the specific rules which give life and meaning to a general principle, hovering between law as it is and law as it ought to be, differ considerably as between these major types of watercourses. This is not to deny that practices and law derived from one type of waterway have frequently been given application to other types of passages with considerable beneficial effect for the progressive development of law and for the harmonization of legal rules dealing with kindred problems.[2]

In the formative years of international law, there was considerable support for the doctrine that there was an inherent right of free passage over an international river. Grotius, for example, expresses the opinion that "rivers, and any part of the sea that has become subject to the ownership

---

[1] See Chapter VI for the technical aspect of freedom of transit.

[2] As, for example, in the development of the analogy between interoceanic canals and international straits in The S.S. Wimbledon, P.C.I.J., ser. A, No. 1, at 28 (1923).

of people, ought to be open to those who, for legitimate reasons, have need
to cross over them" and cites as one of these legitimate reasons the carrying
on of commerce through a waterway.[3] This theme is adopted and elaborated
by Vattel, who shows an awareness of the technical as well as the legal
basis of free transit.[4] But it is quite clear that this principle of free transit
of international rivers was a reflection of a wider right of free passage which
these authorities believed should have application to the seas, to rivers, and
to the land. The fact that no such general right of free passage was recognized in practice at the time and the denial of any such right to unimpeded
transit on land in the international law of the nineteenth and twentieth
centuries justifiably weakened the force of these contentions about a right of
free usage of international rivers.[5] Nevertheless, as late as the controversy
between the United States and Great Britain over free passage of the St.
Lawrence River in the 1820's[6] and in the *Faber* case[7] at the turn of the
century, the venerable authorities were appealed to in support of the right
of nationals of a riparian state to navigate the portion of an international
river falling within the territory of another riparian.

During the seventeenth and eighteenth centuries, a number of treaties
had granted a right to the free passage of certain individual rivers, usually
on a reciprocal basis as between the riparian nations.[8] To the extent that
the treaties of the nineteenth and twentieth centuries provide for a right
of free navigation of the rivers with which they deal, their common ancestry
may be traced to the articles which were adopted at the Congress of Vienna
in 1815. Despite efforts which were made to establish the right of nonriparians to utilize the Rhine and other waterways with which the Congress
was concerned,[9] the principles which were laid down in the Final Act of

---

[3] DE JURE BELLI AC PACIS, Book II, ch. II, §§ xii and xiii (Kelsey transl. 1925).

[4] LE DROIT DES GENS, Book II, ch. ix, §§ 126, 129, ch. x, § 132; Book I, ch. xxii,
§§ 272 and 273 (1758).

[5] H. A. Smith, *The Danube,* [1950] YEAR BOOK OF WORLD AFFAIRS 191, 196.

[6] AMERICAN STATE PAPERS, 6 FOREIGN RELATIONS, 2d ser. 769-75 (1859), 19 BRITISH
AND FOREIGN STATE PAPERS 1067-83 (1834), incorporating the Argument of the American
Plenipotentiary, annexed to the Protocol of the 18th Conference between the British and
American Plenipotentiaries, London, June 19, 1824.

[7] First Opinion of Commissioner Goetsch (the German Commissioner), in Faber
Case (Germany v. Venezuela), RALSTON, VENEZUELAN ARBITRATIONS OF 1903, 600 at
606 (1904).

[8] A number of these are quoted and described in HAJNAL, THE DANUBE: ITS
HISTORICAL, POLITICAL AND ECONOMIC IMPORTANCE 17-25 (1920).

[9] Great Britain, in particular, wished to see the right of free navigation extended to
nonriparians. The view was taken, however, that to limit the right to riparians would

the Congress went no further than to grant rights of navigation to the riparian states.[10] The Articles Concerning the Navigation of the Rhine provided that the entire navigable course of the river "sera entièrement libre, et ne pourra, sous le rapport de commerce, être interdite à personne," but this general statement was qualified by a requirement of compliance with regulations which would operate to discriminate between riparians and nonriparians.[11] It was not until 1868 that the right of free navigation was extended to all nations.[12] A further series of Articles Concerning the Navigation of Rivers which in their Navigable Course Separate or Traverse Different States reflected the same resolution of the problem of free transit as had been worked out in the case of the Rhine. The crucial article provided:

> La navigation dans tout le cours des rivières indiquées dans l'article précédent [those which separated or traversed several states], du point où chacune d'elles devient navigable jusqu'à son embouchure, sera entièrement libre, et ne pourra, sous le rapport de commerce, être interdite à personne, en se conformant toutefois aux règlemens qui seront arrêtés pour sa police d'une manière uniforme pour tous, et aussi favorable que possible au commerce de toutes les nations.[13]

This principle was, however, to be given application in instruments which were to be drawn up by commissioners appointed by the parties, and it was accordingly not self-executing. While this statement was not fully implemented for those rivers of Europe which flowed between two or more states, it had a strong influence upon a number of other arrangements relating

---

be consistent with the principle of article V of the Treaty of Peace between France and Austria and its Allies, signed at Paris, May 30, 1814, 2 MARTENS, N.R. 1 (1818), that navigation of the Rhine was to be free "de telle sorte qu'elle ne puisse être interdite à personne." Article V, notwithstanding the generality of its language, was considered not to have looked to the creation of a right in favor of nonriparians. 3 KLÜBER, ACTEN DES WIENER CONGRESSES 170-71 (1815). Kaeckenbeeck argues that it was actually the purpose of article V of the Treaty of Paris to open the river to all nations. INTERNATIONAL RIVERS 39 (1918).

[10] See CHAMBERLAIN, THE REGIME OF THE INTERNATIONAL RIVERS: DANUBE AND RHINE 190 (1923). This is not to deny that the Articles represented a considerable advance upon previous treaties regulating fluvial navigation.

[11] Article I of the Articles annexed to the Final Act of the Congress of Vienna, signed June 9, 1815, 2 MARTENS, N.R. 416 (1818).

[12] Revised Act for the Navigation of the Rhine, signed at Mannheim, Oct. 17, 1868 (ratifications exchanged on April 17, 1869), art. 1, 20 MARTENS, N.R.G. 355 (1875), 2 RHEINURKUNDEN 80 (1918).

[13] Article II of the Articles annexed to the Final Act of the Congress of Vienna, signed June 9, 1815, 2 MARTENS, N.R. 414 (1818).

to rivers on that continent and elsewhere. The Treaty of 1856 regulating the status of the Danube opened the river to navigation on a basis analogous to that of the Rhine.[14] The provisions of the agreements of 1885 regarding the free navigation of the Niger and Congo rivers in Africa[15] were similarly based on the Final Act of the Congress of Vienna, as were also various treaties relating to the major rivers of South America.[16]

The right of free navigation by nonriparian states was of slower growth. In the *Case relating to the Territorial Jurisdiction of the International Commission of the River Oder*, the Permanent Court of International Justice adverted to the fact that "most previous treaties," prior, that is, to the Treaty of Versailles, had limited the right to riparians alone.[17] Article 332 of that treaty declared certain waterways, which included portions of the Elbe, Oder, Niemen, and Danube, to be international in the sense that the ships and nationals of all states, riparian and nonriparian, were to be treated on a basis of equality. The emphasis which the Court placed on the full internationalization of these rivers is not easily reconciled with the conclusion, following hard on the heels of an analysis of the changes wrought by the Treaty of Versailles and the other peace treaties, that "the Treaty of Versailles adopts the same standpoint as the Act of Vienna and the treaty law which applied and developed the principles of that Act." [18]

The development of commerce during the last century has led to a wider realization of the fact that freedom of navigation is intimately bound up with equality of economic opportunity in the exploitation of a river for the transportation of persons and goods. In the *Oscar Chinn* case,[19] the

---

[14] General Treaty of Peace between Austria, France, Great Britain, Prussia, Russia, Sardinia, and the Ottoman Porte, signed at Paris, March 30, 1856, art. XV, 15 MARTENS, N.R.G. 770 (1857), 46 BRITISH AND FOREIGN STATE PAPERS 8 (1865); cf. State Treaty for the Re-Establishment of an Independent and Democratic Austria, signed at Vienna, May 15, 1955, art. 31, 6 U.S.T. 2369, T.I.A.S. No. 3298.

[15] General Act of the Conference of Berlin, signed at Berlin, Feb. 26, 1885, ch. IV (Act of Navigation of the Congo), art. 13, and ch. V (Act of Navigation of the Niger), art. 26, 10 MARTENS, N.R.G. 2d ser. 414 (1885-86), 76 BRITISH AND FOREIGN STATE PAPERS 4 (1884-85); and see, concerning the indebtedness of these instruments to the Final Act of the Congress of Vienna, FRANCE, DOCUMENTS DIPLOMATIQUES, AFFAIRES DU CONGO ET DE L'AFRIQUE OCCIDENTALE 59 (1885).

[16] E.g., Treaty between the United States and the Argentine Confederation for the Free Navigation of the Rivers Parana and Uruguay, signed at San José, July 10, 1853, 10 Stat. 1001, T.S. No. 3, which contains in article 1 language similar to that of the Final Act of the Congress of Vienna.

[17] P.C.I.J., ser. A, No. 23, at 28 (1929).

[18] *Id.* at 29.

[19] P.C.I.J., ser. A/B, No. 63 (1934).

Belgian Government had subsidized a Belgian shipping company operating on the Congo and other rivers, thus driving out of business Chinn's competing shipping company. The British Government contended that the conduct of Belgium was a violation of the Treaty of St. Germain of September 10, 1919, which superseded the General Act of Berlin of 1885 but maintained the principle of freedom of navigation on the Niger and other rivers.[20] The Permanent Court of International Justice, while denying the British claim, stated that "freedom of navigation implies, as far as the business side of maritime or fluvial transport is concerned, freedom of commerce also."[21] It grounded its rejection of the contention of Great Britain on the fact that the freedom of trade established by the Treaty of St. Germain did not protect British nationals from commercial competition and indicated that its conclusion would have been different if the measures taken by the Belgian Government had constituted "an obstacle to the movement of vessels."[22] The dissenting judges, including Sir Cecil Hurst, Van Eysinga, and Anzilotti, adverted to the close relationship between freedom of navigation and freedom of commerce in this sense.[23] In a more recent period, the concern of the Central Commission for the Navigation of the Rhine with economic questions points to the increasing importance of keeping transport on international rivers free of economic restraints.[24]

Despite the existence of a substantial number of treaties which grant a right of free navigation on international rivers to the vessels of all nations, the practice of opening such waterways to general usage is by no means universal. Treaties which relate to navigable waters forming part of an international boundary between two states normally do no more than to give equality of rights to the nationals of both riparians.[25] To this category of arrangements belongs the Boundary Waters Treaty of 1909 between the

[20] Convention Revising the General Act of Berlin of Feb. 26, 1885, and the General Act and Declaration of Brussels of July 2, 1890, signed at St. Germain-en-Laye, Sept. 10, 1919, art. 5, 8 L.N.T.S. 25.

[21] Oscar Chinn case, *supra* note 19, at 83.

[22] *Ibid.*

[23] In the words of Judge Anzilotti, "It is clear, to begin with, that this Article [art. 5] lays down that navigation is to be free, both as regards movements of shipping, or navigation in the strict sense of the word, and as regards the carriage of passengers and cargo. It is, indeed, in that sense that freedom of fluvial navigation has always been understood in international treaties concerned with the question." *Id.* at 111.

[24] See p. 139 *supra*.

[25] E.g., Convention between Norway and Sweden concerning Common Lakes and Watercourses, signed at Stockholm, Oct. 26, 1905, art. 3, 34 MARTENS, N.R.G. 2d ser. 710 (1907).

United States and Great Britain, which provides that the navigation of all navigable boundary waters is to continue "free and open for the purposes of commerce to the inhabitants and to the ships, vessels, and boats of both countries equally . . ."[26]

The great majority of treaties, whether opening the river to riparians or to all states, contain one or both of two important limitations. One of these is to reserve the right of cabotage, that is, the transport of goods between points within a single state, to nationals of the state concerned.[27] Since the limitation of cabotage to nationals constitutes a qualification of freedom of navigation on an international river, an express reservation of the power to exclude foreign shipping and persons must be made when the river is thrown open to free use.[28]

Warships are seldom expressly mentioned, but language like that of the Articles formulated at the Congress of Vienna, "sous le rapport de commerce," points by implication to the applicability of the Articles to merchant vessels alone.[29] These exclusions of warships from international rivers had their parallel, in the case of the Great Lakes, in the severe limitations placed by the Rush-Bagot Agreement of 1817 on the number of naval vessels allowed

[26] Treaty between the United States and Great Britain relating to Boundary Waters between the United States and Canada, signed at Washington, Jan. 11, 1909, art. 1, 36 Stat. 2448, T.S. No. 548.

[27] E.g., State Treaty for the Re-Establishment of an Independent and Democratic Austria, signed at Vienna, May 15, 1955, art. 31, 6 U.S.T. 2369, T.I.A.S. No. 3298; Convention between the Federal Republic of Germany, the French Republic, and the Grand Duchy of Luxembourg concerning the Canalization of the Moselle, signed at Luxembourg, Oct. 27, 1956, art. 28, 11 VERTRÄGE DER BUNDESREPUBLIK DEUTSCHLAND 34 (1959); LES ACTES DU RHIN 46 (1957).

Restrictions placed by Germany on the participation of foreign vessels in the transport of goods between points in that country, beginning in 1949, were protested by Switzerland and the Netherlands as being in violation of the principle of freedom of navigation on the Rhine. See Fortuin, *The Regime of Navigable Waterways of International Concern and the Statute of Barcelona,* 7 NEDERLANDS TIJDSCHRIFT VOOR INTERNATIONAAL RECHT 125, 133-35 (1960). The case of the Netherlands is presented in VAN DER HOEVEN, DE RIJNVAARTAKTEN EN DE CABOTAGE (1956).

[28] Statute on the Régime of Navigable Waterways of International Concern, annexed to the Convention opened for signature at Barcelona, April 20, 1921, art. 5, para. 1, 7 L.N.T.S. 50, 1 HUDSON, INTERNATIONAL LEGISLATION 645 (1931); Règlement pour la navigation des fleuves internationaux, art. 3, adopted by the Institute of International Law in 1934, [1934] ANNUAIRE DE L'INSTITUT DE DROIT INTERNATIONAL 713, 715.

[29] Articles concerning the Navigation of Rivers which in their Navigable Course Separate or Traverse Different States, art. II, annexed to the Final Act of the Congress of Vienna, signed June 9, 1815, 2 MARTENS, N.R. 414 (1818).

on those waters prior to the Second World War.[30] The only important treaties which expressly provide for navigation by warships are those relating to the Congo River[31] and to certain rivers in South America,[32] which, by reason of running through wild and largely uninhabited territory, occupy a different geographic and political position from that of the rivers of Europe. The absence of any express provision in the Act of the Niger, corresponding to the stipulation as to warships in the Congo Act, led to a diplomatic controversy concerning the interpretation of the agreement. Great Britain contended that the absence of the provision was significant, in view of the explicit statement in the parallel instrument, and that the General Act of Berlin had been patterned on the Final Act of the Congress of Vienna, which by implication excluded warships from the categories of vessels entitled to claim a right of free passage.[33] As a matter of interpretation, it would seem that any language in a treaty which states that the waterway is open for purposes of commerce should be construed to exclude passage by warships.

The increasing number of cases in which rivers have been opened to navigation by the vessels of other states led Westlake to say in 1904, "We may now look back on the history which we have traced and ask whether it does not amount to such an acceptance of that right by the civilised world as makes international law by the consent of states." [34] To this question, which Westlake would have answered in the affirmative, contemporary authority has given a negative answer.[35] As to each river which is opened

[30] Agreement between the United States of America and Great Britain concerning Naval Forces on the American Lakes, signed at Washington, April 28/29, 1817, 8 Stat. 231, T.S. No. 110½.

[31] General Act of the Conference of Berlin, signed at Berlin, Feb. 26, 1885, *supra* note 15, ch. IV (Act of Navigation of the Congo), arts. 21 and 22.

[32] Treaty of Limits between Uruguay and Brazil, signed at Rio de Janeiro, Oct. 30, 1909, art. IX, 5 URUGUAY, MINISTERIO DE RELACIONES EXTERIORES, COLECCIÓN DE TRATADOS, CONVENCIONES Y OTROS PACTOS INTERNACIONALES DE LA REPÚBLICA ORIENTAL DEL URUGUAY 451, 456 (1928).

[33] PILLIAS, LA NAVIGATION INTERNATIONALE DU CONGO ET DU NIGER 85-90 (1900).

[34] 1 WESTLAKE, INTERNATIONAL LAW 157 (1904).

[35] 1 HYDE, INTERNATIONAL LAW, CHIEFLY AS INTERPRETED AND APPLIED BY THE UNITED STATES 563 (2d rev. ed. 1945); 1 OPPENHEIM, INTERNATIONAL LAW 474 (8th ed. Lauterpacht 1955); ROUSSEAU, DROIT INTERNATIONAL PUBLIC 390 (1953); Winiarski, *Principes généraux du droit fluvial international,* 45 HAGUE RECUEIL 77, 159 (1933). Guggenheim regards the right of free passage as conventional in origin, except in the case of rivers connecting interior ports with the sea. 1 TRAITÉ DE DROIT INTERNATIONAL PUBLIC 405 (1953). Professor H. A. Smith stigmatizes the free navigation theory as "in

to international navigation the understanding is that the arrangement is a special one applicable to that waterway alone, constituting particular international law for the river rather than reflecting any principle of the *lex generalis*. The continuing existence of river régimes which accord rights of passage only to riparians, as, for example, in the case of the St. Lawrence,[36] is perhaps the strongest indication that a general right of free transit is not one recognized by international law. Further confirmation of this view is to be found in the fact that the Barcelona Convention of 1921 on the Régime of Navigable Waterways of International Concern,[37] which accorded a right of free navigation over inland "waterways of international concern," has been accepted by a relatively restricted group of states.[38] The Convention requires the contracting states to accord "free exercise of navigation to the vessels flying the flag of any one of the other Contracting States," on international waterways which are navigable from the sea,[39] as more fully defined in article I of the annexed Statute. Although the preamble of the Convention refers to "the principle of Freedom of Navigation" to which it wished to give "fresh confirmation," the principle referred to is akin to that of 1815, which had in the international sphere been received with lethargic deference.[40]

In the *Faber* case,[41] the one important instance in which the question of a right of free passage over a river flowing through two or more states and affording access to the sea has been litigated, a conclusion was reached which appears to support the contention that there is no general right of free navigation. There was in issue in that case the control by Venezuela of the Catatumbo and Zulia rivers, which afforded access from Colombia to the sea. A claim was made by a German national for the losses which

---

flagrant conflict alike with the facts of Nature and with the practice of the world." THE ECONOMIC USES OF INTERNATIONAL RIVERS 9 (1931).

[36] Treaty between the United States and Great Britain relating to Boundary Waters between the United States and Canada, signed at Washington, Jan. 11, 1909, art. I, 36 Stat. 2448, T.S. No. 548.

[37] 7 L.N.T.S. 50, 1 HUDSON, INTERNATIONAL LEGISLATION 645 (1931).

[38] ROUSSEAU, DROIT INTERNATIONAL PUBLIC 393 (1953).

[39] Article 3 of the annexed Statute.

[40] The Statute grants freedom of navigation over "waterways of international concern" to all of the contracting parties, whereas the Final Act of the Congress of Vienna limited the right to riparians.

[41] (Germany v. Venezuela), RALSTON, VENEZUELAN ARBITRATIONS OF 1903, 600 at 620 (1904).

he sustained through the cutting off of his trade as a consequence of the measures taken by Venezuela with respect to the navigation of these rivers in 1901 and 1902. The possibility that hostile forces might enter Venezuela along the Zulia River led the Umpire to conclude that Venezuela had the right to regulate and "if necessary to the peace, safety, or convenience of her own citizens, to prohibit altogether navigation on these rivers."[42] The Umpire also stated, by way of obiter dictum, that if the case turned on the general question of international law, he would be compelled to reach the conclusion that there was no right of free navigation over the river in the absence of a treaty to that effect, even though the waterway linked Colombia with the high seas.[43]

It is consistent with the view that there is no general right of passage over international rivers that proposals have been made after both World Wars that the rivers of Europe should be open to free and unrestricted navigation. When reference has been made to some general principle of "free navigation," the principle has been spoken of as one which is not self-executing and requires implementation through agreements applicable to individual rivers or groups of rivers. At the conclusion of the First World War, the Allied and Associated Powers stated, "It is clearly in accord with the agreed basis of the peace and the established public law of Europe that inland states should have secure access to the sea along navigable rivers flowing through their territory."[44] This "established public law of Europe" is nothing more than the general statement which appeared in the Final Act of the Congress of Vienna. It must also be observed that the only freedom which was referred to was that of inland states to use an international river to reach the high seas. The proposal made by President Truman at the end of the Second World War that the important international waterways, including the Rhine and "all the inland waterways of Europe which border on two or more States," be opened to use by all was clearly based

---

[42] *Id.* at 630.

[43] *Id.* at 630; but cf. Award of the Tribunal of Arbitration constituted under Article I of the Treaty of Arbitration signed at Washington on the 2nd February 1897, between Great Britain and the United States of Venezuela, Oct. 3, 1899, 29 MARTENS, N.R.G. 2d ser. 581, 587 (1903), in which the arbitrators decided that "in times of peace the Rivers Amakuru and Barima shall be open to navigation by the merchantships of all nations."

[44] Reply of the Allied and Associated Powers to the Observations of the German Delegation on the Conditions of Peace, and Ultimatum, June 16, 1919, [1919] 13 FOREIGN REL. U.S. 44, 51 (1947).

upon the assumption that free and unrestricted navigation did not yet exist with respect to these waterways.[45]

The subject of access by landlocked states to the high seas was among the topics considered by the First Geneva Conference of 1958 on the Law of the Sea. The article drafted at that Conference[46] provides that "States having no sea-coast *should have* free access to the sea"[47] but that freedom of transit for the state having no seacoast shall be accorded "by common agreement." The provision, which does not speak in the language of firm legal obligation, was far weaker than the principles adopted by the Pre-liminary Conference of Land-Locked States. These nations had taken the position that there is a right of free access to the sea, deriving from the principle of the freedom of the high seas, and that the "transit of persons and goods from a land-locked country towards the sea and vice versa . . . must be freely accorded."[48] Thus, even in the most compelling case for a right of freedom of transit, the Geneva Conference of 1958, widely repre-sentative of the nations of the world, failed to recognize any immediate and self-executing right of free navigation through an international river.

The opening of international rivers to the ships of other riparians appears to have been dictated to a very considerable extent by the common interest of the riparian states in trade with each other and in the exploitation of the waterway as an avenue of commerce between their territories. The opening of fluvial waterways to nonriparians has come about only gradually over the period of the last century and only collaterally to the main purpose of creat-ing a society of riparians for the more effective utilization of the waterway. As a matter of history, nonriparians' right of free navigation has frequently been of slight significance because of the very small proportion of the river commerce which fell to them.[49] The mixture of indifference and hostility

[45] Radio address of Aug. 9, 1945, 13 Dep't State Bull. 208, 212 (1945).

[46] Convention on the High Seas, signed at Geneva, April 29, 1958, art. 3, 2 United Nations Conference on the Law of the Sea, Off. Rec. 135 (U.N. Doc. No. A/CONF.13/38) (1958); see Jessup, *The United Nations Conference on the Law of the Sea,* 59 Colum. L. Rev. 234, 253-55 (1959).

[47] Emphasis supplied.

[48] Principles I and V of the Principles Enunciated by the Preliminary Conference of Land-Locked States, Annex 7 of the Memorandum Submitted by the Preliminary Conference of Land-Locked States, Feb. 28, 1958, 7 United Nations Conference on the Law of the Sea, Off. Rec. 67, 78 (U.N. Doc. No. A/CONF.13/43) (1958); see Question of Free Access to the Sea of Land-Locked Countries, Memorandum by the Secretariat of the United Nations, 1 *id.* 306, 320-27 (U.N. Doc. No. A/CONF.13/37) (1958).

[49] E.g., in the case of the Rhine. Chamberlain, The Regime of the International

with which the Barcelona Convention was greeted is symptomatic of the continuing reluctance of nations to open their water highways to vessels not belonging to the river community. The success which has been achieved in the maintenance and free employment of the inland waterways of Europe is attributable to the existence of reciprocal rights and obligations which are undertaken by the states bordering on the river. The Permanent Court of International Justice has accurately referred to this as "a community of interest of riparian States."[50] The Court, in its judgment concerning the International Commission of the River Oder, went on to say, "This community of interest in a navigable river becomes the basis of a common legal right, the essential features of which are the perfect equality of all riparian States in the user of the whole course of the river and the exclusion of any preferential privilege of any one riparian State in relation to the others."[51] Each riparian will normally be a territorial sovereign, the operator of the waterway, and a user of the waterway. To the extent that it shares these three interests with other states, it will find it possible to work out arrangements which, by reason of being reciprocal of application, will be regarded as mutually satisfactory and will command the compliance of the signatories. By way of contrast, the territorial sovereign of a canal or strait is very frequently not the primary user of that waterway and may have an essentially adverse interest to that of the state the vessels of which use the waterway. Correspondingly, the users of the strait or canal will have no identity of interest with the sovereign which controls the waterway or operates it, with the possible exception of cases like that of the Panama Canal, where the state which operates the watercourse also supplies the largest use of the Canal. Success in maintaining free passage through international rivers and in the establishment of institutions to control the fluvial watercourses thus does not point to similarly beneficial effects as a necessary consequence of the establishment of similar régimes and institutions for international straits or canals.

As has already been indicated in Chapter I,[52] a strait for legal purposes

---

Rivers: Danube and Rhine 254 (1923). On the other hand, prior to the First World War the British had the largest share of the shipping on the portion of the Danube controlled by the European Commission of the Danube, while the riparian states, Rumania and Bulgaria, supplied a relatively small share of the total tonnage operating on the river. *Id.* at 256.

[50] Territorial Jurisdiction of the International Commission of the River Oder, P.C.I.J., ser. A, No. 23, 27 (1929).

[51] *Ibid.*

[52] See p. 4 *supra.*

constitutes a geographic strait in which the territorial sea of the littoral state or states is such that no area of the high seas is left for free navigation through the waterway. It is of the essence of the strictly legal concept of a strait that passage through a portion of territorial sea should be necessary.[53] The extensions of territorial seas which have in recent years been made by many states should thus have the effect of creating many new international straits, of which the law will be forced to take account. Indeed, if the asserted claims of certain South American states were to be given literal effect under appropriate geographic circumstances, it would be possible to envisage international straits of a breadth of 400 miles, based upon a claim to a territorial sea 200 miles in width.[54] Under such circumstances, the Caribbean Sea, it has been pointed out, would become one great international strait and the very approach to the Panama Canal from the Atlantic would be through a portion of this strait.[55]

Should straits providing a passage between the high seas and the territorial sea of a state, normally within a gulf or bay, be on the same footing as straits connecting two portions of the high seas? If they are, then the dimensions of the right of free passage in the two cases would be the same.

During the war between Israel and Egypt, Egypt denied passage through the Straits of Tiran, which afford access to the Gulf of Aqaba,[56] to Israeli ships and foreign vessels proceeding to and from the Israeli port of Eilat. The Arab case against freedom of passage was grounded in part on the

[53] The provision of the Convention on the Territorial Sea and the Contiguous Zone, signed at Geneva, April 29, 1958, which concerns passage through straits (art. 16, para. 4) is thus treated as a special application of the principle of innocent passage through the territorial sea. 2 UNITED NATIONS CONFERENCE ON THE LAW OF THE SEA, OFF. REC. 132 (U.N. Doc. No. A/CONF.13/38) (1958).

[54] E.g., Political Constitution of El Salvador, 1950, art. 7, in UNITED NATIONS LEGISLATIVE SERIES, LAWS AND REGULATIONS ON THE REGIME OF THE HIGH SEAS 300 (U.N. Doc. No. ST/LEG/SER.B/1) (1951); see Agreement between Chile, Ecuador, and Peru Supplementary to the Declaration of Sovereignty over the Maritime Zone of Two Hundred Miles, signed at Lima, Dec. 4, 1954, UNITED NATIONS LEGISLATIVE SERIES, LAWS AND REGULATIONS ON THE REGIME OF THE TERRITORIAL SEA 729 (U.N. Doc. No. ST/LEG/SER.B/6) (1956).

[55] See the chart in BRITTIN AND WATSON, INTERNATIONAL LAW FOR SEAGOING OFFICERS 55 (2d ed. 1960).

[56] Concerning the geography of the area see BLOOMFIELD, EGYPT, ISRAEL AND THE GULF OF AQABA IN INTERNATIONAL LAW 1-6 (1957); State of Israel, Ministry for Foreign Affairs, Background Paper on the Gulf of Aqaba 1-9 (1956) (mimeographed); Selak, *A Consideration of the Legal Status of the Gulf of Aqaba*, 52 AM. J. INT'L L. 660, 660-67 (1958); Melamid, *Legal Status of the Gulf of Aqaba*, 53 *id.* 412 (1959).

existence of a state of war[57] and in part on the contention that the Gulf consisted of Arab internal waters as a matter of historic right.[58] In the Arab view, the Gulf is bounded only by the states of Egypt (now the United Arab Republic), Jordan, and Saudi Arabia; Israel is unrecognized and to them an interloper. The Gulf ranges in width from three to 17 miles. The entrance is commanded by the islands of Tiran and Sanafir, sovereignty over which has been claimed by both Saudi Arabia and Egypt.[59] The only navigable channel into the Gulf is the Strait of Tiran, measuring four miles in width. Since the coastal states claimed territorial seas of six miles, there was at one time a strip of high seas within the Gulf of Aqaba, if the Arab contention that the Gulf is internal waters is disregarded.[60] In an attempt to consolidate its legal position, Saudi Arabia extended its territorial sea to 12 miles in 1958,[61] so that, assuming the validity of the extension, the Gulf consists entirely of the territorial seas of the littoral states, even if the claim of historical right be rejected.

The position of Israel,[62] in which the United States[63] and a substantial number of other states concurred, was that straits providing access from the territorial sea to the high seas were to be governed by the same rule of free passage as a channel connecting stretches of the high seas. The question of the existence of such a right in the former case came before the International Law Commission in connection with its codification of the law of the sea. The Commission ducked the issue. It referred only to freedom

[57] This question turns on considerations not relevant to the law of international straits.

[58] Hammad, *The Right of Passage in the Gulf of Aqaba*, 15 REVUE ÉGYPTIENNE DE DROIT INTERNATIONAL 118, 123-140 (1959). The Arab case is based in part on the Gulf of Fonseca Case (El Salvador v. Nicaragua), 11 AM. J. INT'L L. 674 (1917), holding that the Gulf, bounded by three states, was "an historic bay . . . possessed of the characteristics of a closed sea." In that case, however, all of the riparians were in accord in regarding the Gulf as closed sea, and the only question was whether the waters outside the territorial sea were held by the riparians jointly or severally.

[59] Selak, *supra* note 56, at 666.

[60] *Id.* at 660, 666-67.

[61] Saudi Arabia, Decree No. 33 defining the Territorial Waters of the Kingdom, Feb. 16, 1958, UNITED NATIONS LEGISLATIVE SERIES, SUPPLEMENT TO LAWS AND REGULATIONS ON THE REGIME OF THE HIGH SEAS (VOLUMES I AND II) AND LAWS CONCERNING THE NATIONALITY OF SHIPS 29 (U.N. Doc. No. ST/LEG/SER.B/8) (1959).

[62] State of Israel, Ministry for Foreign Affairs, Background Paper on the Gulf of Aqaba 9 (1956) (mimeographed).

[63] Aide Mémoire Handed to Israeli Ambassador Eban by Secretary of State Dulles, Feb. 11, 1957, UNITED STATES POLICY IN THE MIDDLE EAST, SEPTEMBER 1956-JUNE 1957, at 290 (Dep't of State Pub. 6505) (1957).

of passage "between two parts of the high seas" [64] and decided that "the question [of passage through the Strait of Tiran] raised by the Israel Government related to an exceptional case which did not lend itself to the formulation of a general rule." [65] The United Nations Conference on the Law of the Sea was bolder. By a narrow vote—a majority of one, if abstentions are disregarded [66]—the committee working on the law of the territorial sea changed the provision on nonsuspension of innocent passage through straits to make it apply to "straits which are used for international navigation between one part of the high seas and another part of the high seas *or the territorial sea of a foreign State*." [67] The Strait of Tiran was not specifically mentioned in the discussion, but the representative of Saudi Arabia, who joined with the representatives of the United Arab Republic and Jordan in voting against the changed language, spoke darkly of the proposal as having "been carefully tailored to promote the claims of one State." [68] In plenary, the revised text was adopted by 62 votes to one, with nine abstentions. [69] The present position of the matter, as reflected in this authoritative pronouncement, is thus that no distinction can legitimately be made between straits connecting portions of the high seas and straits connecting the territorial sea with the high seas. The legitimacy of exercising controls over straits in general, and over the Strait of Tiran in particular, in time of war must be left for later consideration. [70]

So far as the passage of ships through undoubted international straits is concerned, Brüel in his definitive work reached the conclusion that the law had been established before the First World War that "the right of passage of *merchant vessels* in international straits was certain, and *warships* were supposed to have the same right—though not with the same degree of

[64] Articles concerning the Law of the Sea, art. 17, para. 4, in [1956] 2 YEARBOOK OF THE INTERNATIONAL LAW COMMISSION 258 (U.N. Doc. No. A/CN.4/SER.A/1956/Add. 1) (1957).

[65] [1956] 1 YEARBOOK OF THE INTERNATIONAL LAW COMMISSION 203 (U.N. Pub. Sales No. 1956. V. 3, Vol. I).

[66] 3 UNITED NATIONS CONFERENCE ON THE LAW OF THE SEA, OFF. REC. 100 (U.N. Doc. No. A/CONF.13/39) (1958).

[67] Emphasis supplied. Convention on the Territorial Sea and the Contiguous Zone, signed at Geneva, April 29, 1958, art. 16, para. 4, 2 UNITED NATIONS CONFERENCE ON THE LAW OF THE SEA, OFF. REC. 132 (U.N. Doc. No. A/CONF.13/38) (1958).

[68] *Op. cit. supra* note 66, at 96.

[69] *Op. cit. supra* note 67, at 65; the debates in the United Nations are summarized in Gross, *The Geneva Conference on the Law of the Sea and the Right of Innocent Passage through the Gulf of Aqaba,* 53 AM. J. INT'L L. 554, 580-92 (1959).

[70] See p. 209 *infra*.

certainty." [71] The practice of recent years has done nothing to call into question the accuracy of this conclusion as to the passage of merchant vessels.[72] Aside from the case of passage through the Strait of Tiran, which was rendered controversial by the purported assertion of belligerent rights,[73] there appear to have been no instances of late in which states have claimed a right to deny passage to merchant vessels through international straits comprising portions of their territorial sea. The fidelity of nations to this principle is illustrated by the fact that when two ships were detained by Turkey in the Turkish Straits in 1956 and allegations were made that the detention arose out of the vessels' carriage of Russian arms, it was made clear that the actual basis for holding the ships was the circumstance that one of them had run aground and that the other had on a previous voyage damaged an anti-submarine net.[74] The Turkish authorities had thus acted in conformity with article II of the Montreux Convention, which guarantees "freedom of transit and navigation in the Straits" for merchant vessels. It has, of course, been necessary on occasion to close straits to international navigation because of construction work, but these impediments to navigation have been of only temporary duration.[75]

The damaging of several British warships in the Corfu Strait in 1947 by mines which had been laid in the channel with the knowledge of the Albanian authorities offered the occasion for the establishment on a firm basis of the principle, theretofore not altogether undisputed, that vessels of war, like merchant ships, were entitled to free transit through an international strait.[76] On the issue of the right of Great Britain to send its warships through the strait, the International Court of Justice expressed its views in the following terms: "It is, in the opinion of the Court, generally recognized and in accordance with international custom that States in time of peace have a right to send their warships through straits used for international navigation be-

---

[71] 1 BRÜEL, INTERNATIONAL STRAITS 202 (1947).

[72] Indeed, further developments in the law, such as the rules enunciated in the *Corfu Channel Case,* have rendered less tenable the conclusion expressed by Professor H. A. Smith in 1935 that, "Each of these [salt-water straits] has its own history and characteristics, so that it is really impossible to find the common material necessary for a valid generalisation." 2 GREAT BRITAIN AND THE LAW OF NATIONS 255 (1935).

[73] See p. 210 *infra.*

[74] The Times (London), April 9, 1956, p. 7, col. 3.

[75] Canada, Notice to Mariners No. 141 (430), Dec. 8, 1954, in THE BALTIC AND INTERNATIONAL MARITIME CONFERENCE, MONTHLY CIRCULAR, June 1955, No. 159, at 5746.

[76] The Corfu Channel Case, [1949] I.C.J. Rep. 4.

tween two parts of the high seas without the previous authorization of a coastal State, provided that the passage is *innocent*. Unless otherwise prescribed in an international convention, there is no right for a coastal State to prohibit such passage through straits in time of peace."[77]

Albania and Greece were *de facto* in a state of hostility at that time, and it is therefore conceivable that the case has application to passage through international straits in time of war as well as in time of peace. If significance is to be attached to the state of Greco-Albanian relationships at the time, the actual holding of the case that a right of free passage existed for the warships of third states during a period of hostilities would apply a fortiori in time of peace. But it cannot be too strongly emphasized that the Court professed to be dealing with the peacetime situation. Elsewhere in its opinion, the International Court stated that "the North Corfu Channel should be considered as belonging to the class of international highways through which passage cannot be prohibited by a coastal State in time of peace."[78] It was led to place the Corfu Strait in the category of "international highways" by the facts of its convenience for international navigation and the substantial use which had been made of the Channel for that purpose. It specifically rejected the contention that the passage must be a necessary one before a right of free passage might be said to arise.[79]

---

[77] *Id.* at 28. The qualifying words "unless prescribed in an international convention" quite clearly point to the Turkish Straits, which are governed by the Montreux Convention. The statement raises, but does not solve, the problem whether states may by a bilateral or multilateral agreement prohibit the passage of the vessels of third states through a strait previously regarded as an "international highway" and so employed by the ships of nations not parties to the agreement.

[78] *Id.* at 29.

[79] "But in the opinion of the Court, the decisive criterion is rather its [the strait's] geographical situation as connecting two parts of the high seas and the fact of its being used for international navigation. Nor can it be decisive that this Strait is not a necessary route between two parts of the high seas, but only an alternative passage between the Aegean and the Adriatic Seas." *Id.* at 28. This view is criticized by Brüel on the bases that the Corfu Channel is a lateral strait and that passage could be made around the Island of Corfu with little inconvenience. *Some Observations on Two of the Statements concerning the Legal Position of International Straits (Made by the International Court of Justice in its Judgment of April 9th, 1949, in the Corfu Channel Case)*, in GEGENWARTSPROBLEME DES INTERNATIONALEN RECHTES UND DER RECHTSPHILOSOPHIE: FESTSCHRIFT FÜR RUDOLF LAUN 259, 273, 276 (Constantopoulos and Wehberg eds. 1953). This opinion is consistent with Brüel's earlier observation that "only those straits that are of some, not quite inconsiderable importance to the international sea-commerce, enjoy the peculiar legal position accorded to straits." 1 INTERNATIONAL STRAITS 43 (1947).

To the general right of free passage of warships through international straits, as enunciated in the *Corfu Channel Case,* an exception must be made in the case of the Turkish Straits. Passage of ships of war through the Straits must be notified through diplomatic channels eight days in advance of passage.[80] Subject to this requirement of notice, light surface vessels, minor war vessels, and auxiliary vessels of Black Sea and non-Black Sea powers enjoy freedom of transit. However, it is only the Black Sea powers which may send capital ships through the Straits.[81] The requirements concerning passage through the Bosporus and the Dardanelles are obviously corollaries of the limitation on the number and tonnage of war vessels which may be in the Black Sea at any particular time.

Although the Montreux Convention is now subject to denunciation in the discretion of the parties,[82] none of the signatories has yet sought to be released from its obligations nor has there been any formal request for revision of the Convention. Immediately after the close of the Second World War, the United States proposed that the Straits should be opened to transit by vessels of the Black Sea powers at all times and that vessels of non-Black Sea powers should be admitted only with the consent of the littoral states.[83] During the general discussion of this proposal which took place in 1946, there seemed to be general agreement that this régime for the Straits would be a desirable one,[84] but it has never been placed in effect.

The contemporary Russian response to the problem of the Turkish Straits is rooted in both history and in geography.[85] In periods of weakness, it has been the policy of Russia to demand that the Straits should be closed to warships. At times when her power appears to be in the ascendant, Russia has laid demands on Turkey that only Russian warships should be

---

[80] Convention regarding the Régime of the Straits, signed at Montreux, July 20, 1936, art. 13, 173 L.N.T.S. 213, 223.

[81] Articles 10 and 11.

[82] Under article 28, the Convention was to remain in force for twenty years from the date of its coming into force (Nov. 9, 1936), except that the right of free transit for merchant vessels was to continue without limit of time. The Convention may now be denounced on two years' notice.

[83] Note, American Ambassador in Turkey to the Turkish Minister of Foreign Affairs, Nov. 2, 1945, in HOWARD, THE PROBLEM OF THE TURKISH STRAITS 47 (Dep't of State Pub. 2752) (1947).

[84] Tchirkovitch, *La Question de la révision de la Convention de Montreux concernant le régime des Détroits Turcs: Bosphore et Dardanelles,* 56 REVUE GÉNÉRALE DE DROIT INTERNATIONAL PUBLIC 189, 221 (1952).

[85] See pp. 193 and 199 *infra* concerning the position of the Straits in time of war.

allowed through the Straits, that Russia should be allowed to fortify them, and that the Black Sea should become a Russian *mare clausum*.[86] The latest diplomatic negotiations of consequence on this point between Russia and Turkey were initiated by a Russian complaint of large numbers of foreign warships in the Straits.[87] To this Russian protest Turkey replied that under articles 14 and 17 of the Montreux Convention there was full freedom for courtesy visits by warships and that the presence of these vessels arose from the friendly relationships which existed between Turkey and the states despatching naval units to the Straits for these visits.[88] Nothing further appears to have come of the matter, and Russia continues to give notification of the passage of its warships, as in the past.[89] If the law with respect to passage of warships through the Turkish Straits is not consistent with that having general application to straits, this anomaly of the law is attributable to the special position of the Black Sea, rather than to any internal inconsistency in the international law of straits.

If, subject to the special exception made in the case of the Bosporus and the Dardanelles, merchant ships and warships are to be accorded free passage through international straits in time of peace, that right has a certain superficial resemblance to the general right of "innocent passage" through the territorial sea of a state.[90] The logical consequence of the assimilation of the waters of straits and of the territorial sea would be the elimination of any separate body of doctrine relating to straits, for the law respecting passage through territorial waters within a strait would differ in no respect from the principles having application to passage through the territorial sea of a state which does not form part of a strait. The absence of any distinction between the law of straits and the law of the territorial sea would entail as one of its consequences the ability of the coastal state to suspend the right of innocent passage of merchant vessels in a strait, as in other territorial waters, for the protection of its security.[91] The passage of warships through a strait might

---

[86] Bilsel, *The Turkish Straits in the Light of Recent Turkish-Soviet Russian Correspondence*, 41 Am. J. Int'l L. 727 (1947).

[87] Russian note of July 20, 1953, in Howard, *The Development of United States Policy in the Near East, South Asia, and Africa During 1953, Part I*, 30 Dep't State Bull. 274, 278 (1954).

[88] Turkish note of July 24, 1953, *id.* at 278.

[89] N.Y. Times, June 16, 1957, p. 22, col. 5, describes a Russian communication relating to the passage of a cruiser, two destroyers, and three torpedo boats.

[90] See the dissenting opinion of Judge Azevedo in the Corfu Channel Case, [1949] I.C.J. Rep. 4, 105.

[91] As provided with respect to the territorial sea, in Convention on the Territorial

likewise be made subject to the regulations of the coastal state.[92] To recognize such powers in the riparian state would impose needless and undesirable limitations on any "right" of free passage. This consideration undoubtedly accounts for the rejection by the International Court of Justice in the *Corfu Channel Case*[93] of the Albanian contention that British warships were not authorized to proceed through the strait because Albanian territorial waters had been closed to foreign vessels of war.[94]

The articles drafted at the Geneva Conference on the Law of the Sea[95] only obscured the matter of the passage of warships through international straits. Section III (Right of Innocent Passage), subsection A (Rules Applicable to All Ships), article 16, paragraph 3, permits a state to "suspend temporarily in specified areas of its territorial sea the innocent passage of foreign ships if such suspension is essential for the protection of its security." Straits are mentioned specifically only in paragraph 4, which provides: "There shall be no suspension of the innocent passage of foreign ships through straits which are used for international navigation between one part of the high seas and another part of the high seas or the territorial sea of a foreign State." A shadow is cast over this right of free passage of straits by article 23: "If any warship does not comply with the regulations of the coastal State concerning passage through the territorial sea and disregards any request for compliance which is made to it, the coastal State may require the warship to leave the territorial sea." This last is the single "Rule applicable to Warships."

Although there is some authority to the effect that warships do not have a right of "innocent passage,"[96] the better view is that article 16, appearing

---

Sea and the Contiguous Zone, signed at Geneva, April 29, 1958, art. 16, para. 3, 2 UNITED NATIONS CONFERENCE ON THE LAW OF THE SEA, OFF. REC. 132 (U.N. Doc. No. A/CONF.13/38) (1958).

[92] Article 23.

[93] The Corfu Channel Case, [1949] I.C.J. Rep. 4, 27.

[94] See Counter-Memorial of the Albanian Government, 2 CORFU CHANNEL CASE— PLEADINGS, ORAL ARGUMENTS, DOCUMENTS 132 (I.C.J. 1949), and the view of the Court at [1949] I.C.J. Rep. 30.

[95] Convention on the Territorial Sea and the Contiguous Zone, signed at Geneva, April 29, 1958, *supra* note 91.

[96] Statement by Dr. El-Erian (United Arab Republic), U.N. GEN. ASS. OFF. REC. 13th Sess., 6th Comm. 207 (A/AC/SR.590) (1958); Sørenson, *Law of the Sea*, INTERNATIONAL CONCILIATION, No. 520 at 235 (1958). Sørenson admits that the language of the Convention supports the existence of the right but declares that "this was not the intention of the majority of delegations" (*ibid.*).

among the "Rules applicable to All Ships" and not in the "Rules applicable to Merchant Ships," does indeed apply to warships.[97] If this provision does apply to warships, what limits, if any, does article 23 place on that right of free passage? Another rule applicable to all ships[98] requires compliance with national laws and regulations, consistent with international law and the treaty, relating to transport and navigation. Article 23 goes beyond this only by providing a sanction in case of noncompliance by warships with "regulations . . . concerning passage," which must be read as rules of navigation. If this be the case, a consistent reading of the articles would point to a right of free navigation by warships which cannot be suspended but which is qualified by a requirement of compliance with the navigational laws of the coastal state. Article 23 still contains the possibility of mischief for warships exercising a right of innocent passage through straits. It is unfortunate that this question was approached obliquely and that the passage of straits was treated only by way of exception to the law regulating passage through the territorial sea.[99]

The right of free passage through international straits is a product of state practice hardening into customary international law and thence into treaty. The right of free passage through interoceanic canals is a consequence of the opening of each waterway to usage by the international community. It is the origin of the right in a series of individual grants which distinguishes the law relating to canals from the law of straits. The privilege of free passage through the three major interoceanic canals, Suez, Panama, and Kiel, has been created in each case by a treaty to which the territorial sovereign, acting freely

---

Article 24 of the draft prepared by the International Law Commission, dealing with the provision of innocent passage to warships (subject to notification and authorization) could not, even in modified form, command the requisite two-thirds vote and was eliminated from the final Convention. 2 UNITED NATIONS CONFERENCE ON THE LAW OF THE SEA, OFF. REC. 66-68 (U.N. Doc. No. A/CONF.13/38) (1958).

[97] Jessup, *The United Nations Conference on the Law of the Sea,* 59 COLUM. L. REV. 234, 248 (1959).

[98] Article 17.

[99] At the Hague Codification Conference of 1930, there had similarly been no special provision on the passage of straits in the draft Articles concerning the Legal Status of the Territorial Sea, but an observation appended to the article on the passage of warships through the territorial sea stated: "Under no pretext, however, may there be any interference with the passage of warships through straits constituting a route for international maritime traffic between two parts of the high sea." 3 ACTS OF THE CONFERENCE FOR THE CODIFICATION OF INTERNATIONAL LAW 212, 217 (L.N. Doc. No. V. Legal 1930. V. 16) (1930).

or under the pressure of other powers,[100] has been a party. The first of these in time and in importance was the Convention of Constantinople in 1888.[101] Significantly, the treaty opens with a sweeping declaration concerning free usage: "The Suez Maritime Canal shall always be free and open, in time of war as in time of peace, to every vessel of commerce or of war, without distinction of flag."[102] Although subsequent articles place some limitations upon this grant, notably as concerns the régime of the Canal when the security of Egypt is threatened,[103] the position and language of the quoted statement leave no doubt that the grant was comprehensive and to be liberally construed. The régime established for the Canal by the Convention of Constantinople was to be a permanent one, for it was not limited by the duration of the concession of the Compagnie Universelle du Canal Maritime de Suez,[104] which would, in the normal course of events, have expired in 1968. Whatever doubt there may have been regarding Egypt's subjection to the treaty after Egypt was freed from Turkish suzerainty and from British occupation was removed by the Agreement between Great Britain and Egypt regarding the Suez Canal Base, which was signed in 1954.[105] In that treaty, the two governments agreed that the Suez Canal was "a waterway economically, commercially and strategically of international importance" and expressed their "determination to uphold" the Convention of 1888.[106] Throughout the controversy concerning the passage of Israeli vessels and of cargo destined for Israel through the Canal, the United Arab Republic has given no indication that it doubted in any way the continuing force of the

---

[100] As in the case of the Treaty of Versailles and other treaties of peace.

[101] 15 MARTENS, N.R.G. 2d ser. 557 (1890). Prior to this time, the right of free navigation had been dealt with only in the concessions granted to the Suez Canal Company, which declared the Canal and its ports to be "ouverts à toujours, comme passage neutre, à tout *navire de commerce* . . . sans aucune distinction, exclusion ni préférence de personnes ou de nationalités . . ." (emphasis supplied). Concession of Jan. 5, 1856, art. 14, RECUEIL DES ACTES CONSTITUTIFS 6, 9. While the concession does not itself, of course, create international obligations, the introductory language of the article— "Nous déclarons solennellement, pour Nous et Nos successeurs"—might be construed as a unilateral declaration creating international obligations for the declarant state. See OBIETA, THE INTERNATIONAL STATUS OF THE SUEZ CANAL 51-56 (1960), and, concerning unilateral declarations generally, 1 OPPENHEIM, INTERNATIONAL LAW 872-73 (8th ed. Lauterpacht, 1955).

[102] Article I.

[103] Article X.

[104] Article XIV.

[105] Signed at Cairo, Oct. 19, 1954, Great Britain T.S. No. 67 (1955).

[106] Article 8.

Convention of 1888, and in all of the diplomatic interchanges and in the debates of the United Nations the participants framed their arguments in terms of the Convention. The most recent solemn pronouncement on the subject, the Egyptian Declaration of April 24, 1957, contains a reaffirmation of the Convention in the following terms: "It remains the unaltered policy and firm purpose of the Government of Egypt to respect the terms and the spirit of the Constantinople Convention of 1888 and the rights and obligations arising therefrom. The Government of Egypt will continue to respect, observe and implement them."[107]

When Great Britain and the United States reached agreement concerning the construction of a transisthmian canal in 1901, it was natural that the rules governing what was to become the Panama Canal should be based on the Convention of 1888. The rules contained in the Hay-Pauncefote Treaty, which were stated to be "substantially as embodied in the Convention of Constantinople," began with the assertion that "the canal shall be free and open to the vessels of commerce and of war of all nations observing these Rules, on terms of entire equality, so that there shall be no discrimination against any such nation, or its citizens or subjects, in respect of the conditions or charges of traffic, or otherwise."[108] The words "in time of war as in time of peace" which had appeared in the earlier Convention were, it will be noted, omitted from the treaty between the United States and Great Britain. The reason was the desire of the United States to be in a position to defend itself without being fettered by treaty obligations if it should be at war with Great Britain or with any other nation.[109] The omission thus serves approximately the same purpose as article X of the Convention of Constantinople, which secures to Turkey and Egypt the right to take measures necessary for "the defense of Egypt and the maintenance of public order." Without these words, the Hay-Pauncefote Treaty seems to say no more and no less than the Convention of Constantinople. The stipulation that the Canal was to be

[107] Article 1 of the Egyptian Declaration, annexed to letter to the Secretary-General of the United Nations from the Egyptian Minister for Foreign Affairs of the same date (U.N. Doc. No. A/3576, S/3818) (1957).

[108] Treaty between the United States and Great Britain to Facilitate the Construction of a Ship Canal, signed at Washington, Nov. 18, 1901, art. III, 32 Stat. 1903, T.S. No. 401.

[109] History of Amendments Proposed and Considered after the Action of the Senate and which Resulted in the Second Hay-Pauncefote Treaty, Memorandum prepared in the Department of State, in *Diplomatic History of the Panama Canal*, S. Doc. No. 474, 63d Cong., 2d Sess. 64 (1914).

free and open to the ships of "all nations observing these Rules"[110] likewise appears to have effected no real modification of the treaty of 1888, for it could hardly be expected that Turkey or Egypt would have been required to afford transit through the Canal to vessels which refused to respect their neutrality or to abide by the rules of navigation provided for the Canal or to pay the tolls which were required for transit. When in 1903 the United States acquired from Panama a zone through which to build a canal across the Isthmus of Panama, the Hay–Bunau-Varilla Treaty simply adopted by reference the rules governing passage through the canal which had been incorporated in the Hay-Pauncefote Treaty two years before.[111]

To these two great waterways open to the vessels "of all nations" the Kiel Canal was a comparatively late addition. Prior to the Treaty of Versailles, it had not been held out by Germany as an international waterway open to all comers by any formal declaration to that effect, although in fact about half the traffic through the Canal was under foreign flags.[112] That no such declaration was made by Germany is not surprising when it is recalled that the waterway was constructed with the primary purpose of affording freer access to the high seas by Germany's growing navy. The Canal was essentially a military enterprise.[113] With the conclusion of peace at the end of the First World War, the right of free passage through the Canal was established on the same basis as that which had been provided for Suez and Panama.

---

[110] The language was the result of the British insistence that it would be inequitable for Britain to be bound by treaty to respect the neutrality of the Canal, while other users, not being parties to any agreement, would be under no corresponding obligation. The first language proposed was "which shall agree to observe these rules." To this proposal the President of the United States objected, because it suggested that very sort of contractual relationship with third states which the Senate had rejected in declining to give its consent to the ratification of the first Hay-Pauncefote Treaty. The compromise which was reached was "all nations observing these Rules," which suggests a condition rather than a promise. The Marquis of Lansdowne to Lord Pauncefote, Oct. 23, 1901, in *Diplomatic History of the Panama Canal, supra* note 109, at 50; and *id.* at 67.

[111] Convention between the United States of America and the Republic of Panama for the Construction of a Ship Canal to Connect the Waters of the Atlantic and Pacific Oceans, signed at Washington, Nov. 18, 1903, art. XVIII, 33 Stat. 2234, T.S. No. 431. Bunau-Varilla stated that it was his and Secretary of State Hay's desire that the principles of the Convention of Constantinople of 1888 "should become, in a permanent way, the directing principle of the operation of the Panama Canal." BUNAU-VARILLA, PANAMA: THE CREATION, DESTRUCTION, AND RESURRECTION 373 (1914).

[112] THE KIEL CANAL AND HELIGOLAND (Foreign Office Peace Handbook No. 41) at 14 (1920).

[113] *Id.* at 5.

Article 380 of the Treaty of Versailles, in language which reflected its deriva-
tion from the Convention of Constantinople, provided: "The Kiel Canal and
its approaches shall be maintained free and open to the vessels of commerce and of war of all nations at peace with Germany on terms of entire
equality."[114]

No distinction is made in any of these instruments opening the major
interoceanic canals to use by "all nations" between merchant vessels, on the
one hand, and warships and auxiliaries, on the other. In the practice of states,
the right has been recognized to be one enjoyed by both categories of vessels
in time of peace on exactly the same basis. To demand advance notification
of the intended passage of warships is inconsistent with this grant. Although
no such requirement has ever been imposed in the case of the Panama
Canal,[115] there was a portent of the coming war in Germany's insistence in
January 1937 that warships and naval craft of foreign states might pass
through the Kaiser Wilhelm Canal (as the Kiel Canal was then called) only
if advance authorization had been obtained through diplomatic channels.[116]
Similar attempts to require advance notification of the passage of warships
through the Suez Canal were made by Egypt,[117] presumably in the advance-
ment of its contention that the waters of the Canal constituted Egyptian
territorial waters.

The passage of merchant vessels through these waterways in time of
peace has seldom posed any difficulties of principle. From time to time there
have been Russian protests concerning "discriminatory" treatment by the
United States of Soviet vessels passing through the Panama Canal. The
measures complained of, which include the searching of vessels, the placing
of security guards aboard, the denial of shore leave to officers and crew, and
the measurement of ships for the purpose of fixing tolls,[118] are in con-

---

[114] Treaty of Peace between the Allied and Associated Powers and Germany, signed
at Versailles, June 28, 1919, 112 BRITISH AND FOREIGN STATE PAPERS 1 (1919).

[115] When Mexico requested permission in 1917 for the passage of a transport, the
United States replied that no special permission would be required for the transit of the
vessel. Secretary Lansing to the Mexican Ambassador, Oct. 16, 1917, 2 HACKWORTH,
DIGEST OF INTERNATIONAL LAW 790 (1941).

[116] THE TREATY OF VERSAILLES AND AFTER: ANNOTATIONS OF THE TEXT OF THE
TREATY 690 (Dep't of State Pub. 2724) (1947).

[117] In a note of May 28, 1947, the Ministry of Foreign Affairs of Egypt required ten
days' notification of the passage of foreign warships through Egyptian territorial waters
and "par le fait même le transit par le Canal de Suez." Information furnished by the
Suez Canal Company.

[118] A formal note of protest was sent by the Embassy of the U.S.S.R. in Washington
to the Department of State in connection with the passage of the Soviet motor vessel

formity with the regulations governing transit of the Canal and either are necessary to the operation of the waterway or are not in violation of the rights of foreign nations. The United States was on the brink of danger in 1954 when measures were being taken to cut off the supply of arms to communist forces in Guatemala. Had a merchant ship laden with arms for that country presented itself for transit through the Canal, the United States would have had to determine how steadfast its adherence to the terms of the treaties regarding Panama actually was. It was spared this embarrassment, however, and contented itself with requesting permission of the major maritime powers to visit and search their vessels at sea, a suggestion which was not unnaturally rejected with some asperity.[119]

In 1958, the Republic of Panama extended its territorial sea to 12 miles.[120] This measure gave rise to some concern in the United States, because the Hay–Bunau-Varilla Treaty[121] had measured the Canal Zone three miles out into the sea at each end of the Canal and it was feared that the Panamanian decree would therefore box in both ends of the waterway. The United States Department of State took the position that the Panamanian law could not affect the rights of that country,[122] and the Government of the Republic of Panama, which had stated that it would apply the law in conformity with "international treaties in force,"[123] has not for its part given the slightest intimation that it would seek through the broadening of the territorial sea to impede navigation through the Canal. There is thus no ground for supposing that there will be any interference from that source with the free passage of merchant ships or war vessels.

By way of contrast, two other major artificial international waterways have not been formally opened to use by the international community, although

---

Ostrovsky on January 2, 1957 and of other Soviet ships. N.Y. Times, April 10, 1957, p. 1, col. 2, and p. 10, col. 1; see also *id.,* April 28, 1962, p. 39M, col. 4.

[119] N.Y. Times, June 18, 1954, p. 1, col. 3; June 19, 1954, p. 1, col. 8; June 22, 1954, p. 3, col. 8; June 23, 1954, p. 3, col. 5; June 24, 1954, p. 3, col. 7.

[120] Law No. 58, Dec. 18, 1958, 55 REPUBLIC OF PANAMA, GACETA OFICIAL, No. 13,720, p. 1 (1958). At least six nations protested the extension; the protests were rejected by Panama. N.Y. Times, March 14, 1959, p. 46M, col. 5.

[121] Convention between the United States of America and the Republic of Panama for the Construction of a Ship Canal to Connect the Waters of the Atlantic and Pacific Oceans, signed at Washington, Nov. 18, 1903, art. II, 33 Stat. 2234, T.S. No. 431.

[122] 40 DEP'T STATE BULL. 128 (1959), pointing out that article XXIV of the Hay–Bunau-Varilla Treaty states that no change in the law of Panama can affect the rights of the United States.

[123] Article 2, para. c, of the Panamanian law.

both are freely employed by vessels under the flags of states other than the territorial sovereigns. The first of these is the Corinth Canal, which runs entirely through Greek territory and has never been the subject of an international agreement or other form of grant. The other is the St. Lawrence Seaway, which, by reason of being in fact nonnavigable in its natural state, might be placed in the category of artificial waterways. It is probably the circumstance that it constitutes a boundary river and offers access, not to another section of the high seas, but to inland lakes, that accounts for the fact that it is by treaty open as a matter of right only to nationals of the United States and of Canada.[124] It is safe to infer from these two instances that foreign vessels do not have a legal right to use an international artificial waterway, even though it may like Corinth connect two stretches of the high seas, unless there has been an express grant of this privilege.

The question of the right of free passage is relatively uncomplicated when transit is claimed for a vessel by a state which is a party to the treaty which has thrown the waterway open. To the Convention of Constantinople of 1888, nine states—Great Britain, Germany, Austria-Hungary, Spain, France, Italy, the Netherlands, Russia, and Turkey—were parties.[125] The treaties regarding free passage through the Panama Canal were concluded by the United States only with Great Britain and with Panama. In the case of the Convention of Constantinople, the High Contracting Parties undertook to bring the treaty to the attention of states which had not signed it and to invite them to accede to it,[126] but no other states did so accede. A similar provision in the first text of the Hay-Pauncefote Treaty was rejected by the United

---

[124] Treaty between the United States and Great Britain relating to Boundary Waters between the United States and Canada, signed at Washington, Jan. 11, 1909, art. I, 36 Stat. 2448, T.S. No. 548.

[125] Great Britain did not become a party to the Convention until 1904; see Declaration between Great Britain and France respecting Egypt and Morocco, together with the Secret Articles, signed at London, April 8, 1904, art. VI, 101 BRITISH AND FOREIGN STATE PAPERS 1053 (1912).

Difficult problems of succession would arise if an attempt were made in this day to convene a conference of the signatories to the Convention of 1888. The Austro-Hungarian monarchy has been dissolved and Germany is divided in two portions. The Soviet Union maintained at the First London Conference of 1956 that the German Democratic Republic as well as the Federal Republic of Germany should have been invited and that Austria, Hungary, Czechoslovakia, and Yugoslavia were entitled to attend as successor states to the Austro-Hungarian monarchy. THE SUEZ CANAL PROBLEM, JULY 26-SEPTEMBER 22, 1956, at 59 (Dep't of State Pub. 6392) (1956); see HOYT, THE UNANIMITY RULE IN THE REVISION OF TREATIES: A RE-EXAMINATION 234-44 (1959).

[126] Article XVI.

States Senate on the ground that the United States should not be contractually bound by an agreement which would create rights and duties as between the United States and third states.[127] The Kiel Canal is similarly governed by a treaty to which, in numerical terms, only a minority of the international community is privy. No question about the rights of states not parties to the agreement arose in the case of *The S.S. Wimbledon,* [128] since Great Britain, France, Italy, and Japan, which jointly instituted proceedings in the Permanent Court of International Justice, and Germany, the respondent state, were all parties to the Treaty of Versailles.

Thus in the case of all three major interoceanic canals, a great many of the user nations are not parties to the agreements opening the waterways to use by the vessels of "all nations." The same problem arises with respect to the instruments which have thrown international rivers open to general use.[129]

If the treaty stipulation in favor of "all nations" is violated by the obligor to the detriment of a nonsignatory to the treaty, there can be no doubt that a party to the treaty has standing to complain of the violation. The controversial matter is whether the nonsignatory may itself assert the right to freedom of passage and whether it acquires a right which is proof against modification or termination of the treaty by the parties thereto—one which is independent of the treaty.[130]

[127] History of Amendments Proposed and Considered after the Action of the Senate and which Resulted in the Second Hay-Pauncefote Treaty, Memorandum prepared in the Department of State, in *Diplomatic History of the Panama Canal,* S. Doc. No. 474, 63d Cong., 2d Sess. 61, 64, 67 (1914); 3 MOORE, DIGEST OF INTERNATIONAL LAW 210, 216 (1906).

The Senate report on the first Hay-Pauncefote Treaty includes an enigmatic statement: "Special treaties for the neutrality, impartiality, freedom, and innocent use of the two canals that are to be the eastern and western gateways of commerce between the great oceans are not in keeping with the magnitude and universality of the blessing they must confer upon mankind. The subject rather belongs to the domain of international law." S. Doc. No. 268, 56th Cong., 1st Sess. (1900), in 8 COMPILATION OF REPORTS OF THE COMMITTEE ON FOREIGN RELATIONS UNITED STATES SENATE, 1789-1901, S. Doc. No. 231, 56th Cong., 2d Sess. 636, 645 (1901).

[128] P.C.I.J., ser. A, No. 1 at 15 (1923).

[129] In 1887, the United States claimed rights of navigation on the Congo, even though it was not a party to the General Act of Berlin, opening the river to navigation by the ships of all nations. The Secretary of State to the Minister to Belgium, Dec. 19, 1887, [1887] 1 FOREIGN REL. U.S. 27 (1889), 1 MOORE, DIGEST OF INTERNATIONAL LAW 652 (1906).

[130] See Jiménez de Aréchaga, *Treaty Stipulations in Favor of Third States,* 50 AM. J. INT'L L. 338, 340 (1956).

Since the United States is the territorial sovereign and operator of the Panama Canal and one of the principal users of the Suez Canal, its position and practice might be expected to shed some light on the question of the rights of third parties. In the first role, assertions of right by nations other than Great Britain and Panama might be made against it, while in the second instance, the United States, a nonsignatory to the Convention of 1888, would be a claimant of rights under the Convention. With respect to the Hay-Pauncefote Treaty, Secretary of State Hughes wrote to President Harding that "other nations . . . not being parties to the treaty have no rights under it."[131] During the Suez Canal crisis, the highest officials of the United States Government maintained, consistently with the views of Secretary Hughes, that the situations of the two canals were quite different. In a press conference on August 8, 1956, President Eisenhower remarked, "It [the Suez Canal] is completely unlike the Panama Canal, for example, which is strictly, was a national undertaking carried out under bilateral treaty."[132] Later in the same month, in the performance of the same sort of rite, the Secretary of State, Mr. Dulles, elaborated on this theme: "And there is no international treaty giving other countries any rights at all in the Panama Canal except for a treaty with the United Kingdom which provides that it has the right to have the same tolls for its vessels as for ours."[133] The other basis upon which he found the statuses of Panama and Suez to be "totally dissimilar" was that Suez had been "internationalized," while Panama had not. To this it seems proper to reply that if the United States denies to states not parties to the treaties opening the Panama Canal to the vessels of "all nations" any legal right to employ the Canal, it has no basis for the assertion that ships of the United States, including its warships, have a legal right to transit Suez. The mere declaration that the two waterways are "completely unlike" and "totally dissimilar" is not legally persuasive. It is quite true that a provision inviting other nations to become parties to the Hay-Pauncefote Treaty was eliminated from the agreement at the demand of the Senate, but on the other hand, the express invitation of the Convention of Con-

[131] 5 HACKWORTH, DIGEST OF INTERNATIONAL LAW 221-22 (1943).

[132] THE SUEZ CANAL PROBLEM, JULY 26-SEPTEMBER 22, 1956, at 45 (Dep't of State Pub. 6392) (1956).

[133] News Conference of Aug. 28, 1956, 35 DEP'T STATE BULL. 411 (1956). A statement on Aug. 29, 1956, by the President that the Suez Canal was an "internationalized" waterway brought a prompt protest from President Nasser of Egypt. N.Y. Times, Aug. 31, 1956, p. 1, col. 1. The President of the United States later explained that he was only referring to the status of the Canal under the Convention of 1888. N.Y. Times, Sept. 1, 1956, p. 1, col. 2.

stantinople has never been accepted by the United States or others of the nonsignatory states. Furthermore, if free passage for the ships of all nations is a right which can be actually asserted only by signatories to the treaty, the United States could not of its own motion protest any denial of free passage to its vessels by the proprietor of the Suez Canal.[134] It is quite clear that any argument directed against the rights of third parties with respect to Panama cuts the ground from under the United States and other nonsignatory users of the Suez Canal. The effect of the United States position is to limit to the actual signatories of the agreement legal rights of passage over rivers and canals declared open to the vessels of all nations.

The extent to which the international community employs the great international waterways and to which the territorial sovereign and operators of the waterways have acquiesced in such usage is in itself sufficient refutation of the contention that rights of free navigation may be acquired only by treaty. If such be the case, it then becomes necessary to determine the foundation upon which the rights of nonsignatories may be based. The question may be approached from a number of different legal perspectives:

1. The right of passage has by some been asserted to constitute an "international servitude."[135] The doctrine of servitudes occupies at best, however, a questionable position in international law,[136] and the Permanent Court of International Justice, when offered the opportunity in *The S.S. Wimbledon*

---

[134] The correct position is stated by Sir Gerald Fitzmaurice, who writes with respect to the conventions creating a right of free navigation for the shipping of all countries through certain canals and other waterways, that "although only certain countries are actually parties to these Conventions, there is little doubt that the benefit of them could be directly claimed by any country; or to put the matter in another way, that any country would have a direct right of complaint if the free navigation of its shipping through those waterways was impeded." *The Juridical Clauses of the Peace Treaties,* 73 Hague Recueil 259, 360 (1948).

[135] Reid refers to such rights of passage as "interoceanic transit servitudes." International Servitudes in Law and Practice 131 (1932).

[136] The tribunal in the North Atlantic Coast Fisheries Arbitration said that the doctrine was "but little suited to the principle of sovereignty which prevails in States under a system of constitutional government such as Great Britain and the United States, and to the present international relations of sovereign States," and had therefore "found little, if any, support from modern publicists." S. Doc. No. 870, 61st Cong., 3d Sess., vol. 1 at 76 (1912); and see the weighty criticism of the doctrine by McNair in *So-Called State Servitudes,* 6 Brit. Y.B. Int'l L. 111 (1925). But cf. Lauterpacht, Private Law Sources and Analogies of International Law 238 (1927), who maintains the view that the tribunal "did not reject the conception of servitudes, but it demanded a strict proof that a servitude has been created." That burden could probably not be sustained in the case of any major waterway as to which any third state claims rights.

to decide the question of free passage in these terms, expressly declined
to do so.[137] And Váli, whose thorough study of servitudes in international
law makes his views particularly deserving of attention, sees the right of
free passage through an international canal as a general restriction on territorial sovereignty which has no relevance to international servitudes.[138]

2. By analogy to municipal law, it might be asserted that states not parties
to the instrument throwing the waterway open to general use are third-party
beneficiaries of the treaty and may in their own right assert the rights so
conferred. But, as is true of the doctrine of international servitudes as well,
the doctrine of third-party beneficiaries has received an unenthusiastic reception from the Permanent Court, notably in the *Case concerning Certain
German Interests in Polish Upper Silesia*[139] and the *Chorzów Factory* case.[140]
On the other hand, in the *Case of the Free Zones of Upper Savoy and the
District of Gex,* the same court, while professedly refraining from dealing
with the question of the extent to which international law recognizes "stipulations in favour of third Parties,"[141] seems actually to have grounded its
decision on that doctrine.[142]

In the writings of publicists, the doctrine has received a mixed reception.[143]
The Harvard Research in its draft on the Law of Treaties recognized that a
treaty might contain a stipulation which is expressly for the benefit of a
state not a party to the treaty[144] but excluded from that category treaties
"which, without the parties so intending, do incidentally secure advantages
or benefits to third States."[145] In this latter category was placed article 380 of

---

[137] P.C.I.J., ser. A, No. 1, at 24 (1923).

[138] SERVITUDES OF INTERNATIONAL LAW 52-54 (2d ed. 1958). Hostie envisages the
right of free passage as an absolute one good *erga omnes* and therefore not capable of
analysis in terms of servitude, which looks to a duty running to a specific piece of
land or to a certain person or persons. *Notes on the International Statute of the Suez
Canal,* 31 TUL. L. REV. 397, 418-19 (1957).

[139] P.C.I.J., ser. A, No. 7, at 28-29 (1926).

[140] P.C.I.J., ser. A, No. 17, at 45 (1928).

[141] Order of Aug. 19, 1929, P.C.I.J., ser. A, No. 22, at 20 (1929).

[142] LAUTERPACHT, THE DEVELOPMENT OF INTERNATIONAL LAW BY THE INTERNATIONAL
COURT 306-08 (1958); Jiménez de Aréchaga, *Treaty Stipulations in Favor of Third
States,* 50 AM. J. INT'L L. 338, 342-43 (1956).

[143] Compare the writings referred to in the preceding note with ROXBURGH, IN-
TERNATIONAL CONVENTIONS AND THIRD STATES (1917). Roxburgh concludes (at 69-70)
that while the Hay-Pauncefote Treaty did not create rights in third parties, it might
in the course of time become the basis of a rule of customary international law.

[144] Article 18(b), 29 AM. J. INT'L L., Sp. Supp. 924 (1935).

[145] *Id.* at 926.

the Treaty of Versailles, relating to the free transit of the Kiel Canal. In the view of the draftsmen of the Harvard Research, the treaty could not create a legal right for states other than the contracting parties, although they admitted that with the passage of time a customary right might come into existence.[146] This conclusion is consistent with the view advanced below that both a grant and a period of reliance are necessary to the creation of any right in nonsignatories.

In one of his reports to the International Law Commission, Sir Gerald Fitzmaurice expressed a view consistent with that of the Harvard Research but did not elaborate upon the treaties regulating passage through interoceanic canals: "Where a treaty expressly confers rights or benefits on, or makes provision for the exercise of rights or faculties, or for the enjoyment of facilities or benefits by a third State, in such a way as to indicate that the parties meant to create legal rights for the third State, or to bind themselves to grant them, or to create a legal relationship between themselves and the third State, the third State concerned thereby acquires a legal right to claim the benefit of the provisions in question."[147]

As observed above, the presence of an article in the Convention of Constantinople of 1888[148] whereby the signatories undertook to bring the Convention to the attention of other states and to invite them to accede to it is a complicating element. It is possible to contend that such an accession clause provides the only route whatsoever by which a third state might acquire any rights under the treaty.[149] According to a literal construction of this article, the requirement of an invitation from a signatory cannot be reconciled with an unqualified right of accession by third states; no such invitations were extended. Even if third states could accede without an invitation, the act of accession should be taken as an express assumption of all of the obligations of the convention, rather than as a necessary condition to be fulfilled before the right of free navigation could be enjoyed.[150] And in the practice of states, a

---

[146] *Id.* at 926. Jiménez de Aréchaga points out, however, that article 386 gave to "any interested power" the right to appeal to a jurisdiction instituted for the purpose by the League of Nations and that means of enforcement by third parties suggests the existence of rights running to them. *Supra* note 142, at 350.

[147] Fifth Report on the Law of Treaties (Treaties and Third States), art. 20(1), p. 42 (U.N. Doc. No. A/CN.4/130) (1960).

[148] Convention of Constantinople, signed Oct. 29, 1888, art. XVI, 15 MARTENS, N.R.G. 2d ser. 557 (1891).

[149] ROXBURGH, *op. cit. supra* note 143, at 49-51.

[150] Hostie, *Notes on the International Statute of the Suez Canal*, 31 TUL. L. REV. 397, 415-16 (1957).

provision for accession has never been set up as a defense to a claim by a nonacceding state that it is entitled to rights under the treaty. In the dispute between the United States and Belgium over navigation on the Congo, the accession clause in the General Act of Berlin did not enter into the controversy.[151]

Even should the doctrine of third-party beneficiaries be considered to be applicable to treaties opening international waterways to general usage, the principle would not be a satisfactory basis for the creation of rights in the international community on a permanent footing. The most weighty objection is that the rights of all states would be tied to the treaty itself.[152] The view does not guard against possible denunciation of the treaty or its termination by mutual assent of the parties with the possible destruction of the rights of third parties. In the event of the coming into existence of a new state controlling the area through which a waterway runs or the cession of that territory, difficult problems of succession to treaties would arise, and it is not inconceivable that the rights of signatories and nonsignatories might be recognized to have been terminated by the change in the status of the territory. Further complications would be caused by the fact that a treaty concerning an international river or canal may be suspended or terminated should war break out between parties to the treaty. Such suspension or termination might, according to the generally recognized principles concerning the construction of treaties, be considered to affect the "rights" of third parties not engaged in hostilities. Over and above these objections based upon the mutability of treaties, the third-party beneficiary theory overlooks the possibility that the international status of the canal may be created by unilateral declaration rather than by treaty. To the extent that Egypt assumed certain obligations with respect to the Suez Canal by its declaration of April 24, 1957,[153] the international community cannot be said to have acquired any rights by virtue of a treaty. Finally, the doctrine of third-party beneficiaries takes no account of the place of reliance on the existence of the canal, without which a claim to rights in the waterway can hardly evoke sympathetic consideration.

3. Somewhat more satisfactory is the view that treaties which open rivers

---

[151] See p. 175, n. 129 *supra*.

[152] Harvard Research, Draft Convention on the Law of Treaties, art. 18(b), 29 Am. J. Int'l L., Sp. Supp. at 924 (1935).

[153] Annexed to letter to the Secretary-General of the United Nations from the Egyptian Minister for Foreign Affairs, April 24, 1957 (U.N. Doc. No. A/3576, S/3818) (1957).

and canals to general or limited use by the vessels of states other than the parties to the instrument in question are dispositive in nature.[154] Such dispositive treaties create real rights which are attached to a territory and are therefore not dependent upon the treaty creating them.[155] Although such real rights are theoretically related to international servitudes,[156] it does not appear that adherence to the doctrine of "real rights" necessarily entails acknowledgment of the existence of servitudes in international law. The demilitarization of the Aaland Islands was recognized by a committee of jurists appointed by the League of Nations to create "a special international status" not grounded on the existence of an international servitude.[157] The terms in which this "special international status" was discussed by the committee suggest that it regarded the obligation as a "real" one, having its origin in a dispositive treaty.

So far as problems of succession are involved, the theory of dispositive treaties and of real rights gives some degree of permanency to the régime established for an international waterway.[158] The recognition of a real right in the case of the Kiel Canal would make it unnecessary to consider what

[154] Concerning this concept see McNair, The Law of Treaties, 1961, at 255-59 (1961); O'Connell, The Law of State Succession 49-63 (1956); Udina, *La Succession des états quant aux obligations internationales autres que les dettes publiques*, 44 Hague Recueil 669, 742-48 (1933); Fitzmaurice, *The Juridical Clauses of the Peace Treaties*, 73 *id*. 259 at 260, 360 (1948).

[155] In the words of Lord McNair, ". . . the treaties belonging to this category create, or transfer, or recognize the existence of, certain permanent rights, which thereupon acquire or retain an existence and validity *independent of the treaties which created or transferred them*." The Law of Treaties, 1961, at 256 (1961). See also his separate opinion in the advisory opinion on the International Status of South-West Africa, [1950] I.C.J. Rep. 128 at 155-57.

[156] O'Connell regards "international servitudes" as a species of "real right." *Op. cit. supra* note 154, at 51.

[157] *Report of the International Commission of Jurists entrusted by the Council of the League of Nations with the task of giving an advisory opinion upon the legal aspects of the Aaland Island question*, League of Nations Off. J., Sp. Supp. No. 3, at 18 and 19 (1920). The Committee stated, ". . . such a settlement cannot be abolished or modified either by the acts of one particular Power or by conventions between some few of the Powers which signed the provisions of 1856, and are still parties to the Treaty" (at 18). It would appear from this assertion that the demilitarized status of the Islands could be altered with the consent of all of the parties to the treaty which had originally demilitarized them. If such be the case, the opinion does not lend support to the view that dispositive treaties freeing international waterways to navigation by all may not, if once relied upon by the international community, be modified or terminated by the joint action of all states parties to the original treaty.

[158] O'Connell, *op. cit. supra* note 154, at 51.

continuing effect the Treaty of Versailles may have. The status of the water-way, like an international boundary laid down by the terms of a treaty, is not altered by the vicissitudes which the treaty creating it may undergo. A clear attempt to create a right which would be independent of territorial and political changes is discernible in article IV of the Hay-Pauncefote Treaty, in which it is agreed that "no change of territorial sovereignty or of the inter-national relations of the country or countries traversed by the before-mentioned canal shall affect the general principle of neutralization or the obligation of the High Contracting Parties under the present Treaty."[159] While problems of state succession are avoided by the adoption of this theory, it is not altogether clear whether this view satisfactorily accounts for the legal privileges which may be claimed by nations that are not parties to the dispositive treaty and are not successor states. The possibility that the parties to the original instrument creating the real right may by a subsequent treaty destroy that right would also make the theory of dispositive treaties subject to the same objection as those of international servitudes or third-party bene-ficiaries. The principle of dispositive treaties, furthermore, fails to take ac-count of an additional element which, in the view of the writer, is of sub-stantial importance. This factor is reliance on the waterway by those to whom access is afforded. In view of these weighty objections, the theory of dispositive treaties must be rejected as a basis for a permanent right of transit through international waterways which have been opened to the ships of all nations.

4. The preferable theory concerning the rights of nonsignatories is that a state may, in whole or in part, dedicate a waterway to international use, which dedication, if relied upon, creates legally enforceable rights in favor of the shipping of the international community.[160] A treaty, a unilateral declaration—perhaps even a concession[161]—may be the instrument whereby

---

[159] Treaty between the United States and Great Britain to Facilitate the Construction of a Ship Canal, signed at Washington, Nov. 18, 1901, 32 Stat. 1903, T.S. No. 401.

[160] Cf. the theory of Guillien, who considers that the international status of the Suez Canal is a consequence of the fact that the world received the Convention of Constantinople from the hands of plenipotentiaries and accepted it without discussion. He is led to conclude that the convention was in fact legislation. *Un Cas de dédouble-ment fonctionnel et de législation de fait internationale: Le Statut du Canal de Suez,* 2 LA TECHNIQUE ET LES PRINCIPES DU DROIT PUBLIC: ETUDES EN L'HONNEUR DE GEORGES SCELLE 735, 751 (1950). But this tacit acceptance is not far removed from the theory of tacit adhesion from which Guillien rightly dissents (at 749). Tacit adhesions and ac-ceptances belong to the realm of fiction; reliance is real and measurable.

[161] Several authorities take the view that the source of the right of free passage through the Suez Canal, at least in so far as passage by merchant ships in time of peace

the dedication is effected.[162] Its form is not important; what is important is that it speaks to the entire world or to a group of states who are to be the beneficiaries of the right of free passage.

This theory of dedication, but without the element of reliance, is that of the Permanent Court of International Justice in *The S.S. Wimbledon,* wherein the court spoke of Kiel as "an artificial waterway connecting two open seas [which] has been permanently dedicated to the use of the whole world."[163] Justice does not demand that third states acquire any rights until there has been actual international use of the waterway. The harm caused by the dislocation of established international channels of communication is the real reason why the barring of free passage is considered to be an international wrong. In the case of states having treaty rights in the waterway, no reliance on the use of the highway is needed as a basis for a complaint of discrimination or exclusion. Such treaties have a dispositive character, and successor states thus acquire the same obligations and rights as their predecessors. But third states should not gain rights unless the dedication has induced them to make some measurable use of the canal or of the river and to make it one of their shipping routes.

Is it necessary that this reliance be by the particular state claiming rights of transit or by international shipping generally? In the case of a waterway which has been opened only to the ships of stipulated third states—and it must be emphasized that grants of free passage to a limited number of states are normally effected by treaties to which all of such states are parties—the reliance would have to be by the vessels of such of those nations as were

---

is concerned, is the concessions of 1854 and 1866. Obieta does so on the basis that the concession constituted a unilateral declaration tacitly accepted by the international community. THE INTERNATIONAL STATUS OF THE SUEZ CANAL 39, 57 (1960). Hostie sees the concessions as "elements of the statute of the Suez Canal" in public international law. *Notes on the International Statute of the Suez Canal,* 31 TUL. L. REV. 397, 420-22 (1957). In view of later developments, the question whether the Canal was internationalized by the concessions or by the Convention of 1888 is of only academic interest.

[162] The dedication of the Suez Canal to international usage may be seen as having been effected in the first instance by the concessions and in the second by the Convention of Constantinople. The dedication was then reaffirmed by the Egyptian Declaration of 1957.

Concerning the doctrine of dedication with respect to interoceanic canals, see Farran, *The Right of Passage through International Maritime Canals: A Comparative Study,* [1959] SUDAN LAW JOURNAL AND REPORTS 140, 143-59.

[163] P.C.I.J., ser. A, No. 1 at 28 (1923).

not signatories of the treaty. If, however, as is true of Suez, Panama, and Kiel, the waterway is, subject to some exceptions, thrown open to use by the ships of all nations by the state through the territory of which it flows, it would appear that reliance by the shipping of the international community in general suffices. It follows that the shipping of a state whose vessels had not hitherto used the canal would have a right to such passage if there had previously been substantial reliance on the waterway by international shipping generally. Such a principle avoids difficulties about the extent of use by individual states and takes into account the creation of new states, or new merchant fleets, and of new trade routes. To require substantial reliance on an individual basis, state by state, would take no account of change and would make it possible for the littoral state to discriminate between nations by excluding the ships of those countries which might be just beginning their utilization of the waterway.

If third states acquire rights under the treaties internationalizing waterways, there remains the question of the source of their obligations with respect to the territorial sovereign or the operator or supervisor of the waterway. These obligations are in actuality conditions on the grant of the right of passage. If the vessels of a state will not or cannot assume these obligations, and thereby fulfill the conditions which have been imposed, the operator of the waterway has no obligation to accord freedom of transit. A ship has, for example, no right of free transit unless tolls are paid. A warship has no right of transit unless it is prepared to respect the neutrality of the waterway and to refrain from acts of hostility within it. A vessel undertakes as one of the conditions of passage compliance with the rules of navigation which have been promulgated for the waterway. In short, a state which accepts the benefit of transit assumes with it the burdens which condition that benefit.

The rights of free passage which exist in rivers, canals, and straits, considered as separate categories of waterways, are of diverse origin. It is probably correct to say that the right of free passage through international straits is but one aspect of the freedom of the seas. Without the right to pass from one portion of the high seas to another through a strait, the freedom of the seas would be much impaired, for wide detours would be caused and whole stretches of ocean cut off by a state's control of a single strait. What rights littoral states have under customary international law are accordingly derogations from a more general principle of freedom. On the other hand, rivers and canals are, in the absence of express provisions to the contrary, governed by principles which go to the opposite extreme. Subservient as they are to the land masses through which they flow, international rivers might properly be regarded as nothing more than fluvial highways, as fully within the sovereignty of a state as the roads which it has built. Practical considerations of commerce have impelled states to open many of these rivers to wider use, first by the other riparians, later by the ships of all nations. As in the case of rivers, canals do not by the mere fact of their construction automatically become international highways for the vessels of all nations. If a state has no obligation to construct an interoceanic canal in the first place, there is no reason in principle or in policy why it should be required, once the waterway is in existence, to make it available to all. Again, as is true of rivers as well, the opening of artificial interoceanic waterways has come about as the result of grants by the territorial sovereign or operator. In the absence of other arrangements, political geography keeps even such rivers as are international closed to the vessels of other states. What geography demands for rivers is, in the case of canals, a consequence of artificiality. Neither type of waterway is in its natural state open to all comers.

If rights of transit through the three types of waterways have differing sources, the results of the opening of these waterways present a far different pattern. Each river has been recognized to be opened individually, so that it remains difficult to construct any body of principles applicable to all international rivers, at least in so far as navigation is concerned. A number of international rivers remain closed, as a matter of law, to general use. On the other hand, the opening of the three major interoceanic canals by grants which are virtually identical in their terms has meant the emergence of a common law of international canals, the existence of which was by implication recognized by the Permanent Court of International Justice in *The S.S. Wimbledon*.[164] Were a fourth major interoceanic canal to be constructed, it is quite possible that it would be regarded as being outside this international canal-law, unless it were opened on terms similar to those provided for Suez, Panama, and Kiel. The development of international canal-law is thus largely a consequence of the coincidence that all of the class have been subjected to similar régimes of transit. Historical parallelism has counted for far more in the development of the law of canals than has any theoretical consideration like the freedom of the seas as applied to straits.

As the law stands today, canals and straits bear a close affinity as regards freedom of transit in time of peace. A series of grants in the one case and

[164] P.C.I.J., ser. A, No. 1 (1923).

freedom of the seas in the other have worked to bring about the same result. The establishment of a right of free passage common to both categories of waterways has been aided by the functional analogy resulting from the fact that both offer access from one stretch of the high seas to another. The characterization of a canal as an "artificial strait"[165] is a highly accurate one, provided it is understood that a canal may achieve that status but it is not necessarily born to it. It is a consequence of this similarity of function that in time of peace the right of transit through both types of waterway remains untrammeled. By contrast, international rivers remain, in a legal sense, the products of the instruments internationalizing them. The right of free passage exists only to the extent that it has been created for a particular river; it has not passed into customary international law.

[165] This analogy was drawn for a limited purpose in The S.S. Wimbledon, *supra* note 163, at 28, wherein it was stated: "Moreover they [the precedents afforded by Suez and Panama] are merely illustrations of the general opinion according to which when an artificial waterway connecting two open seas has been permanently dedicated to the use of the whole world, such waterway is assimilated to natural straits in the sense that even the passage of a belligerent man-of-war does not compromise the neutrality of the sovereign State under whose jurisdiction the waters in question lie."

CHAPTER IV

PASSAGE OF SHIPS THROUGH INTERNATIONAL WATER' WAYS IN TIME OF WAR

T HE EXISTENCE of a war to which either the state through whose territory the waterway passes or a user[1] of the waterway is a party causes
conflicting demands of a particularly compelling nature. If a user state but
not the littoral state is at war, the state which sends its vessels through the
waterway will naturally be anxious to continue that use in time of war. With
this interest must be reconciled the desire of the littoral state to maintain its
neutrality by precluding the commission of hostile acts and to keep the waterway open for the use of other states. When the littoral state is itself at war,
one of its primary concerns must necessarily be for its own safety and for that
of the waterway. Its enemy will in all probability be anxious to seize the
waterway or interdict its use or may, on the contrary, desire to continue to
draw some of the sustenance for its military activities through it. Opposed to
these interests will be those of neutral users of the waterway whose demands
for the free and unfettered use of the waterway can be expected to conflict to
a greater or less degree with the littoral state's legitimate interest in defending
itself and in waging economic warfare against the enemy.

International law has for the most part not attempted to reconcile these
conflicting interests with any precision. It is only exceptionally that treaties
which open rivers to navigation by the ships of all nations make any reference
to the régime which is to exist in time of war. Article 49 of the Convention
Instituting the Statute of Navigation of the Elbe, for example, contents itself
with stating that "the provisions of the present Convention continue valid in
time of war to the fullest extent compatible with the rights and duties of

---

[1] The fact that belligerent states normally exercise strict control over their shipping
in time of war makes it particularly appropriate to speak of states, rather than shipping
companies, as the users of international waterways in time of war.

belligerents and neutrals."[2] The Convention of Mannheim, regulating the status of the most important of European waterways, the Rhine, contains an even more equivocal reference to the "measures prescribed for the maintenance of the general security."[3] This absence of provisions having application to periods of war can probably be attributed to the considerations that rivers fall within the domain of land warfare and that their control is not an aspect of maritime strategy. Involved as they are in battle and occupation, it would be unrealistic to suppose that they can be accorded a status differing to any marked degree from that of the shores between which they run.

Since the legal status of only a small minority of the straits used for international communication is determined by treaty, the principle of the free navigation of international straits must find its origin in the practice of states which has gained acceptance as customary international law. For the most part such treaty provisions as exist are imprecise about passage in time of war. The Montreux Convention, a number of the articles of which contain detailed provisions concerning the passage of merchant ships and warships when Turkey is at war, is the one notable exception.[4] The uselessness of the few treaties as guides to day-to-day conduct and the fact that free passage is in most cases the creation of customary international law combine to make the practice of states of particular significance in dealing with practical problems relating to the navigation of these waterways by warships and merchant ships.

Freedom of navigation through interoceanic canals which have been "permanently dedicated to the use of the whole world"[5] is uniformly established through international agreement or by the unilateral act of the pro-

---

[2] Signed at Dresden, Feb. 22, 1922, 26 L.N.T.S. 221, 241.

[3] Revised Convention for the Navigation of the Rhine, signed at Mannheim, Oct. 17, 1868, art. 1, 20 MARTENS, N.R.G. 355 (1875), 2 RHEINURKUNDEN 80 (1918). The principle of business before battle in early nineteenth century practice is illustrated by the Convention between France and Germany on the Octroi of Navigation of the Rhine, signed at Paris, Aug. 15, 1804, 8 MARTENS, RECUEIL 261, 292 (1835), RHEINURKUNDEN 6 (1918), article CXXXI of which provides: "S'il arrivait (ce qu'à Dieu ne plaise) que la guerre vînt à avoir lieu entre quelques uns des états situés sur le Rhin ou même entre les deux empires, la perception du droit d'octroi continuera à se faire librement sans qu'il y soit apporté d'obstacle de part ni d'autre.

"Les embarcations et personnes employées au service de l'octroi jouiront de tous privilèges de la neutralité . . ."

[4] Convention regarding the Régime of the Straits, signed at Montreux, July 20, 1936, arts. 4-6 and 19-21, 173 L.N.T.S. 213, 7 HUDSON, INTERNATIONAL LEGISLATION 386 (1941).

[5] The S.S. Wimbledon, P.C.I.J., ser. A, No. 1, at 28 (1923).

prietor of the canal. The Convention of Constantinople of 1888 provides that "the Suez Maritime Canal shall always be free and open, in time of war as in time of peace, to every vessel of commerce or of war, without distinction of flag."[6] Subsequent articles contain provisions designed to secure the neutrality of the Canal in time of war. The extent of the power of the United Arab Republic to take measures for its defense is, however, left unclear by related provisions that Turkey, to the rights of which the United Arab Republic succeeded, may take measures which it "might find it necessary to take to assure by their own forces the defense of Egypt and the maintenance of public order,"[7] but that such acts "shall not interfere with the free use of the Canal."[8] The Hay-Pauncefote Treaty of 1901,[9] which regulates, at least vis-à-vis Great Britain, the status of the Panama Canal, is hardly more definite. The Canal is to be "free and open to the vessels of commerce and of war of all nations observing these Rules."[10] The prohibition of blockade and of acts of hostility within it seem to be directed to users of the waterway; the Treaty is silent with respect to the measures which may be taken by the United States for the defense of the Canal or in time of war.[11] Significantly, the phrase "in time of war as in time of peace," which appears in the Convention of Constantinople, was not repeated in the Hay-Pauncefote Treaty, which otherwise bears an intentional similarity to the earlier convention. In establishing the status of the third major interoceanic canal of international concern, the Treaty of Versailles requires that "the Kiel Canal and its approaches shall be maintained free and open to the vessels of commerce and of war of all nations at peace with Germany on terms of entire equality."[12] In none of these three important international agreements are the rights and duties in time of war of the operator or supervisor of the canal, the territorial sovereign, and the users of the canal made plain.

[6] Signed at Constantinople, Oct. 29, 1888, art. 1, 15 Martens, N.R.G. 2d ser. 557 (1891).

[7] Article X.

[8] Article XI.

[9] Treaty between the United States and Great Britain to Facilitate the Construction of a Ship Canal, signed at Washington, Nov. 18, 1901, 32 Stat. 1903, T.S. No. 401.

[10] Article III, rule 1.

[11] Article III, rule 2, which corresponds in its first sentence to article IV of the Convention of Constantinople, provides: "The canal shall never be blockaded, nor shall any right of war be exercised nor any act of hostility be committed within it. The United States, however, shall be at liberty to maintain such military police along the canal as may be necessary to protect it against lawlessness and disorder."

[12] Treaty of Peace between the Allied and Associated Powers and Germany, signed at Versailles, June 28, 1919, art. 380, 112 British and Foreign State Papers 1 (1919).

It may thus be observed that the law relating to international waterways, largely conventional in origin as it is, has to a very large extent failed to concern itself with the common problem of assuring the free passage of such waterways during war, while permitting the defense of the legitimate interests of the state through the territory of which the river or strait or canal in question runs. The answer to this question must therefore be sought in "international custom, as evidence of a general practice accepted as law."[13] The recent practice of states is of primary importance to this inquiry, since ancient precedents[14] lose their relevance with the immense changes which have taken place in the grim art of waging war.

When the Littoral or Riparian State is not at War

During a war to which the state through which an international waterway passes is not a party, such a state rests under certain obligations which are the consequence of its neutral status. Conscious of the responsibilities imposed upon them by the existence of the passage and solicitous of their own defense, littoral states have in practice generally discharged their duties as neutrals in conformity with law. In so far as warships are concerned, the primary concern of the riparian states must normally be to prevent the commission of hostile acts within or in the vicinity of the waterway, while permitting the free passage of all ships, whether merchant vessels or warships, belligerent or neutral. In exceptional cases, dictated either by the facts of geography or of politics, the responsibility of the littoral state may, however, extend to denying passage to certain categories of ships.

Warships

The very scarcity of state practice[15] and of jurisprudence relating to the passage of warships through international straits, far from suggesting any

---

[13] Article 38, para. 1(b), of the Statute of the International Court of Justice.

[14] To this latter category probably belongs the incident which took place in the Suez Canal in 1870. On August 15 of that year, the birthday of the Emperor, a French warship, which had dressed ship for the occasion, encountered a German warship in Lake Timsah, midway in the Canal. Notwithstanding the existence of a state of war between the two countries, the German vessel saluted the French warship. Bahon, Le Libre Usage du Canal de Suez et sa "neutralité," Communication faite à l'Académie de Marine, April 23, 1936, Ms. on file in the records of the Compagnie Financière de Suez, Paris.

[15] I Brüel, International Straits III (1947).

lacuna in the law of nations relating to this subject, indicates that restrictions are placed on such passage only as an exceptional measure. During the Second World War, the Strait of Gibraltar was freely used as an avenue of approach for warships during the invasion of North Africa.[16] When the Strait was blockaded during an early stage of the war, Prime Minister Churchill had insisted that Spanish territorial waters should not be violated in the course of this process.[17] In the previous World War, Chile had declared the whole of the Strait of Magellan to be within its territorial waters[18] but did not interdict the passage of warships.[19] The act of a British warship in capturing within the Strait the *Bangor,* a Norwegian merchant vessel chartered to carry supplies to a German cruiser, was of doubtful compatibility with the Treaty of 1881 between Argentina and Chile, which had neutralized the Strait of Magellan and declared the Strait open to the shipping of all nations.[20] In prize proceedings for the recovery of the vessel, the court quite properly held that any possible violation of Chilean neutrality resulting from the capture in neutral waters could be asserted only by Chile and could not be the basis for the restoration of the ship to its owners.[21] Prior to the First World War, Denmark and Sweden allowed warships to pass freely through the Danish Straits between the Baltic and the North Sea during the Crimean War, the Franco-Prussian War of 1870, the Russo-Turkish War, and the Russo-Japanese War.[22] Denmark refused to accede to a British request during the First World War for the closure of the Straits

[16] MORISON, HISTORY OF UNITED STATES NAVAL OPERATIONS IN WORLD WAR II, VOL. II: OPERATIONS IN NORTH AFRICAN WATERS, OCTOBER 1942-JUNE 1943, at 191 (1947). There was considerable fear about the ability of Spain, in a military sense, to close the Straits. *Id*. at 187.

[17] Prime Minister to General Ismay for C.O.S. Committee, Oct. 13, 1940, in CHURCHILL, THEIR FINEST HOUR 501, 502 (Boston, 1949).

[18] Decree of Dec. 15, 1914, on 83 BOLETIN DE LAS LEYES I DECRETOS DEL GOBIERNO 1660 (1914); French translation in 23 REVUE GÉNÉRALE DE DROIT INTERNATIONAL PUBLIC, Documents at 13 (1916).

[19] Both British and German warships passed through the Strait. 2 BRÜEL, *op. cit. supra* note 15, at 249.

[20] Boundary Treaty between Argentina and Chile, signed at Buenos Aires, July 23, 1881, art. V, 12 MARTENS, N.R.G. 2d ser. 491 (1887).

[21] The Bangor, [1916] P. 181, 185. The case is discussed in 2 BRÜEL, *op. cit. supra* note 15, at 246-49; see also Mathieu, *The Neutrality of Chile during the European War,* 14 AM. J. INT'L L. 319, 327 (1920).

[22] 2 H. A. SMITH, GREAT BRITAIN AND THE LAW OF NATIONS 262-63 (1935); 2 BRÜEL, *op. cit. supra* note 15, at 55-56. It is difficult to agree with Brüel's conclusion (at 58) that the practice of Denmark and Sweden created only a "presumption" that the Straits should be left open to passage by warships in time of war.

to belligerent ships of war.[23] The neutrality rules of the two countries, promulgated in 1912, reserved from the areas which might be closed to foreign warships in time of war those portions of the Straits necessary to the passage of ships.[24] Under pressure from Germany and in order to protect herself from belligerent activities which might be conducted in close proximity to Danish territory by reason of the strategic importance of this passage to the Baltic, Denmark mined the three passages of the Straits (the Sound, the Great Belt, and the Little Belt) early in the First World War. The passage of warships was prohibited, and none in fact passed through the mine fields which had been laid. Merchant vessels were permitted through the mine fields by day and under pilotage.[25] Prior to the Second World War, Denmark and Sweden adopted, pursuant to an agreement between the Scandinavian states,[26] rules of neutrality, which authorized belligerent vessels to traverse "the natural routes of traffic between the North Sea and the Baltic Sea."[27] With the outbreak of war and the occupation of Denmark by German forces, the Straits became the scene of vigorous efforts at interdiction through the laying of a net barrage and mines—activities in which both German and Allied forces participated.[28]

The complete closure of straits by a littoral state has been effected in other instances, particularly when the geographic nature of a strait has been such as to require its closing to protect the adjacent territory of a neutral state from the hazards of warfare. The Straits of Messina were thus completely closed

[23] 2 SMITH, *op. cit. supra* note 22, at 262.

[24] Denmark, Royal Order No. 293 Concerning the Neutrality of Denmark in Case of War between Foreign Powers, Dec. 20, 1912, para. 1, [1912] LOVTIDENDE FOR KONGERIGET DANMARK 1342, 1 DEÁK AND JESSUP, A COLLECTION OF NEUTRALITY LAWS, REGULATIONS AND TREATIES OF VARIOUS COUNTRIES 476 (1939); Sweden, Royal Ordinance No. 346 Concerning the Neutrality of Sweden in Case of War between Foreign Powers, Dec. 20, 1912, [1912] SVENSK FÖRFATTNINGSSAMLING 853, 2 DEÁK AND JESSUP, *op. cit. supra,* at 963.

[25] 2 BRÜEL, *op. cit. supra* note 15, at 59-78.

[26] Declaration between Denmark, Finland, Iceland, Norway and Sweden for the Purpose of Establishing Similar Rules of Neutrality, signed at Stockholm, May 27, 1938, 188 L.N.T.S. 295.

[27] Denmark, Royal Order No. 209, May 31, 1938, [1938] LOVTIDENDE FOR KONGERIGET DANMARK 1064, 188 L.N.T.S. 297; Sweden, Royal Ordinance No. 187, May 27, 1938, [1938] SVENSK FÖRFATTNINGSSAMLING 373, 188 L.N.T.S. 323.

[28] ROSKILL, THE WAR AT SEA, 1939-1945, VOL. 1: THE DEFENSIVE 390 (1954); TUCKER, THE LAW OF WAR AND NEUTRALITY AT SEA, 50 U.S. NAVAL WAR COLLEGE, INTERNATIONAL LAW STUDIES, 1955, at 297, n. 35 (1957); The Times (London), April 13, 1940, p. 5, col. 2; April 30, 1943, p. 4, col. 6. On the general law concerning the laying of mines at sea, see STONE, LEGAL CONTROLS OF INTERNATIONAL CONFLICT 583-85 (1954).

to warships by Italy during the period of that country's neutrality in the First World War.[29] Considerations of history have, moreover, combined with those of geography to place the Turkish Straits in a peculiar position which has no exact counterpart in the case of other waterways. Some survival of the ancient concept of the Black Sea as a *mare clausum* is to be found in those provisions of the Montreux Convention which forbid the warships of belligerents to pass through the Turkish Straits in time of war when Turkey is not a belligerent. Exceptions are made in those instances in which the vessels are assisting in the defense of a state linked with Turkey in a treaty of mutual assistance, are carrying out the obligations of a state under the Covenant of the League of Nations, or are returning to bases from which they have become separated.[30] The basis for this prohibition of transit in time of war lies in the considerations that the Turkish Straits could hardly fail to become the scene of active hostilities if warships were allowed to pass freely through them and that a right of free passage for the warships of Black Sea powers alone would transform the Black Sea into a sanctuary whence no vessels of war of a nonriparian state might follow. In this latter event, strategic necessities could induce a nonriparian state to embroil Turkey in the war either as an enemy or as an ally of the nation seeking access to the Black Sea.[31] While the prohibition has certain defensive advantages for Russia, that nation has, as its strength has increased in recent years, made a number of efforts to alter the terms of the Montreux Convention in such a way as to give it free access through the Turkish Straits in time of war and to allow it to participate in the defense of the Straits themselves.[32]

[29] Avis du ministre de la marine sur la navigation dans le détroit de Messine, May 30, 1915, JOURNAL OFFICIEL DE LA RÉPUBLIQUE FRANÇAISE, June 1, 1915, at 3518, 22 REVUE GÉNÉRALE DE DROIT INTERNATIONAL PUBLIC, Documents at 216 (1915).

[30] Convention concerning the Régime of the Straits, signed at Montreux, July 20, 1936, arts. 19 and 25, 173 L.N.T.S. 213, 7 HUDSON, INTERNATIONAL LEGISLATION 386 (1941).

[31] Ahmed Sükrü Esmer, *The Straits: Crux of World Politics*, 25 FOREIGN AFFAIRS 290, 301-02 (1946-47).

[32] Both at the conference between Churchill and Stalin at Moscow in October 1944 and at the Yalta Conference in 1945, Stalin pressed for the revision of the Montreux Convention. At the earlier meeting he had specifically sought "modification for the free passage of Russian warships." FOREIGN REL. U.S., THE CONFERENCES AT MALTA AND YALTA, 1945, at 328 and 903-04 (1955).

A note from the Soviet Chargé at Washington to the U.S. Acting Secretary of State, Aug. 7, 1946, in HOWARD, THE PROBLEM OF THE TURKISH STRAITS 47-49 (Dep't of State Pub. 2752) (1947), proposed the joint defense of the Straits by Russia and Turkey in order to prevent their utilization by "other countries for aims hostile to the Black

The ability of Turkey to enforce the provisions of the Montreux Convention relating to passage of ships through the Straits in time of war followed in large measure from the capacity of Turkey to maintain its neutrality through the larger portion of the Second World War.[33] The desire of the belligerents to permit warships to pass through the Straits was, indeed, one of the motives for attempting to induce Turkey to alter the firm position of neutrality which it had adopted. When plans were being laid for the Yalta Conference, Mr. Churchill pointed out to President Roosevelt that one of the difficulties of holding the projected meeting at that spot would be that warships would not be able to pass through the Dardanelles and Bosporus, since Turkey was not yet at war with Germany. He suggested that Turkey might be persuaded either to enter the war or to waive the provisions of the Treaty of Montreux.[34] The arrangements ultimately arrived at involved the passage of four minesweepers and two noncombatant auxiliary vessels, which were to furnish communications and other services to the conference. To preclude any embarrassment in the passage of these vessels through the Straits, prior notice had been given to the Turkish Government and its permission had presumably been obtained.[35] The lawfulness of the passage of the Straits by these vessels is dependent upon the characterization of those and other ships as "vessels of war" or as belonging, on the other hand, to other categories under the Montreux Convention. Despite the clarity of the provisions of that Convention which govern the passage of both warships and merchant ships in time of war, attempts to evade the prohibitions of the agreement and certain lacunae in the categorization of vessels in Annex II to the Treaty[36] were

---

Sea powers." The territorial claims of Russia were withdrawn in a new Soviet note to Turkey on May 30, 1953, in Howard, *The Development of United States Policy in the Near East, South Asia, and Africa during 1953, Part I*, 30 DEP'T STATE BULL. 274, 277-78 (1954); see also W. B. SMITH, MY THREE YEARS IN MOSCOW 53 (1950), regarding Stalin's desire for bases in the Dardanelles, and Bilsel, *The Turkish Straits in the Light of Recent Turkish-Soviet Russian Correspondence*, 41 AM. J. INT'L L. 727, 743 (1947).

[33] See generally, Bilsel, *International Law in Turkey*, 38 AM. J. INT'L L. 546 (1944).

[34] Cable from Prime Minister Churchill to President Roosevelt, Oct. 23, 1944, in FOREIGN REL. U.S., THE CONFERENCES AT MALTA AND YALTA, 1945, at 10-11 (1955).

[35] Memorandum from the President's Naval Aide to the Chief of Staff to the Commander in Chief of the United States Fleet, Jan. 3, 1945, *id.* at 27; memorandum from the President's Chief of Staff to the President, Jan. 13, 1945, *id.* at 35; cable from Prime Minister Churchill to President Roosevelt, Jan. 13, 1945, *id.* at 34.

[36] 173 L.N.T.S. 235. The wording is taken from the Treaty for the Limitation of Naval Armament, signed at London, March 25, 1936, art. 1, 50 Stat. 1363, T.S. No. 919.

the occasion, as in the case of the vessels which entered the Black Sea in connection with the Yalta Conference, of much diplomatic consultation. These problems of characterization must be left for separate consideration later in this chapter.[37] If these difficult questions are left aside, there appears to be little doubt that Turkey administered the provisions of the Montreux Convention with due regard for its dual role as a neutral and as the guardian of the Straits.[38] The mining of these waters undoubtedly had considerable influence in persuading the belligerents that it would not be to their advantage to attempt to force a passage through the Straits.[39]

The legal analogies which have been drawn between straits used as avenues of maritime communication and interoceanic canals[40] suggest that there should be some similarity in the legal régime and in the resulting practice of states as regards their passage by warships when the littoral nation is neutral.[41] Of the existing interoceanic canals, that of Suez, which was opened to traffic in 1869, has by reason of its age and strategic importance been productive of a larger body of precedent than any other maritime canal. The concessions which had been granted to the Compagnie Universelle du Canal Maritime de Suez by the Viceroy of Egypt had stipulated for the Canal a neutral status but, as for its passage by shipping, referred only to merchant vessels.[42] Nevertheless, the ships of the belligerent powers passed

---

[37] See p. 239 *infra*.

[38] With regard to the application of the Montreux Convention by the Turkish Government, Mr. Bevin stated in the House of Commons on October 22, 1946, that His Majesty's Government were of the view that "on the whole, its terms had been conscientiously observed." 427 H.C. Deb. (5th ser.) 1945 (1945-46). But compare the view of Tchirkovitch in *La Question de la révision de la Convention de Montreux concernant le régime des Détroits Turcs: Bosphore et Dardanelles,* 56 Revue générale de droit international public 189, 202 (1952), that the application of the Convention gave no satisfaction to the Allies during the Second World War.

[39] The Times (London), March 4, 1941, p. 4, col. 2.

[40] The logic of having this discussion of passage through the Suez and Panama canals follow closely on the heels of a consideration of the Turkish Straits is borne out by the fact that a United States briefing paper prepared in anticipation of the Yalta Conference stated, "'Internationalization' of the [Turkish] Straits is not a practical solution at this time because, if that is done, the Suez Canal and the Panama Canal logically should receive the same treatment." Foreign Rel. U.S., The Conferences at Malta and Yalta, 1945, at 329 (1955).

[41] See The S.S. Wimbledon, P.C.I.J., ser. A, No. 1 at 28 (1923), assimilating canals to natural straits.

[42] First Act of Concession, Nov. 30, 1854, in Recueil des actes constitutifs 2 (1950); Firman of Concession, Jan. 5, 1856, *id.* at 6.

freely through the Canal during the Franco-Prussian War.[43] With the adoption of the Convention of Constantinople of 1888, which required that the Canal be kept open "in time of war as in time of peace, to every vessel of commerce or of war" and bound the signatories to refrain from acts of hostility within and about the Canal,[44] a somewhat firmer foundation was laid for free transit of the Canal by warships. Subsequent practice has given some indication that states not parties to the Convention of Constantinople may also avail themselves of the privilege of sending their warships through the Canal,[45] but it is not altogether clear to what extent the agreement created duties incumbent upon nonsignatories. There is sound basis for the view that a state which takes advantage of the privilege of passing through the waterway thereby assumes the correlative obligations of the Convention regarding abstention from belligerent acts in the Canal and its approaches. In later wars in which Egypt and Great Britain remained neutral, the Canal was freely used by the warships of nations at war. The Spanish reserve fleet passed through in inglorious fashion in 1898,[46] and in the Russo-Japanese War, Great Britain allowed Russian warships to proceed through the Canal in order to engage in hostilities with Japan, then allied with Great Britain.[47] Italian warships passed through the Canal in 1911 during that country's war with Turkey, in spite of Turkey's position vis-à-vis Egypt.[48] During the two World Wars, Egyptian territory was itself of a belligerent status, and measures, especially of a military nature, were taken to ensure that enemy ships did not pass through the Canal. No restrictions appear to have been placed on the passage

---

[43] HALLBERG, THE SUEZ CANAL: ITS HISTORY AND DIPLOMATIC IMPORTANCE 281 (1931).

[44] Convention of Constantinople, signed Oct. 29, 1888, arts. I and IV, 15 MARTENS, N.R.G. 2d ser. 557 (1891).

[45] In 1898, the United States, which was not a party to the Convention of Constantinople, caused inquiries to be made of the British Government concerning its attitude toward the use of the Suez Canal for the passage of United States warships. The United States Ambassador reported to his Government: "The attitude of the British Government is that we are unquestionably entitled to the use of the canal for war ships." The Department of State replied that it had not been disposed to rely upon, or formally to appeal to, the Convention, since the United States was not one of the signatory powers. [1898] FOREIGN REL. U.S. 982 (1901).

[46] 2 CHADWICK, THE RELATIONS OF THE UNITED STATES AND SPAIN: THE SPANISH-AMERICAN WAR 388 (1911); Le Fur, *Chronique des faits internationaux, Espagne et Etats-Unis,* 6 REVUE GÉNÉRALE DE DROIT INTERNATIONAL PUBLIC 169, 196, 216-19 (1899).

[47] HERSHEY, THE INTERNATIONAL LAW AND DIPLOMACY OF THE RUSSO-JAPANESE WAR 189, 191-92 (1906); Rapisardi-Mirabelli, *La Guerre italo-turque et le droit des gens,* 15 REVUE DE DROIT INTERNATIONAL ET DE LÉGISLATION COMPARÉE, 2d ser. 85, 115 (1913).

[48] HALLBERG, *op. cit. supra* note 43, at 301.

of warships during the conflict in Korea or in the Italo-Abyssinian hostilities prior to the Second World War.[49] In recent years there have been newspaper reports that, on the one hand, President Nasser had given assurances to Premier Khrushchev that the Canal would be kept open in the event of a world conflict and, on the other hand, that the United Arab Republic would prohibit the passage of Dutch warships if hostilities should break out between the Netherlands and Indonesia.[50] Fortunately no occasion has arisen to put to the test what the policy of the United Arab Republic might be.[51]

For the period of both World Wars during which the United States was a neutral, warships of the belligerents were permitted to use the Panama Canal. The United States Proclamation of November 13, 1914,[52] gave effect to the provisions of the Hay-Pauncefote Treaty by establishing rules limiting the number of belligerent warships[53] in the Canal, prescribing their order of departure, and prohibiting the furnishing of supplies,[54] in order to preclude

[49] See CHURCHILL, THE GATHERING STORM 170 (Boston, 1948). The question of the passage of Italian shipping to assist in Italy's aggression against Ethiopia more properly belongs to the consideration of the closing of international waterways as a sanction against an aggressor state, discussed at p. 237 *infra*.

[50] N.Y. Times, May 6, 1958, p. 1, col. 6; 202 THE ECONOMIST 209 (1962).

[51] A report that a vessel of the Portuguese navy had been denied passage through the Suez Canal during the Indian invasion of the Portuguese colony of Goa (N.Y. Times, Dec. 21, 1961, p. 19, col. 1) was subsequently shown to have been incorrect. Letter to the author, dated Jan. 4, 1963, from H.E. Dr. Pedro Theotonio Pereira, Ambassador of Portugal to the United States, and letter to the author, dated Jan. 15, 1963, from Mr. Mohamed Habib, Press Counselor, Embassy of the United Arab Republic, Washington, D.C.

[52] 38 Stat. 2039.

[53] Auxiliary vessels were dealt with separately in the Proclamation of 1914, but for most purposes the two categories of warships and auxiliary vessels were, while separately identified, treated in the same way. An auxiliary vessel was not referred to as such but fell within a class defined as a ship, not being a vessel of war, which "is employed by a belligerent Power as a transport or fleet auxiliary or in any other way for the direct purpose of prosecuting or aiding hostilities . . ." Proclamation of Nov. 13, 1914, rules 1 and 2, 38 Stat. 2039. However, in the case of the new proclamation on the Panama Canal issued in 1939, the general neutrality proclamation of the United States (Proclamation No. 2348, Sept. 5, 1939, 54 Stat. 2629, 2634, 4 Fed. Reg. 3809, 3811 (1939)) had provided: "The provisions of this proclamation pertaining to ships of war shall apply equally to any vessel operating under public control for hostile or military purposes." The neutrality proclamation dealing specifically with the Canal Zone was designed to modify Proclamation No. 2348 in its application to the Panama Canal, and references to "belligerent ships of war" in the Canal Zone instrument must therefore be read in light of the definition in the general neutrality proclamation.

[54] These regulations are obviously based on the corresponding provisions relating to "ports" and "roadsteads" of articles 12-20 of Convention No. XIII of The Hague

the commission of hostilities in the Canal and thus to guarantee that the waterway would "remain free and open on terms of complete equality to vessels of commerce and of war." A similar proclamation was issued at the outbreak of the Second World War.[55] Both regulations required that the total number of vessels of war of each belligerent and its allies should not exceed three and limited to six the total number of warships which were permitted within the Canal and the territorial waters of the Canal Zone at any one time.[56] Commanding officers of public vessels of both belligerent and neutral nations were required, during the Second World War, to give written assurance that the rules, regulations, and treaties of the United States would be faithfully observed.[57] Since prizes were assimilated to vessels of war by the Hay-Pauncefote Treaty and by the proclamations of neutrality, no difficulty was presented by the passage of a German merchant vessel, the *Dusseldorf*, under a British prize crew in December 1939.[58]

Although armed public vessels have been employed on international rivers, no significant practice with regard to their passage through neutral riparian states in time of war appears to have been reported. The use of neutral stretches of international rivers by warships and auxiliaries of the belligerents would in the first place seem to fall afoul of article 2 of Convention No. V of The Hague of October 18, 1907, respecting the Rights and Duties of Neutral Powers and Persons in Case of War on Land,[59] which forbids belligerents to move "troops or convoys of either munitions of war or supplies" across the "territory" of a neutral state. This prohibition is no more than a reflection of the customary international law on the subject. Secondly, many of the principal instruments opening international rivers to free navigation by the ships of all states make reference only to the use of the rivers for commercial purposes. The very document which established the European policy

---

concerning the Rights and Duties of Neutral Powers in Naval War, signed Oct. 18, 1907, 36 Stat. 2415, T.S. No. 545.

[55] Proclamation No. 2350, 54 Stat. 2638, 4 Fed. Reg. 3821 (1939). Concerning the desirability of establishing rules of neutrality for the Panama Canal separate and apart from those generally applicable in the United States, see [1936] 1 FOREIGN REL. U.S. 174-75, 178 (1953).

[56] Proclamation of Nov. 13, 1914, rule 10, 38 Stat. 2039; Proclamation No. 2350, para. 2, 54 Stat. 2638.

[57] Exec. Order No. 8234, Sept. 5, 1939, 4 Fed. Reg. 3823 (1939).

[58] PADELFORD, THE PANAMA CANAL IN PEACE AND WAR 161-62 (1943).

[59] 36 Stat. 2310, 2322, T.S. No. 540; see Accioly, *Freedom of River Navigation in Time of War,* 19 IOWA L. REV. 231, 233 (1934).

of opening rivers to free navigation by the ships of all nations, the Articles concerning the Navigation of Rivers, adopted by the Congress of Vienna, contains the qualifying phrase "sous le rapport de commerce."[60]

Merchant Ships

The restrictions which have been placed by neutral riparian states upon the navigation of international waterways by the merchant ships of belligerent powers have been considerably less stringent than those imposed upon vessels of war and auxiliaries, in all probability because of the lesser capacity of these vessels to engage in belligerent acts while in transit. Despite the mining of the Danish Straits during the First World War and their prohibition to vessels of war, some 32,084 merchant ships were escorted through the mine fields laid in the Great Belt.[61] The closing of the Straits thus had application to warships alone. The action of Denmark in thus permitting the passage of merchant vessels was fully consistent with the policy which had been followed by Denmark and Sweden during the wars of the previous seventy years in which the two countries had remained neutral.[62]

Article 4 of the Montreux Convention places upon a conventional basis the right of merchant vessels to enjoy freedom of transit and navigation in the Turkish Straits in time of war in which Turkey is not a belligerent. Mining and other defensive measures undertaken by that country during the Second World War required, however, that, notwithstanding the provision of the Convention that pilotage remains optional,[63] the use of Turkish pilots be

[60] Articles concerning the Navigation of Rivers Which in Their Navigable Course Separate or Traverse Different States, art. II, annexed to the Final Act of the Congress of Vienna, signed June 9, 1815, 2 MARTENS, N.R. 414 (1818). And see article 327 of the Treaty of Versailles, *supra* note 12, which refers to equality of treatment in the use of German inland waterways in connection with the transport of goods and passengers.

Cf. Treaty between Brazil and Colombia regarding Frontiers and Inland Navigation, signed at Rio de Janeiro, Nov. 15, 1928, art. 6, 129 BRITISH AND FOREIGN STATE PAPERS 262 (1928), which permits "vessels and transports of war" of the one state to navigate freely on that portion of the waters of common rivers under the jurisdiction of the other state. Concerning the passage of a Colombian flotilla over the Amazon, pursuant to the Treaty, during the Leticia dispute, see Woolsey, *The Leticia Dispute between Colombia and Peru,* 27 AM. J. INT'L L. 317, 318 (1933).

[61] 2 BRÜEL, INTERNATIONAL STRAITS 74 (1947).

[62] 1 *id.* at 111.

[63] Convention concerning the Régime of the Straits, signed at Montreux, July 20, 1936, art. 2, 173 L.N.T.S. 213, 7 HUDSON, INTERNATIONAL LEGISLATION 386 (1941).

made mandatory.[64] The measures taken in conjunction with the sanitary inspection required by the Convention[65] made it possible for Turkey to inspect ships applying for passage in order to ascertain whether they were indeed merchantmen.[66] The amount of Axis shipping which used the passage was substantial, but underwent sharp variations as the fortunes of war shifted.[67] As the result of control of the Aegean by Germany, Italy, Hungary, and Rumania, no Allied shipping was seen in the Turkish Straits until 1945, when such traffic was resumed,[68] and there is reason to suppose that the use of the Straits by only one of the belligerent parties during any given period had much to do with simplifying the role of Turkey as the guardian of the waterway.

The slight difficulty which has been encountered in securing the free movement of warships through artificial waterways traversing neutral territory correctly suggests that the passage of merchant vessels should present even less of a problem. The Suez Canal Convention holds the waterway open to "every vessel of commerce or of war . . . in time of war as in time of peace,"[69] but the corresponding provision of the Hay-Pauncefote Treaty[70] omits the phrase referring to times of war and peace. Nevertheless, there appears to have been no denial of transit to belligerent merchant vessels during periods when either the United Arab Republic or Egypt or the United States has been neutral. While no quantitative restrictions have been placed

[64] The Times (London), March 4, 1941, p. 4, col. 2. In a Soviet note to the Turkish Ministry of Foreign Affairs of Sept. 24-25, 1946, reference is made to notices to mariners dated Feb. 25 and May 6, 1941, and June 27, 1942. Howard, The Problem of the Turkish Straits 55-58 (Dep't of State Pub. 2752) (1947). The denial of pilotage through mined waters was tantamount to a refusal to permit the passage of the vessel, as the *Tarvisio* affair (see p. 240 *infra*) made clear. See The Times (London), Aug. 18, 1941, p. 3, col. 4.

[65] Montreux Convention, *supra* note 63, art. 3.

[66] The Times (London), June 16, 1944, p. 4, col. 5; June 17, 1944, p. 4, col. 5. See the remarks of Mr. Eden in the House of Commons, June 21, 1944, in 401 H.C. Deb. (5th ser.) 167 (1943-44).

[67] The Times (London), May 22, 1941, p. 3, col. 3; Oct. 11, 1943, p. 4, col. 7; Nov. 27, 1943, p. 3, col. 5; Jan. 14, 1944, p. 3, col. 4; Jan. 10, 1944, p. 4, col. 5; March 15, 1944, p. 3, col. 4.

[68] Bilsel, *International Law in Turkey*, 38 Am. J. Int'l L. 546, 553 (1944); The Times (London), Jan. 18, 1945, p. 3, col. 5.

[69] Convention of Constantinople, signed Oct. 29, 1888, art. 1, 15 Martens, N.R.G. 2d ser. 557 (1891).

[70] Treaty between the United States and Great Britain to Facilitate the Construction of a Ship Canal, signed at Washington, Nov. 18, 1901, art. III, rule 1, 32 Stat. 1903, T.S. No. 401.

on the number of such ships which may pass through the Panama Canal, the United States has taken certain affirmative measures for the protection of the waterway. These have included the requirement that cameras be sealed,[71] a prohibition on the use of wireless sets,[72] the placing of armed guards on merchant vessels in transit through the Canal,[73] and the granting of authority to take possession and control of vessels if necessary to prevent damage or injury to ships or the Canal or "to secure the observance of the rights and obligations of the United States."[74] Under regulations of this nature, belligerent merchantmen have continued to use the Canal freely while the United States has been neutral.[75]

Since the case of *The S.S. Wimbledon*,[76] decided by the Permanent Court of International Justice in 1923, the obligation of a neutral littoral state to allow a merchant ship carrying contraband of war to a belligerent state to transit an artificial interoceanic waterway lying within the territory of the neutral state has been undisputed. The *Wimbledon* was registered in Great Britain, a neutral state, but was carrying munitions to the Polish naval base in Danzig for use in the Russo-Polish War, in which Germany was also a neutral. Although the case thus dealt with the passage of a neutral ship carrying contraband rather than the transit of a merchant ship of one of the belligerents, it would appear that no logical distinction could be drawn between the two categories of vessels. No such differentiation is to be inferred from article 380 of the Treaty of Versailles, the interpretation of which was in issue in the case, for the treaty declares the Kiel Canal "free and open to the vessels of commerce and of war of all nations at peace with Germany on terms of entire equality." In determining that Germany was obliged to permit the *Wimbledon* to proceed through the Canal, the Court placed heavy reliance on the fact that both belligerent warships and merchant ships of such powers were authorized, and had in fact been allowed, to pass through the Suez and Panama canals when Egypt and the United States were not at war. It also alluded to the fact that it had "never been alleged that the neu-

---

[71] Exec. Order No. 8382, March 25, 1940, 5 Fed. Reg. 1185 (1940).

[72] Exec. Order No. 8715, March 18, 1941, 6 Fed. Reg. 1531 (1941).

[73] Regulation of the Acting Governor of the Panama Canal, "Inspection and Control of Vessels in Canal Zone Waters," approved by the President on July 8, 1941, 6 Fed. Reg. 3407 (1941).

[74] Exec. Order No. 10226, March 23, 1951, 16 Fed. Reg. 2673 (1951). These measures seem to be responsive to the emergency precipitated by the Korean conflict.

[75] PADELFORD, THE PANAMA CANAL IN PEACE AND WAR 133, 169 (1943).

[76] P.C.I.J., ser. A, No. 1 (1923).

trality of the United States, before their entry into the war [World War I], was in any way compromised by the fact that the Panama Canal was used by belligerent men-of-war or by belligerent or neutral merchant vessels carrying contraband of war."[77] This reliance upon precedents furnished by other artificial waterways had the sound effect of indicating that there exists a common body of canal law, which has application to all of the major interoceanic canals. In another sense as well, the judgment of the Court has a wider significance than its facts alone would indicate, for the Court assimilated international canals to natural straits and emphasized that even the passage of a belligerent warship through such waterways would not compromise the neutrality of the state in whose jurisdiction the waters lie. There is nothing, however, in the judgment of the Permanent Court of International Justice which would suggest that a state is precluded from taking reasonable measures for the protection of the waterway and of its own neutrality, even though these may impose some burdens upon shipping using the waterway.

If the practice with respect to the transit of belligerent merchantmen through canals and straits, and a fortiori with regard to the passage of neutral merchantmen carrying contraband, seems to indicate a simple rule of free passage, the little conventional law and slight practice which exists with respect to the use of international rivers by vessels of commerce of the belligerents suggests that a contrary rule may apply to such waterways. Although, as hitherto observed,[78] the treaties normally stipulate only that the rivers will be open to transit by vessels of commerce, a significant minority expressly provide that the general right of free passage does not extend to the transportation of materials destined for a belligerent and characterized as contraband of war.[79] In only one instance has the transport of goods to a belligerent caused any great difficulty. The shipment of sand, gravel, coal, minerals,

---

[77] *Id.* at 28.

[78] See p. 154 *supra.*

[79] Article 25 of the Act of Navigation of the Congo, annexed to General Act of the Conference of Berlin, signed at Berlin, Feb. 26, 1885, 10 MARTENS, N.R.G. 2d ser. 414 (1885-86), 76 BRITISH AND FOREIGN STATE PAPERS 4 (1884-85) is typical: "Il ne sera apporté d'exception à ce principe qu'en ce qui concerne le transport des objets destinés à un belligérant et considérés, en vertu du droit des gens, comme articles de contrabande de guerre." A similar stipulation exists in the same instrument with respect to the Niger (art. 33), and in the Treaty between Argentina and Brazil, signed at Paraná, March 7, 1856, art. 19, COLECCION DE TRATADOS CELEBRADOS POR LA REPUBLICA ARGENTINA CON LAS NACIONES ESTRANGERAS 347, 356 (1863), as regards the River Plata. See Accioly, *Freedom of River Navigation in Time of War,* 19 IOWA L. REV. 231, 234 (1934).

and metals on the Rhine between Germany and occupied Belgium during the First World War placed the Netherlands in a difficult position, beset as it was by the conflicting demands of Germany, which desired to employ this route of transport, and of Great Britain, which protested against the use of neutral territory for this purpose.[80] In this controversy, which centered about the interpretation of article 2 of Convention No. V of The Hague of 1907, prohibiting the movement of convoys of munitions of war or supplies across the territory of a neutral power,[81] reference was also made to article 2 of the Act of Mannheim of 1868.[82] The contention was made by both the Netherlands and Germany that that provision, which authorized a free choice of routes through the Netherlands from the Rhine to the sea, justified the passage of boats carrying the products mentioned. However, the agreement of 1868 must be read in the context of its authorization of free passage for ships engaged in commerce, as had historically been the case with regard to navigation on the Rhine.[83] Indeed, it would seem that freedom of navigation on the Rhine is limited, both in war and in peace, to commercial shipping. Retorsion by the British Government ultimately lead to the placing of restrictions on the shipments by the Netherlands, and it is not possible to say that the incident can form the basis for any definitive interpretation of the Act of Mannheim.

From the foregoing survey, which has necessarily been of a summary nature, it is possible to draw certain conclusions regarding the right of warships and merchant ships to pass through international waterways when the littoral or riparian state is neutral. Warships have the right to pass through interoceanic straits and canals dedicated to public use and do not thereby compromise the neutrality of the littoral state. However, the littoral state may take necessary and reasonable measures for the protection of the waterway and of its neutrality, including the prohibition of belligerent acts within the waterway itself. The success of littoral states in keeping interoceanic canals open to transit by belligerent warships suggests that neutral states have on

[80] Correspondence respecting the Transit Traffic Across Holland of Materials Susceptible of Employment as Military Supplies, Misc. No. 17 (1917) (Cd. 8693); De Visscher, *Chronique des faits internationaux,* 26 Revue générale de droit international public 142 (1919).

[81] 36 Stat. 2310, T.S. No. 540.

[82] Revised Convention for the Navigation of the Rhine, signed at Mannheim, Oct. 17, 1868, 20 Martens, N.R.G. 355 (1875), 2 Rheinurkunden 80 (1918).

[83] See Articles concerning the Navigation of the Rhine, annexed to the Final Act of the Congress of Vienna, signed June 9, 1815, 2 Martens, N.R. 416 (1818), especially article I.

occasion acted unreasonably, and perhaps unlawfully, in placing a complete prohibition on the passage of straits by belligerent warships. The practice of states, confirmed by international agreements and by judicial precedent, also establishes that merchant ships of belligerent states or neutral shipping carrying contraband for the use of such states may in time of war pass through international canals or straits running through neutral territory, without prejudice to the right of the neutral littoral state to impose reasonable regulations for its protection. Is this conclusion compatible with the prohibition of the "suspension of the innocent passage of foreign ships through straits which are used for international navigation" laid down in the Geneva Convention of 1958 on the Territorial Sea and the Contiguous Zone?[84] Although on its face this provision might seem to have application to periods when the littoral state is a neutral and a belligerent desires to have its warships and merchant ships utilize the waterway, it is doubtful that the law of war and neutrality was taken into account by the architects of the Geneva Conventions of 1958. The absence of any reference in the Conventions to the consequences of war or of armed conflict at points at which these events might be expected to qualify the rules laid down in the treaties makes it logical to assume that the treaties were to apply only to normal peacetime situations. The International Law Commission made it clear that its draft, which was the basis of the work of the Geneva Conference of 1958, had reference to the law of peace only.[85] The Conventions simply have no application to the situation in which the littoral or user state is at war or engaged in armed conflict.

As the great majority of international rivers have been opened only to commercial use, an exception to the general principle of free passage through neutral international waterways in time of war must be made in regard to such inland watercourses. It is highly questionable whether the portion of such waterways lying within neutral territory may be used in time of war for other than ordinary commercial traffic, which, under modern conditions, may disappear altogether in time of war. In this respect, such waterways appear to be governed by the law of neutrality in land warfare,[86] rather than

[84] Signed at Geneva, April 29, 1958, art. 16, para. 4, 2 UNITED NATIONS CONFERENCE ON THE LAW OF THE SEA, OFF. REC. 132 (U.N. Doc. No. A/CONF.13/38) (1958).

[85] *Report of the International Law Commission Covering the Work of Its Eighth Session, 23 April-4 July 1956,* [1956] 2 YEARBOOK OF THE INTERNATIONAL LAW COMMISSION 253, 256 (U.N. Doc. No. A/CN.4/SER.A/1956/Add.1) (1956).

[86] Convention No. V of The Hague respecting the Rights and Duties of Neutral Powers and Persons in Case of War on Land, signed Oct. 18, 1907, 36 Stat. 2310, T.S. No. 540.

by any body of principles having application to waterways connecting stretches of the high seas.

When the Littoral or Riparian State is at War

The outbreak of war, to which a state through whose territory an international waterway runs is a party, is productive of conflicting demands of a practical order. On the one hand, the state in question must, as a minimum, be accorded the right to take measures necessary for its self-defense. It may also demand legal authority to take offensive measures against enemy shipping or neutral shipping carrying contraband or performing unneutral service which finds its way into the waterway. Opposed to these requirements is the desire of neutral states to continue to utilize the waterway for the passage of both vessels of war and merchant ships, notwithstanding the existence of the war in which the littoral state is engaged. In this respect, controversy about the use of an international waterway may be no more than a reflection of the unending struggle between belligerents and neutrals to maintain and extend the spheres of action within which they may respectively have freedom to pursue their own interests. In conformity with modern tendencies in the conduct of war, the practice relating to passage through international waterways lying within belligerent territory appears to take increasing account of the need of the territorial sovereign to take defensive measures.

Straits

Straits possess a strategic significance which frequently makes them a scene of active hostilities, whether on land[87] or at sea.[88] This circumstance and the significance which they have as part of the line of communications of belligerents make it altogether unrealistic to suppose that a belligerent littoral state is required to allow passage to enemy warships and other vessels bent on hostile missions. It is for this reason that states rarely find it necessary to

[87] E.g., the Straits of Messina in the invasion of Italy during the Second World War. See Morison, History of United States Naval Operations in World War II, vol. IX: Sicily-Salerno-Anzio, January 1943-June 1944, at 201-24 (1954).

[88] E.g., the Battle of Sunda Strait. See Morison, History of United States Naval Operations in World War II, vol. III: The Rising Sun in the Pacific, 1931-April 1942, at 363-70 (1948).

promulgate any legal instrument expressly closing a waterway of this nature to enemy warships.[89]

For the purposes of the law concerning captures and prize on the high seas, it is necessary to make a distinction between enemy merchant vessels, neutral merchant vessels carrying contraband of war or engaged in unneutral service, and other neutral vessels. Similar differentiations must be made when questions arise concerning the passage of merchantmen through such international waterways as straits while the littoral state is at war. Attempted passage by the first category of shipping is, during a period of active hostilities, not of great practical consequence. Belligerent states will have no reason to risk the capture of their merchant vessels by permitting them to venture unescorted in, or within the vicinity of, an international waterway controlled by the enemy. If belligerent merchant vessels should during the course of hostilities enter a strait controlled by an enemy, they would be subject to seizure and adjudication in prize as if captured on the high seas.[90] A more pressing problem, which continues for the duration of the hostilities, is the matter of inspecting neutral ships to ascertain whether they carry contraband.[91] Because of the facility with which control may be exercised over neutral shipping at a strait connecting two sections of the high seas, such location is ideal for the control of contraband. During the Second World War, contraband-control bases were established by Great Britain at Gibraltar and at Aden,[92] the latter being of significance because of its proximity to the Strait of Bab al Mandab at the mouth of the Red Sea. Though the United States was still a neutral in 1940, the measures of control exercised at Gibraltar were the subject of sharp protests by the United States, which complained of prolonged detention of its shipping at that point and drew unfavorable comparisons between the time required to clear United States and Italian vessels.[93] Under the Montreux Convention, when Turkey

[89] In Royal Decree No. 595, June 6, 1940, [1940] 2 Raccolta ufficiale delle leggi e dei decreti del Regno d'Italia 1551, Italian territorial waters, which would include the Straits of Messina, were closed to all foreign vessels of war, including apparently the warships of neutrals as well.

[90] Concerning which see Colombos, A Treatise on the Law of Prize 48 (3d ed. 1949).

[91] Notwithstanding the fact that the measures of contraband control used in modern warfare emphasize control at the source and rely to a much diminished extent on the instrument of visit and search. See Fitzmaurice, *Some Aspects of Modern Contraband Control and the Law of Prize*, 22 Brit. Y.B. Int'l L. 73 (1945).

[92] 1 Medlicott, The Economic Blockade 73-74 (1952).

[93] An aide-mémoire handed to the British Ambassador at the United States De-

is at war, the Turkish Straits are to remain open to merchant vessels "not belonging to a country at war with Turkey" on the condition they do not in any way assist the enemy.[94] If a neutral ship carrying contraband were to present itself for passage through the Straits, it would seem that it would be assisting the enemy and might therefore be denied passage and be subject to seizure in prize. There can be no doubt that the right of visit and search may be exercised by a belligerent in a portion of its territorial sea forming part of a strait in exactly the same manner as on the high seas. The real question is that of the legal authority of the belligerent to prohibit passage altogether, even though the merchant ships thus denied transit carry no contraband and do not aid the enemy in any way. Recorded precedents regarding the passage of vessels of commerce through straits when the littoral state has been at war have been so extremely scarce, during both the Second World War and earlier periods,[95] that it is difficult to say that any clear rule of law has been established.

The commonest method for interdicting passage through natural straits, the laying of mines, is merely a special application of the device of mining areas of the high seas, to which belligerents have increasingly resorted in recent wars.[96] Indeed, the laying of extensive minefields in the approaches to straits diminishes the practical importance of the question of the blocking of the straits themselves by this means. By way of example, not only the Danish Sound was mined in the Second World War but the whole of the Skagerrak and Kattegat, with the exception of a channel twenty miles in breadth and the territorial waters of Sweden.[97] Reports of postwar mine-

---

partment of State on Jan. 20, 1940, stated: ". . . [I]t [the United States] now regrets the necessity of being forced to observe not only that British interference, carried out under the theory of contraband control, has worked a wholly unwarrantable delay on American shipping to and from the Mediterranean area; but also that the effect of such action appears to have been discriminatory." 2 DEP'T STATE BULL. 93 (1940); see 1 MEDLICOTT, *op. cit. supra* note 92, at 353-57.

[94] Montreux Convention, *supra* note 63, art. 5.

[95] 1 BRÜEL, INTERNATIONAL STRAITS 108-09 (1947).

[96] STONE, LEGAL CONTROLS OF INTERNATIONAL CONFLICT 583-85 (1954); SPAIGHT, AIR POWER AND WAR RIGHTS 492-95 (3d ed. 1947); TUCKER, THE LAW OF WAR AND NEUTRALITY AT SEA, 50 NAVAL WAR COLLEGE, INTERNATIONAL LAW STUDIES 299-300, 304-05 (1957).

[97] The Times (London), April 13, 1940, p. 4, col. 2; STONE, *op. cit. supra* note 96, at 573, n. 11. The German forces laid a net barrage in the Great Belt but supplied no pilots to assist ships in entering the narrow passage through it. The Times (London), May 1, 1940, p. 6, col. 3.

lifting operations indicate to what a tremendous extent avenues of international commerce and communication, such as the Strait of Dover, had been blocked.[98] Such measures as these naturally have the effect of prohibiting passage not only to enemy warships and merchant shipping serving the enemy but to neutral vessels of commerce and warships as well. The widespread use of mines, especially in narrow waterways, suggests possible practical difficulties in giving effect to the principle enunciated by the International Court of Justice in the *Corfu Channel Case*.[99] The Court took account of the fact that Greece considered herself to be technically in a state of war with Albania and that disputed claims existed with respect to territory bordering on the Corfu Channel, and to this extent may have grounded its conclusions on the assumption that Albania may have been in a *de facto* state of war. If this is so, the conclusion of the Court that Albania would have been justified in issuing regulations regarding the passage of warships through the Strait, "but not in prohibiting such passage or in subjecting it to the requirement of special authorization," [100] might justify the conclusion that it is unlawful for a belligerent littoral state to block passage through a strait by the laying of mines, even if the possibility is held out that authorized ships would be escorted through the mine field. It might, however, be more reasonable to suppose that the Court was announcing that the restrictions adopted must bear some reasonable relation to the danger to be averted and that the nature of the war may determine the extent to which mines may be laid and passage prohibited.[101] There is some basis for concluding that a belligerent is under an obligation to provide passage to neutral vessels, subject to reasonable measures of security and control such as compulsory pilotage and navigation by day, and that it may completely block

[98] Third Interim Report by the International Central Board (1st October, 1946, to 30th June, 1947) in 5 CORFU CHANNEL CASE—PLEADINGS, ORAL ARGUMENTS, DOCUMENTS 9 (I.C.J. 1949). This report refers to the lifting of mines in the Skagerrak, the Kattegat, the Baltic Straits, the English Channel, the Kerch Strait, and the Irbensky Strait, among others.

[99] Judgment of April 9, 1949, [1949] I.C.J. Rep. 4.

[100] *Id.* at 29.

[101] Thus the Great Belt and Skagerrak were opened to free usage by all nations shortly after the termination of hostilities in the Second World War, notwithstanding the continuation of a technical state of war. See the statement by Mr. Bevin in the House of Commons on Feb. 21, 1946, 419 H.C. DEB. (5th ser.) 1356 (1945-46).

It was reported that as of September 1957 the Corfu Channel was still not cleared of mines and that negotiations between Greece and Albania on the clearance of the strait had been in progress in Moscow for more than a year. N.Y. Times, Sept. 17, 1957, p. 5, col. 6.

passage of a strait only as a last resort in the most urgent and compelling of circumstances.

The question of the right of shipping to pass through a strait controlled by a belligerent has arisen in its most dramatic form in the controversy about the passage of ships through the Straits of Tiran, which afford access to the Gulf of Aqaba.[102] The solutions which were arrived at in this instance afford a somewhat questionable foundation for any rules to be applied in the future, since they more often had their origin in political and policy considerations than in legal principle. Furthermore, the juridical status of the waterway was affected by the various stages through which the relations between Egypt and Israel passed—the period of armed hostilities; suspension of hostilities under an armistice; the period of the armistice following the resolution of the Security Council of September 1, 1951, which placed further limitations on the powers of the belligerents; the active hostilities during the conflict of 1956; and the present era in which a degree of tranquillity has been restored.

In order to deal meaningfully with the essential issues regarding passage through the Straits of Tiran, it is necessary to make certain assumptions about matters which are for present purposes extraneous to the problem of free transit. The first of these is that a state of war existed between Egypt and Israel, notwithstanding the fact that both parties indulged in fictions of non-recognition.[103] It will also be taken for granted that Egypt exercised sovereignty over the Sinai Peninsula,[104] and that Egypt was validly in occupation

---

[102] The most important studies dealing with this question are BLOOMFIELD, EGYPT, ISRAEL AND THE GULF OF AQABA IN INTERNATIONAL LAW (1957); PORTER, THE GULF OF AQABA: AN INTERNATIONAL WATERWAY (1957); Gross, *Right of Passage through the Strait of Tiran and the Gulf of Aqaba*, 1 FLETCHER ALUMNI REVIEW 8 (1957); Selak, *A Consideration of the Legal Status of the Gulf of Aqaba*, 52 AM. J. INT'L L. 660 (1958); Gross, *The Geneva Conference on the Law of the Sea and the Right of Innocent Passage through the Gulf of Aqaba*, 53 AM. J. INT'L L. 564 (1959); and *Gaza and the Gulf: What International Law Has to Say*, The Times (London), March 8, 1957, p. 11, col. 6.

See p. 160 *supra* concerning passage through the Straits of Tiran as a channel affording access between a gulf and the high seas.

[103] See Baxter, *The Definition of War*, 16 REVUE ÉGYPTIENNE DE DROIT INTERNATIONAL 1, 4-5 (1960); but cf. Letter from the Permanent Representative of Israel to the United Nations, addressed to the Secretary-General, Jan. 25, 1957 (U.N. Doc. No. A/3527) (1957), and FEINBERG, THE LEGALITY OF A "STATE OF WAR" AFTER THE CESSATION OF HOSTILITIES (1961).

[104] Although an argument to the contrary is made by BLOOMFIELD, *op. cit. supra* note 102, at 98-143.

of the islands of Tiran and Sanafir under the arrangements which were made between that country and Saudi Arabia in 1949 or prior to that time.[105] A ship entering the Gulf of Aqaba from the Red Sea must accordingly pass through the territorial waters of Arab states.

It does not appear that the Straits of Tiran were exploited for the control they exercised over the entrance to the Gulf of Aqaba until after the conclusion of the armistice of 1949 between Egypt and Israel.[106] While hostilities were still in progress, there would have been no legal impropriety in refusing access to enemy warships and merchantmen and in exercising the right of visit and search in order to ascertain whether neutral shipping was carrying contraband through the Gulf to the Israeli port of Eilat at the head of the Gulf. The conclusion of the armistice between Egypt and Israel interjected a new element in the problem. According to the traditional law regarding the effect of an armistice, the conclusion of this agreement between the belligerents did not, in the absence of any express provision to this effect, limit Egypt's power to exercise the right of visit and search and to condemn vessels and cargo as prize. The question of the interpretation of the armistice in this respect must, however, be left for consideration in connection with the matter of similar Egyptian controls in the Suez Canal.[107] It must be emphasized that the measures of control which Egypt exercised were carried on not only at sea but from the shore as well, the batteries mounted there lending silent, and occasionally noisy, emphasis to the demands of the Egyptian authorities that ships be subjected to control at this point.[108] There is admittedly a certain air of unreality in making a distinction between the régime applicable on the high seas and that pertaining to land warfare, for in this case at least the two types of measures were part of the same pattern.

Discussions between the United States and Egypt in early 1950 concerning the right of free passage through the Straits of Tiran led to an Egyptian aide-mémoire in which assurances were given that the occupation of the

[105] An Israeli document uses the 1949 date (State of Israel, Ministry for Foreign Affairs, Background Paper on the Gulf of Aqaba, May 1956, at 8 (mimeographed)), but Egypt maintained that it had first placed military installations on the islands of Sanafir and Tiran during the Second World War. U.N. SECURITY COUNCIL OFF. REC. 9th year, 659th meeting 10 (S/PV.659) (1954).

[106] Egyptian-Israeli General Armistice Agreement, signed at Rhodes, Feb. 24, 1949, 42 U.N.T.S. 251.

[107] See p. 227 *infra*.

[108] As in the *Empire Roach* episode, in which a British steamship was stopped by an Egyptian navy corvette in waters commanded by the Egyptian shore batteries. See text at note 111, *infra*.

islands of Tiran and Sanafir was not effected for the purpose of hindering "innocent passage" through the Straits and that passage would remain free "in conformity with international practice and recognized principles of the law of nations." [109] In light of what international practice had hitherto been established, this statement did not constitute the assumption of any responsibility to grant passage to enemy vessels or to permit the free passage of vessels carrying contraband.[110] The detention of the British steamer *Empire Roach* in the Straits on July 1, 1951, indicated that the Egyptian authorities were maintaining their right to see that contraband did not flow through this narrow waterway into the Gulf of Aqaba and so to Israel.[111] Following this incident, which Great Britain protested to the Egyptian Government,[112] arrangements were worked out between the two countries whereby British ships would be cleared at either Suez or Adabia before entering the Straits,[113] thus foreclosing, it was hoped, further episodes of this sort. Nevertheless, there were further instances of firing upon British vessels in 1955.[114]

What might otherwise have been a lawful control of neutral commerce through the Gulf of Aqaba was called sharply in question by the resolution of the United Nations Security Council of September 1, 1951,[115] which was directly concerned with Egyptian interference with free transit through the Suez Canal. The resolution makes no express reference to Aqaba at all, but the preamble of the instrument asserts that "neither party can reasonably assert that it is actively a belligerent or requires to exercise the right of visit, search, and seizure for any legitimate purpose of self-defence." The binding force of the resolution as regards Aqaba is clouded by the fact that the operative portion of the resolution refers only to Suez. On balance, it would appear that the preambular language about the right of visit and

[109] This statement was made to the United States by the Egyptian Ministry of Foreign Affairs on Jan. 28, 1950. See Aide-Mémoire Handed to Israeli Ambassador Eban by Secretary of State Dulles, Feb. 11, 1957, in UNITED STATES POLICY IN THE MIDDLE EAST, SEPTEMBER 1956-JUNE 1957, at 291 (Dep't of State Pub. 6505) (1957).

[110] It was the Egyptian contention that the passage of contraband through the territorial sea does not constitute "innocent passage." U.N. SECURITY COUNCIL OFF. REC. 9th year, 661st meeting 19 (S/PV.661) (1954).

[111] 490 H.C. DEB. (5th ser.) 424-28, 835-40 (1950-51).

[112] *Id.* at 1408-09.

[113] Exchange of notes between Great Britain and Egypt, signed July 29/30, 1951, U.N. SECURITY COUNCIL OFF. REC. 9th year, 661st meeting 19-20 (S/PV.661) (1954).

[114] A British vessel was fired upon on April 10 (543 H.C. DEB. (5th ser.) 1915 (1955-56)), and another vessel was actually hit by fire from Egyptian shore batteries on July 3 (544 *id.* at 23-24).

[115] U.N. Doc. No. S/2322 (1951).

search constitutes a valid political, if not necessarily legally correct, determination of the Aqaba question in favor of the position upheld by Israel. The Egyptian Government continued to give a narrow construction to the armistice agreement and to the resolution of the Security Council, with the result that Israel filed a further protest with the Security Council in early 1954, in which it claimed that Egypt was not complying with the Security Council's action as to either Suez or Aqaba.[116] A Russian veto on March 29, 1954, foreclosed the adoption of a draft resolution submitted by New Zealand which would have referred the complaint regarding Aqaba to the Mixed Armistice Commission established under the armistice agreement between Egypt and Israel.[117] Although the United States subsequently maintained the position that "the Gulf of Aqaba comprehends international waters and that no nation has the right to prevent free and innocent passage in the Gulf and through the Straits giving access thereto,"[118] it nevertheless circulated a Notice to Mariners in 1955, which called attention to the requirement imposed by the Egyptian authorities that notification of the passage of vessels be given seventy-two hours in advance of that event and that vessels might be required to halt for inspection by the customs authorities of that country.[119]

The resumption of hostilities in 1956 had at least the temporary effect of throwing Israel and Egypt into what was, for all practical purposes, virtually the same legal relationship as before the conclusion of the General Armistice Agreement in 1949. Israeli forces occupied the Sharm el Sheikh area and the islands commanding the entrance to the Gulf, to be replaced by the troops of the United Nations Expeditionary Force only after prolonged and delicate negotiations with Israel. It was recognized that the

[116] Letter of Jan. 28, 1954 from the Permanent Representative of Israel to the President of the Security Council (U.N. Doc. No. S/3168) (1954).

[117] U.N. SECURITY COUNCIL OFF. REC. 9th year, 664th meeting 12 (S/PV.664) (1954).

[118] Department of State Circular, June 5, 1957, in 37 DEP'T STATE BULL. 112 (1957).

[119] Notice to Mariners No. 44, Oct. 29, 1955, *id.* at 113. The Egyptian regulations are incorporated in *Government of Egypt, Ports and Lighthouses Administration Circular to Shipping No. 4 of 1955, Sept. 5, 1955, ibid.;* see also Journal du commerce et de la marine (Alexandria), Aug. 29, 1955.

Great Britain also acquiesced in this procedure, but made it plain to the Egyptian Government that Her Majesty's Government "do not recognise the legality of the blockade of Israel nor the right of the Egyptian Government to grant or withhold permission to ships to use the international channel at the mouth of the Gulf of Aqaba." 545 H.C. DEB. (5th ser.) 1450-51 (1955-56) and 546 *id.* at 2287-88.

Straits of Tiran and Suez together constituted two of the greatest strategic assets at stake in the struggle for power between Egypt and Israel. At issue was not only the question of who was to control the territory in question but also the larger problem of whether free and unimpeded passage was to be allowed through the waterway. Israel refused to withdraw its troops until this principle, which had formed one of its objects in the resumption of hostilities, was firmly established.[120]

On February 11, 1957, the United States stated that it was, "on behalf of vessels of United States registry . . . prepared to exercise the right of free and innocent passage and to join with others to secure general recognition of this right," but cautioned that the "enjoyment of a right of free and innocent passage by Israel" would depend upon the withdrawal of its forces.[121] This declaration was accepted by Israel on March 1 in a statement by the Foreign Minister to the United Nations General Assembly, in which she made it plain that Israel was resolved that vessels of Israeli registry should enjoy a right of free and innocent passage and that her country would regard any attack on vessels of Israeli registry engaged in such transit as an attack entitling Israel to exercise its inherent right of self-defense under article 51 of the United Nations Charter.[122] These references to a right of "free and innocent passage" are consistent with the establish-

---

[120] Aide mémoire on the Israeli position transmitted by the Permanent Representative of Israel, Jan. 23, 1957 (U.N. Doc. No. A/3511) (1957); UNITED STATES POLICY IN THE MIDDLE EAST, SEPTEMBER 1956-JUNE 1957, at 255 (Dep't of State Pub. 6505) (1957).

[121] Aide-Mémoire Handed to Israeli Ambassador Eban by Secretary of State Dulles, Feb. 11, 1957, in UNITED STATES POLICY IN THE MIDDLE EAST, *op. cit. supra* note 120, at 290.

The views of the Secretary-General of the United Nations underwent an interesting metamorphosis between his two reports of January 15 and January 24, 1957. In the first of these, he concluded that "the international significance of the Gulf of Aqaba may be considered to justify the right of innocent passage through the Straits of Tiran and the Gulf in accordance with recognized rules of international law," but in the second report, he was forced to concede that "a legal controversy exists as to the extent of the right of innocent passage through these waters." Note by the Secretary-General on compliance with General Assembly resolutions calling for withdrawal of troops and other measures, Jan. 15, 1957, at 5 (A/3500) (1957), and Report by the Secretary-General in pursuance of the resolution of the General Assembly of 19 January 1957 (A/RES/453), Jan. 24, 1957, at 8 (A/3512) (1957).

[122] Statement in the United Nations General Assembly by Israeli Foreign Minister Golda Meir, March 1, 1957, U.N. GEN. ASS. OFF. REC. 11th Sess., Plenary 1275 (A/PV. 666) (1957), UNITED STATES POLICY IN THE MIDDLE EAST, SEPTEMBER 1956-JUNE 1957, at 328 (Dep't of State Pub. 6505) (1957).

ment of such a régime for the Straits of Tiran as would exist in time of peace[123] and tend to exclude an interpretation which would allow Egypt to maintain that ships carrying contraband to Israel were not in innocent passage. The statements which were made by the representatives of the United States, who took an important part in the negotiations culminating in the withdrawal of Israeli forces, did little to clarify the legal position.[124] The United States representative in the General Assembly, Mr. Henry Cabot Lodge, maintained that there was a right of free and innocent passage through the Straits of Tiran in accordance with the principles adopted by the International Law Commission.[125] This assertion appears, with all respect, to be internally inconsistent in view of the fact that the International Law Commission, mindful of the political implication involved in laying down any principle applicable to Aqaba, had expressly refused to deal with the "legal position of straits forming part of the territorial sea of one or more States and constituting the sole means of access to a port of another State."[126]

The establishment of a right of free passage through the Straits of Tiran and the Gulf of Aqaba was symbolized by the passage of a tanker flying the American flag on April 6, 1957, thus bringing to Eilat the first cargo of crude oil ever to reach that port.[127] Israeli vessels shortly thereafter joined the steady procession of ships passing through the Straits and trading with the port of Eilat.[128] Israel remained silent. Egypt and the other Arab states

---

[123] The pertinent law in this respect is probably that of the Corfu Channel Case, Judgment of April 9, 1949, [1949] I.C.J. Rep. 4 at 28.

[124] In his press conference of February 19, 1957, Secretary of State Dulles parried questions about the characterization of the carriage of what would otherwise be contraband as "innocent passage" with a statement that "innocent passage" is a "phrase which is a conventional phrase used in international law which has a meaning, although the authorities, I think, differ slightly as to the precise implications of it." 36 DEP'T STATE BULL. 400, 401 (1957).

[125] U.N. GEN ASS. OFF. REC. 11th Sess., Plenary 1278 (A/PV.666) (1957).

[126] *Report of the International Law Commission Covering the Work of Its Eighth Session, 23 April-4 July 1956,* [1956] 2 YEARBOOK OF THE INTERNATIONAL LAW COMMISSION 253, 273 (U.N. Doc. No. A/CN.4/SER.A/1956/Add.1) (1956). The Commission had also decided that a question raised by the Government of Israel about the status of the Straits of Tiran as "territorial sea" related "to an exceptional case which did not lend itself to the formulation of a general rule." 1 *id.* at 203.

[127] N.Y. Times, April 7, 1957, p. 1, col. 8.

[128] N.Y. Times, June 16, 1957, sec. 4, p. 4E, col. 3. The United States Department of State requested that it be notified of any denial of free and innocent passage to vessels of United States registry. Department Circular, 37 DEP'T STATE BULL. 112 (1957).

protested, but they lacked the means and the resolution to take forceful measures against the shipping.[129] The legal situation which prevailed constituted a re-establishment of the régime which was established *de jure* by the resolution of the Security Council of September 1, 1951, but not hithertc implemented.[130]

The conclusions which may be drawn from this body of experience do not form an important contribution to international straits law. There can be no serious question—and this principle had already been established before the Aqaba problem arose—that a belligerent may exploit a strait which it controls in order to cut off contraband from the enemy and to deny passage to enemy shipping in time of active hostilities. It matters not whether this control be exercised within the straits or off the straits by way of a blockade.

According to the traditional law, this right would continue in existence even after an armistice had been concluded, provided the armistice made no reference to a suspension of the right of visit and search. The resolution adopted by the Security Council, political in its inspiration, had the effect of constituting, if not declaring, a right of free passage through the Straits of Tiran for all shipping, including vessels carrying war materials to Israel. Since the exercise of the right of visit and search during the period of an armistice may very well lead to firing and bloodshed if exercised with respect to belligerent or neutral vessels which resist visit and search, there is some basis for the contention that the general prohibition of belligerent activities in an armistice ought to terminate the belligerent right of visit and search. The issue may be a relatively unimportant one in so far as the international law of the future is concerned, for a repetition of the problem may be avoided in future armistices by an express reference to the right of visit and search on the high seas and within narrow waterways. Seen in

---

[129] N.Y. Times, July 8, 1957, p. 1, col. 8. It was reported that President Eisenhower had written to King Saud that the right of passage upon which the United States was insisting applied only to "innocent passage," thus by implication ruling out the passage of Israeli warships. N.Y. Times, July 19, 1957, p. 1, col. 3.

[130] One of the subsidiary questions which was, of course, posed in connection with the negotiations between Israel and the Secretary-General of the United Nations regarding the withdrawal of Israeli forces from the Gaza Strip and the Sharm el Sheikh area was whether the General Armistice Agreement of 1949 remained in effect despite the revival of hostilities. One of the objectives of the Secretary-General was to bring about a return to the state of affairs envisaged in the Armistice Agreement. See Report by the Secretary-General in pursuance of the resolution of the General Assembly of 19 January 1957 (A/RES/453), Jan. 24, 1957, at 9 (A/3512) (1957).

perspective, the whole Aqaba controversy may boil down for legal purposes to nothing more than an illustration of drafting problems in armistice agreements.

Canals

The strategic importance of canals has meant that they too, like straits, have frequently been the scene of hostilities. Despite the neutralized status of the Suez and Panama canals, these waterways have been attacked by enemies and have been defended by those states for whom these channels of communication possess special importance. During the First World War shipping bound for the Suez Canal was attacked by German submarines in the Mediterranean, and traffic to and from the Far East had, as a result, to be diverted to the Cape of Good Hope route.[131] A land attack by Turkish forces was repelled.[132] At the beginning of the Second World War, British warships denied access to the Canal to Italian vessels of war and merchant shipping.[133] For the most part, the enemy's attack upon the Canal was confined to aerial bombardment and the dropping of mines in the Canal.[134] The Canal once more became the scene of international conflict when Great Britain, France, and Israel mounted an attack upon Egypt in October and November of 1956. While both Great Britain and France justified the military measures which were taken on the grounds of the necessity of extending protection to the Suez Canal and of maintaining the right of free passage, the ultimatum which was issued to Egypt left no doubt that the measures contemplated, far from being a defense of the Canal in cooperation with Egypt, were rather intended to displace Egypt as defender of the Canal and to substitute a temporary military occupation by Great Britain and France.[135] Under the circumstances, it was inevitable that Egypt should regard the joint Franco-British action as an infringement of Egyptian sovereignty, to be resisted with force.

Although the Panama Canal, in the literal language of the Hay-Paunce-

---

[131] Hallberg, The Suez Canal: Its History and Diplomatic Importance 345-48 (1931).

[132] Douin, L'Attaque du Canal de Suez (3 février 1915) (1922).

[133] The Times (London), June 13, 1940, p. 7, col. 6.

[134] Compagnie Universelle, The Suez Canal 53-62 (1952); *Compagnie Universelle du Canal de Suez*, in Collection économie du monde 75-77 (1947).

[135] See statement by Prime Minister Eden in the House of Commons on Oct. 30, 1956, 558 H.C. Deb. (5th ser.) 1274-75 (1956), and statement in the French National

fote Treaty, is free and open to the vessels of all nations in time of war, neutralized, and immune from attack by any belligerent,[136] it has proved impossible to give effect to these provisions in so far as they may confer rights upon an enemy of the United States.[137] Proclamations of the President of the United States have expressly denied access to the Canal to vessels of war, auxiliary vessels, and private vessels of enemies of the United States.[138] United States naval forces were deployed about the Canal during the Second World War in order to protect its sea approaches from attack by either Japanese or German warships but were not successful in the early stages of the war in preventing frequent sinkings of merchant vessels by enemy submarines operating in the vicinity of the waterway.[139] In practice, the measures which were set in motion to prevent attack on the Canal by enemy warships were taken, not within the Canal or its territorial waters, but at considerable distances out on the high seas. The provisions of the Treaty which might extend a right of free passage to enemy ships and which give the Canal a neutralized status must accordingly be considered to have yielded, in legal as well as in practical effect, to the realities of the strategic importance of interoceanic canals and of the manner in which modern wars are fought. It is not surprising, therefore, that Germany, in 1937, conscious of the importance of the Kiel Canal to its projected military adventures, had closed that canal to the warships and naval craft of all foreign states except when permission of passage had been obtained through diplomatic chan-

---

Assembly by Premier Mollet, Oct. 30, 1956, in UNITED STATES POLICY IN THE MIDDLE EAST, SEPTEMBER 1956-JUNE 1957, at 139 (Dep't of State Pub. 6505) (1957).

[136] Treaty between the United States and Great Britain to Facilitate the Construction of a Ship Canal, signed at Washington, Nov. 18, 1901, art. III, rules 1, 2, and 6, 32 Stat. 1903, T.S. No. 401; see ARIAS, THE PANAMA CANAL: A STUDY IN INTERNATIONAL LAW AND DIPLOMACY 126 (1911).

[137] Knapp takes the view that the Hay-Pauncefote Treaty would be suspended if the United States should be at war with Great Britain, and that other powers would not have a right of passage because the United States has no treaty with them. *The Real Status of the Panama Canal as Regards Neutralization,* 4 AM. J. INT'L L. 314, 358 (1910).

[138] Proclamation of May 23, 1917, rule 15, 40 Stat. 1667, the provisions of which are still in force. 35 C.F.R. § 4.176. Vessels of other belligerents are subjected to certain restrictions while passing through the Canal, and their commanding officers must give written assurances to the Canal authorities that the relevant rules and regulations will be observed. 35 C.F.R. § 4.164.

[139] MORISON, HISTORY OF UNITED STATES NAVAL OPERATIONS IN WORLD WAR II, VOL. I: THE BATTLE OF THE ATLANTIC, SEPTEMBER 1939-MAY 1943, at 148-54 (1948).

nels.[140] One is forced to agree with Professor Siegfried that the Panama and Suez canals in particular are no longer neutral, that they are defended by the interested states in their own interest, and that full freedom of passage ceases to exist in time of war.[141]

The closing of canals to enemy warships is only a matter of common sense. No less so are naval measures designed to keep such vessels out of the vicinity of a canal altogether. The substantial legal problem is that of a right of free passage for merchant vessels when the state through the territory of which the canal runs is at war.[142] Bound up with this question is the exercise of the right of visit and search and of seizure of contraband within the waterway itself. These matters have been litigated principally in connection with the position of the Suez Canal during the First and Second World Wars.

It will be recalled that article IV of the Convention of Constantinople of 1888[143] requires that "no right of war, act of hostility, or act having for its purpose to interfere with the free navigation of the Canal, shall be committed in the Canal," its ports, and within a radius of three miles from such ports. While Egypt was technically a neutral during the First World War, its territory was under British occupation, and the Canal must accordingly be considered to have been within belligerent territory during the period of the war.[144] Furthermore, Egypt granted to Great Britain the power to take belligerent measures, in the following terms: "Les forces navales et militaires de Sa Majesté Britannique pourront exercer tout droit de guerre dans les ports et territoire égyptiens, et les vaisseaux de guerre, les navires marchands et les marchandises capturés dans les ports ou territoire égyptiens pourront être déférés en jugement devant un tribunal des prises britannique."[145] At the outbreak of the war, a number of enemy merchant ships were in the Canal or entered it in ignorance of the outbreak of war or

[140] UNITED STATES DEPARTMENT OF STATE, THE TREATY OF VERSAILLES AND AFTER: ANNOTATIONS OF THE TEXT OF THE TREATY 690 (Dep't of State Pub. 2724) (1947).

[141] Siegfried, *Les Canaux internationaux et les grandes routes maritimes mondiales,* 74 HAGUE RECUEIL 5, 36, 60 (1949).

[142] See text at note 158 *infra* concerning freedom of passage during "limited war," e.g., the Korean conflict.

[143] 15 MARTENS, N.R.G. 2d. ser. 557 (1891).

[144] WHITTUCK, INTERNATIONAL CANALS (Foreign Office, Peace Handbook No. 150) at 77 (1920).

[145] Décision du Conseil des Ministres, Aug. 5, 1914, art. 13, JOURNAL OFFICIEL DU GOUVERNEMENT EGYPTIEN, No. 98, Aug. 6, 1914. See also the special provisions of the Decision relating to the Suez Canal in article 20.

purposely entered the neutral area of the Canal in search of refuge. Some of these were seized because they had already committed or were about to commit hostile acts. Orders were issued that those which sought to make the Canal and its ports places of refuge were to be escorted outside the Canal and the three-mile zone surrounding its ports.[146] The enemy vessels were conducted to British warships waiting outside the three-mile limit, which then seized the vessels as prize. When the claimants contended that these actions were inconsistent with article IV of the Convention of 1888, the Privy Council advised that the Convention had no application to ships using the Canal not for passage but "as a neutral port in which to seclude themselves for an indefinite time, in order to defeat belligerents' rights of capture."[147] In the early stage of the war, all inspections of neutral ships were carried on outside the waters of the Canal and its ports, but this process soon proved to be both troublesome and the cause of allowing some vessels to slip through. The solution was the conduct of inspection within the three-mile limit for the ostensible purpose of ascertaining whether the vessels were engaged in any undertaking likely to affect the free navigation and safety of the Canal. If contraband or enemy cargo were found aboard, the ship would be allowed to pass through the Canal and would not be captured until it had passed outside the three-mile limit.[148]

During a substantial part of the Second World War, Egypt was not a party to the conflict. By virtue of the powers which it enjoyed under the Treaty of 1936,[149] Great Britain took measures for the protection of the

[146] Great Britain, Notification relative to Enemy Ships in the Suez Canal, Oct. 23, 1914, 108 BRITISH AND FOREIGN STATE PAPERS 154 (1914).

[147] The Pindos, The Helgoland, The Rostock, [1916] 2 A.C. 193, 196 (P.C.). See also The Concadoro, [1916] 2 A.C. 199 (P.C.); The Sudmark, [1917] A.C. 620 (P.C.); The Gutenfels (1915), 1 B. & C. P. C. 102. In the last of the cited cases, the British Prize Court for Egypt spoke of the possibility that a German ship would request passage as an "unlikely event," but stated that the British authorities would be required to let the ship pass through the Canal. In 1916 certain lighters and tugs belonging to a German company carrying on business along the Canal were, at the direction of the Procurator, held by the liquidator of the company at the disposal of the Crown. The Privy Council held that valid seizure had been effected by this order and that there had been no exercise of any right of war in the Canal in violation of article IV of the Suez Canal Convention. The "de facto tranquillity" which the Convention envisages "was fully respected," in the view of Lord Sumner. His Majesty's Procurator in Egypt v. Deutsches Kohlen Depot Gesellschaft, [1919] A.C. 291 (P.C.). The case is somewhat difficult to reconcile on theoretical grounds with the reluctance of the British authorities to perform any act which might call in question the neutral status of the Canal.

[148] WHITTUCK, *op. cit. supra* note 144, at 82-84.

[149] Treaty of Alliance between His Majesty, in respect of the United Kingdom, and

Canal against hostile action.[150] A cooperative Egyptian Government promulgated a series of decrees establishing an inspection service for ships passing through the Canal and prescribing special regulations for the navigation
of the Canal, dealing with such matters as blackout and use of radio.[151]
The seizure of ships carrying contraband was conducted according to the
same principles applied during the First World War. Inspection services
were set up at Port Said and Suez, but no ships were detained until they
had left the waters of the Canal.[152] In this fashion the formal requirements
of article IV of the Convention were met.

Despite the fact that the régime of the Panama Canal bears a close similarity to that of the Suez Canal, the United States actually seized six German ships which were in Canal Zone waters upon the entry of the United
States into the First World War.[153] However, two Dutch merchant ships
which were "requisitioned" in the Canal were released on the ground that
the treaty obligations of the United States prohibited the exercise of belligerent authority within the Canal Zone.[154] In June 1941, the Italian liner
Conte Biancamano was taken into the possession of the United States on
the ground that there was evidence of a plan to sabotage the vessel and that
the safety of the vessel and of the Canal was thereby endangered.[155] In both
World Wars, the restrictions which had been placed on neutral shipping
during the time that the United States was neutral were maintained in

His Majesty the King of Egypt, signed at London, Aug. 26, 1936, art. 8, Great Britain
T.S. No. 6 (1937).

[150] EL-HEFNAOUI, LES PROBLÈMES CONTEMPORAINS POSÉS PAR LE CANAL DE SUEZ
182-86 (1951). The author's conclusion is that "la situation . . . n'a donné à personne
l'occasion de respecter le traité de 1888" (at 186).

[151] Proclamation No. 4 sur l'inspection des navires à Port-Saïd et à Suez, Sept. 3,
1939, JOURNAL OFFICIEL DU GOUVERNEMENT EGYPTIEN, No. 92, Sept. 4, 1939, at 2;
Arrêté No. 2 of 1939, by the Military Governor, Canal Zone, Sept. 14, 1939, *id*. No.
102, Sept. 10, 1939, at 4; Military Arrêté No. 3/1939 re Control of Wireless in Merchant
Vessels, Sept. 19, 1939, *id*. No. 106, Sept. 28, 1939, at 5; Military Arrêté No. 5/1939 re
Control of Shipping in the Suez Canal, Sept. 25, 1939, *id*. No. 106, Sept. 28, 1939, at 5;
Arrêté No. 8 de 1939 portant règlement d'exécution du Service d'Inspection des bateaux
à Port-Saïd et Suez, Bureau du Gouverneur Militaire de la Zone du Canal, Oct. 15,
1939, *id*. No. 119, Oct. 19, 1939, at 5.

[152] See *Conclusions du Gouvernement Egyptien au sujet des plaintes des Gouvernements étrangers quant à la visite des navires neutres et la saisie des objets de contrebande
dans les ports égyptiens*, 7 REVUE ÉGYPTIENNE DE DROIT INTERNATIONAL 235, 248 (1951).

[153] PADELFORD, THE PANAMA CANAL IN PEACE AND WAR 144 (1943).

[154] [1918] FOREIGN REL. U.S., Supp. I, 1433, 1536 (1933).

[155] PADELFORD, *op. cit. supra* note 153, at 177.

force and supplemented when that country became a belligerent.[156] The waters surrounding the Canal were declared restricted even before the United States became a belligerent late in 1941, and ships were forbidden to enter them without instructions.[157] During the Korean conflict, no enemy merchant ships passed through the Canal, and no vessel or contraband cargo was seized by the Canal authorities.[158]

The right of free passage through the Suez Canal was put to its most severe tests in the decade of open and of suspended hostilities which followed the establishment of the State of Israel.[159] As in the case of the Gulf of Aqaba, the legal situation has not remained uniform throughout this period and has been materially affected by considerations extrinsic to the abstract principle of free transit through the Canal. Among these must be numbered the General Armistice Agreement between Israel and Egypt, the various dealings of the United Nations with the troubled Near Eastern situation, the resumption of hostilities in 1956, and the subsequent pacification of the area through the ministrations of the United Nations.

Shortly after the beginning of hostilities in Palestine, the Egyptian Government instituted an inspection of vessels in the ports of Alexandria, Port Said, and Suez[160] with the purpose of precluding the shipment of munitions and other goods, directly or indirectly, to persons in Palestine.[161] As in past wars, the inspections were to be conducted by customs officials. A later proclamation established the Prize Court in Alexandria, prescribed its procedure, and stipulated that goods having "pour objet d'intensifier l'effort de guerre des sionistes" would be considered as prize if captured in Egyptian ports, in Egyptian or Palestinian territorial waters, or on the high seas.[162]

---

[156] Proclamation of May 23, 1917, 40 Stat. 1667.

[157] Regulation of the Acting Governor of the Panama Canal, "Inspection and Control of Vessels in Canal Zone Waters," approved by the President on July 8, 1941, 6 Fed. Reg. 3407 (1941).

[158] Letter to the writer, dated Dec. 14, 1954, from Executive Secretary, Office of the Governor, Canal Zone Government.

[159] Extended discussion of the measures taken by Egypt during its hostilities with Israel are to be found in Dinitz, *The Legal Aspects of the Egyptian Blockade of the Suez Canal,* 45 GEO. L.J. 169 (1956-57), and Gross, *Passage through the Suez Canal of Israel-Bound Cargo and Israel Ships,* 51 AM. J. INT'L L. 530 (1957).

[160] Proclamation No. 5 relative à l'inspection des navires dans les ports égyptiens, May 15, 1948, JOURNAL OFFICIEL DU GOUVERNEMENT EGYPTIEN, No. 51, May 15, 1948, at 2.

[161] Proclamation No. 13 relative à l'inspection des navires dans les ports égyptiens, May 18, 1948, *id.* No. 55, May 19, 1948, at 1.

[162] Proclamation No. 38 instituant un Conseil des prises, July 8, 1948, *id.* No. 93, July 8, 1948, at 1; concerning the jurisprudence and procedure of the Prize Court, see

Although Egypt did not recognize the State of Israel, the Egyptian Prize Court adopted what is submitted to be the correct view, namely that such nonrecognition did not preclude Egypt from treating Israel as a belligerent.[163]

The view has been widely entertained[164] that the action taken by the Egyptian authorities to prevent the shipment of contraband to the enemy through the Suez Canal was in contravention of the Convention of Constantinople of 1888, which provides that the Canal is always to be "free and open, in time of war as in time of peace, to every vessel of commerce or of war, without distinction of flag."[165] Egypt justified the inspections and seizures which it effected in the area of the Canal by resort to article X of the Convention, the first paragraph of which provides:

Likewise, the provisions of Articles IV [dealing with the passage of vessels of war of belligerents], V [dealing with the embarkation and disembarkation of troops in the Canal area], VII [prohibiting the stationing of vessels of war in the Canal itself], and VIII [charging the agents in Egypt of the signatory powers with supervision of the execution of the treaty] shall not stand in the way of any measures which His Majesty the Sultan and His Highness the Khedive in the

---

Safwat Bey, *The Egyptian Prize Court: Organization and Procedure*, 5 REVUE ÉGYPTIENNE DE DROIT INTERNATIONAL 28 (1949).

[163] The Fjeld, Prize Court of Alexandria, Nov. 4, 1950, 7 REVUE ÉGYPTIENNE DE DROIT INTERNATIONAL 121, 125 (1951), [1950] Int'l L. Rep. 345, 347 (No. 108). The contention, not infrequently advanced, that the United Nations Charter has abolished war and that belligerents are accordingly without legal authority to take belligerent measures under the law of war in effect demands that a body of law which is dedicated to placing limits on the violence which may be employed by belligerents should cease to exist before the human institution with which it deals has been effectively outlawed. See Baxter, *The Rôle of Law in Modern War*, [1953] PROC. AM. SOC'Y INT'L L. 90. Frequent reliance by the parties to the conflict on the principles of the law of war, the very conclusion of an armistice, and countless references to the "war" between Israel and Egypt bespeak a realistic view of the conflict which is inconsistent with the theoretical objection based upon the United Nations Charter.

[164] This was a recurring theme of the complaints of Israel against restrictions on passage through the Canal (e.g., letters of the Permanent Representative of Israel to the President of the Security Council, dated July 11, 1951 (S/2241) (1951), March 17, 1959 (S/4173) (1959), and Aug. 31, 1959 (S/4211) (1959)), although Israel placed much greater emphasis in the debates on the point that the Armistice of 1949 had been violated by the Egyptian measures. See also to like effect AVRAM, THE EVOLUTION OF THE SUEZ CANAL STATUS FROM 1869 UP TO 1956, at 119-22 (1958); OBIETA, THE INTERNATIONAL STATUS OF THE SUEZ CANAL 85 (1960); LAWYERS COMMITTEE ON BLOCKADES, THE UNITED NATIONS AND THE EGYPTIAN BLOCKADE OF THE SUEZ CANAL 8 (1953); Dinitz, *supra* note 159, at 175; Gross, *supra* note 159, at 549-50.

[165] Article I, 15 MARTENS, N.R.G. 2d ser. 557 (1891).

name of His Imperial Majesty, and within the limits of the Firmans granted, might find it necessary to take to assure by their own forces the defense of Egypt and the maintenance of public order.

To this the reply is made by the proponents of the Israeli position that article XI requires that "the measures taken in the cases provided for by Articles IX and X of the present Treaty shall not interfere with the free use of the Canal ..." Taken in their totality, the provisions of the Convention seem to say that free passage is guaranteed to all vessels, but that the United Arab Republic may take defensive measures notwithstanding this right of free passage (in time of war under article IV), but that the measures taken must not interfere with free passage. What is on the surface an internally inconsistent rule must be broken into its component parts and given an interpretation which gives effect to all of the articles of the Convention. If this is to be accomplished, article X must be read as authorizing some interferences with the right of free navigation,[166] for the United Arab Republic could hardly be defended by action which stopped short abruptly where the waters of the Canal lapped the sands of the desert. It does not seem open to any serious question that the United Arab Republic would be authorized to close the Canal to enemy warships, since this action would be quite necessary to "the defense of Egypt." No more could it be expected that the United Arab Republic would be required to allow Israeli merchant vessels and neutral merchant shipping carrying materials of war to employ the Canal as a channel through which Israel might be supplied in its war with the United Arab Republic and the other Arab States. The Egyptian Government pointed out, in defense of its prohibition of the passage of contraband to Israel, that similar measures had been taken by the British Government in past wars, both as to ships within the Canal and as to vessels apprehended outside the territorial waters of the Canal, and that the lack of a large navy, like that available to Great Britain, prevented visit

---

[166] The Flying Trader, Prize Court of Alexandria, Dec. 2, 1950, 7 REVUE ÉGYPTIENNE DE DROIT INTERNATIONAL 127, 133 (1951), [1950] Int'l L. Rep. 440 (No. 149); *Conclusions du Gouvernement Egyptien au sujet des plaintes des Gouvernements étrangers quant à la visite des navires neutres et la saisie des objets de contrebande dans les ports égyptiens,* 7 id. 235, 245 (1951).

The provision in article I of the Convention of Constantinople that the Canal is never to be "subject to the exercise of the right of blockade" appears to refer to blockade by hostile forces directed against the United Arab Republic and the Canal, rather than to restrictive measures taken by that country. When the "blockade" of the Canal by Egypt or the United Arab Republic is referred to, the term is presumably used in its nontechnical sense, as contrasted with the legal concept of blockade.

and search of vessels upon the high seas.[167] Far from limiting the measures which were taken to those designed to preclude the transport of goods which might do injury to the Canal, the British authorities had during the Second World War utilized inspections for the purpose of checking for contraband. With considerable justice, Egypt pointed out that it was as difficult in principle as it had been in practice to separate inspection to discover dangerous cargo from inspection to discover contraband.[168] Aside from these valid considerations of a technical order, it would be wrong to expect that Egypt or the United Arab Republic would stand idly by and watch such vital munitions of war as oil pass unimpeded to Israel. No more could the United States be required to suffer the transport of munitions of war to its enemies through the Panama Canal in time of war. From a military and strategic standpoint, it would have been utterly unrealistic to require that Egypt's right to control vessels passing through the Suez Canal be exercised only in case of an actual or imminent attack on Egyptian territory,[169] for to draw the line in this way would have denied that country the means of cutting off military supplies to the enemy until the attack which the logistical build-up had permitted had actually taken place. Viewed in the light of the Convention of Constantinople of 1888, the past practice of belligerents, and the equities of the affair, the measures taken by Egypt prior to 1949 are not open to valid legal objection.

A General Armistice Agreement between Egypt and Israel was signed at Rhodes on February 24, 1949.[170] It recited that its provisions were adopted "with a view to promoting the return to permanent peace in Palestine." The agreement stipulated that "no aggressive action by the armed forces—land, sea, or air—of either Party shall be undertaken, planned, or threatened against the people or the armed forces of the other . . ."[171] and that "no element of the land, sea or air military or para-military forces of either Party, including non-regular forces, shall commit any warlike or hostile act against the military or para-military forces of the other Party, or against civilians in territory under the control of that Party . . ."[172] Notwithstanding this sweeping renunciation of warlike acts, the Egyptian government maintained in force its restrictions on the passage of contraband to Israel. However, it

[167] *Conclusions du Gouvernement Egyptien, supra* note 166, at 247-51.

[168] *Id.* at 249.

[169] As suggested in Note, *The Security Council and the Suez Canal,* 1 INT'L & COMP. L.Q. 85, 88-89 (1952).

[170] 42 U.N.T.S. 251.

[171] Article I, para. 2.

[172] Article II, para. 2.

mitigated their rigor by reducing the inspection to a single one by customs authorities instead of the dual inspection by military and customs authorities previously required, by reducing the categories of contraband, and by limiting the circumstances under which an enemy destination might be presumed.[173] The contraband list, which was subsequently incorporated in a royal decree of February 6, 1950, consolidating the various directives which had hitherto been issued,[174] was confined to arms and munitions of war, materials for chemical warfare, combustibles, aircraft and boats, trucks and automobiles for military purposes, materials for chemical warfare, and money and specie, provided such goods were destined directly or indirectly for the enemy.[175] This list was more restricted than the contraband lists initially promulgated by Great Britain and Germany at the beginning of the Second World War.[176] The control of contraband met a continuing chorus of protest, not only from Israel, but from a host of neutral nations as well—Great Britain, the United States, the Netherlands, Turkey, France, Denmark, Italy, Norway, New Zealand, and Australia.[177] To these objections, Egypt replied with statistics showing that, for example, out of over 11,000 ships which passed through the Canal from May 1948 to May 1949, 754 had been visited, and only 96 had been found to be carrying contraband.[178] In March 1954, an Egyptian representative announced that subsequent to the adoption of the resolution of the Security Council of September 1, 1951, "only 55 suspected ships have been subjected to the inspection procedure out of 32,047 ships passing through the Suez Canal." [179] What these figures concealed was the fact that no Israeli vessels at all were passing through the Canal, that some neutral vessels were in fact being subjected to delays in transit, and that the

---

[173] Instructions of July 21, 1949, and Sept. 19, 1949. *Conclusions du Gouvernement Egyptien, supra* note 166, at 252.

[174] Décret réglementant la procédure relative à l'inspection des navires et des avions et la capture des prises de la guerre palestinienne, Feb. 6, 1950, JOURNAL OFFICIEL DU GOUVERNEMENT EGYPTIEN, No. 30, April 5, 1951, at 1.

[175] Article 10 of the decree.

[176] 7 HACKWORTH, DIGEST OF INTERNATIONAL LAW 24-26 (1943).

[177] U.N. SECURITY COUNCIL OFF. REC. 6th year, 553d meeting 24 (S/PV.553) (1951) (The Netherlands, Turkey, Great Britain, France, United States); The Times (London), Dec. 13, 1950, p. 5, col. 4 (Denmark); Oct. 10, 1949, p. 4, col. 1 (Italy); Dec. 14, 1950, p. 5, col. 1 (Norway); Aug. 8, 1951, p. 4, col. 5 (New Zealand); July 2, 1951, p. 5, col. 3 (Australia). A number of these protests appear to have been the result of coordinated action.

[178] U.N. SECURITY COUNCIL OFF. REC. 6th year, 549th meeting 20 (S/PV.549) (1951); *Conclusions du Gouvernement Egyptien, supra* note 166, at 250.

[179] U.N. SECURITY COUNCIL OFF. REC. 9th year, 661st meeting 17 (S/PV.661) (1954).

effectiveness of the Egyptian measures of control meant that what might be regarded as contraband by those authorities was frightened away from the Canal. Particularly vexatious to the British Government was the fact that the interdiction of shipments of oil destined for Israel had necessitated the closing or restricted operation of the refineries at Haifa.[180] The resulting cost to the United Kingdom was considerable, and one may speculate that the financial loss may have bulked somewhat larger in the minds of Her Majesty's Government than the abstract principle of free passage through the waterway.

Several other practices resorted to by the Egyptian Government also proved to be vexatious to neutral shipping. One of these was the demanding of certificates of ultimate destination as to cargo passing through the Canal. These instruments constituted guarantees that cargo would not be reshipped to Israel from the neutral country to which it was consigned.[181] The other related device was the blacklisting of ships which had previously carried cargo to Israel.[182] Such blacklisting was taken into account in determining whether goods were contraband[183] and also for the purpose of denying bunkerage and other facilities to shipping so blacklisted.[184] Both of these techniques

---

[180] The refineries' not operating was estimated in March 1950 to be costing the United Kingdom about $50,000,000 per year. 473 H.C. Deb. (5th ser.) 310 (1950); see also 485 *id.* at 2370 (1950-51). The Egyptian Government maintained that oil constituted contraband and that the products of the refineries would be used by Israel for warlike purposes. *Conclusions du Gouvernement Egyptien, supra* note 166, at 253, quoting an Egyptian note of Sept. 22, 1949.

[181] An Egyptian identic note of June 18, 1950, to the powers which had protested the Egyptian restrictions stated: "Pour pouvoir profiter de ces facilités, la destination innocente de la cargaison doit être évidemment attestée par un document faisant foi, émanant soit des Consuls étrangers ou des Capitaines de navires. C'est la raison d'être des certificats exigés par les décisions du 18 juin dernier, lesquels certificats devront attester que les cargaisons de pétrole, ayant traversée le Canal, ont été bel et bien déchargées dans un pays neutre pour la consommation locale, et non point pour une destination ennemie ultérieure à peine déguisée." *Conclusions du Gouvernement Egyptien, supra* note 166, at 256; see The Times (London), May 31, 1949, p. 4, col. 7; Dec. 11, 1950, p. 3, col. 4, concerning British protests.

[182] Letter dated 13 October 1956 from the Representative of Israel Addressed to the President of the Security Council at 6 (U.N. Doc. No. S/3673) (1956).

[183] E.g., decree of Feb. 6, 1950, art. 11, para. 2(g), Journal officiel du Gouvernement Egyptien, No. 30, April 5, 1951. One of the ameliorations of the situation wrought by this decree was that this circumstance was required to be accompanied by one of the other conditions listed in the decree before there could be a condemnation in prize. This fact is not taken into account in Gross, *Passage through the Suez Canal of Israel-Bound Cargo and Israel Ships,* 51 Am. J. Int'l L. 530, 539, n. 35 (1957).

[184] Avis du Gouverneur militaire général concernant l'exécution de la Proclamation No. 13 et de l'avis complémentaire du 6 juin 1948, relatifs à l'inspection des navires,

of economic warfare were adaptations to the war between Egypt and Israel of devices utilized by Great Britain during the Second World War.[185]

When the matter of Egyptian restrictions on shipping through Suez was placed on the agenda of the Security Council as the result of a protest by Israel, the burden of the case made by Mr. Eban, the representative of that country, was that the Egyptian measures were in contravention of the General Armistice Agreement.[186] He argued that the Armistice was not a mere suspension of hostilities in the traditional sense but provided for a definitive end to fighting and that Egypt accordingly had no right to continue the visit and search of vessels passing through the Canal.[187] This contention, which was grounded in what was alleged to be the distinctive nature of the General Armistice Agreement, was not supported by the established law relating to the effect of armistice on the right of visit and search. In the words of Oppenheim-Lauterpacht, "since the exercise of the right of visitation is not an act of warfare, it may be exercised during the time of a partial or general armistice."[188] The pronouncements of the authorities, which appear not to have been based upon any intensive consideration of the point, support this view in the main, although there is some argument to the contrary.[189] As-

---

June 28, 1948, art. 6, J.O., No. 89, June 28, 1948; see 485 H.C. Deb. (5th ser.) 2404 (1950-51).

[185] 1 Medlicott, The Economic Blockade 81-83, 443-48 (1952).

[186] Signed at Rhodes, Feb. 24, 1949, 42 U.N.T.S. 251.

[187] U.N. Security Council Off. Rec. 6th year, 549th meeting 2-24 (S/PV.549) (1951).

[188] 2 Oppenheim, International Law 848-49 (7th ed. Lauterpacht 1952).

[189] Castrén, The Present Law of War and Neutrality 130 (1954); 2 Rolin, Le Droit moderne de la guerre 294 (1920); but cf. Colombos, A Treatise on the Law of Prize 182 (3d ed. 1949). The Institute of International Law has taken inconsistent positions on this point. In its Règlement international des prises maritimes, adopted in 1882 and 1883, it concluded: "Le droit de prise . . . cesse durant l'armistice et avec les préliminaires de la paix" (§ 5). Institut de Droit International, Tableau général des résolutions (1873-1956) 209, 210 (1957). In the so-called Oxford Manual (Manuel des lois de la guerre maritime dans les rapports entre belligérants), adopted at the session of 1913, article 92, dealing with armistices, provides: "Le droit de visite continue à pouvoir être exercé. Le droit de capture cesse hormis les cas où ce droit existerait à l'égard des navires neutres." *Id.* at 253.

There is only an insignificant amount of case law on this subject. A capture of a neutral vessel carrying German cargo during the period of the armistice following the hostilities of the First World War and the subsequent condemnation of ship and cargo were upheld in The Rannveig, [1922] 1 A.C. 97, 3 B. & C. P. C. 1013 (P.C.). One of the bases adduced for the exercise of the right of capture was that "the armistice was an armistice only and quite consistent with the maintenance of the German organiza-

sertions of a continuing right of blockade are not particularly helpful in this respect, since the measures taken by Egypt do not fall within this category of belligerent activity in its technical sense. If a right of visit and search and of blockade is held to survive the conclusion of an armistice in the absence of any provision to the contrary in the instrument, there is a lurking danger of a revival of a "shooting war," for these belligerent measures are made effective through a threat of force exercised by the warships of the belligerent power.[190] A forced passage through the Canal by an Israeli merchant vessel or warship might have led to gunfire by Egypt in contravention of the terms of the General Armistice Agreement. However, since ships cannot transit the Canal unassisted, it might be maintained that the mere denial of facilities would be sufficient to block the Canal. An enemy could, therefore, force its way through the Canal only by its own use of violence, which would inevitably disrupt the normal operation of the Canal. The possibility that resort to visit and search and blockade may be the spark setting off the war again may be a persuasive argument against the present state of customary law relating to armistices. The law, however, must be taken as it is.

To this possibly conservative interpretation of the law, the reply was made[191] that the purpose of the Armistice had been to bring to an end all warlike acts by both belligerents.

A special character might be said to attach to armistices concluded under the auspices of the United Nations, the ends of which include the maintenance of international peace and security.[192] New canons of interpretation might therefore be applied in place of principles which owed their genesis to an earlier era and to different conceptions of international order. The whole purpose of the sweeping provisions of the Armistice relating to the activities of land, sea, and air forces had been to bring the hostilities be-

---

tions in view of a possible renewal of hostilities." [1922] 1 A.C. 104. The same consideration would apply to the suspension of hostilities between Israel and Egypt. And see the cases decided by German, Italian, and French courts in COLOMBOS, *op. cit. supra* at 183-84.

[190] See Levie, *The Nature and Scope of the Armistice Agreement,* 50 AM. J. INT'L L. 880, 904 (1956).

[191] See, for example, the views expressed by the representatives of Great Britain, France, and the United States in the Security Council. U.N. SECURITY COUNCIL OFF. REC. 6th year, 552d meeting 2, 7, 10 (S/PV.552) (1951). See also ROSENNE, ISRAEL'S ARMISTICE AGREEMENTS WITH THE ARAB STATES 84 (1951), for a thoughtful analysis of the Armistice Agreement in this aspect.

[192] ROSENNE, *op. cit. supra* note 191, at 84-85.

tween Egypt and Israel to a final halt. In a resolution of August 11, 1949,[193] the Security Council had found that the "armistice agreements constitute an important step toward the establishment of permanent peace in Palestine" and had borne in mind that "the several armistice agreements include firm pledges against any further acts of hostility between the parties." The eloquent voice of Mr. Bunche had been heard to say in the Security Council that "the entire heritage of restrictions which developed out of the undeclared war should be done away with," and that "no vestiges of the wartime blockade should be allowed to remain as they are inconsistent with both the letter and the spirit of the Armistice Agreements."[194] In the debate which took place in the Security Council during the month of August 1951, a large number of countries contended that the Egyptian measures could not be reconciled with these understandings of the effect of the Armistices.[195]

The letter of the Agreement is easily dealt with. The Chief of Staff of the Truce Supervision Organization had been faced with this problem before the matter went to the Security Council, and he had been constrained to conclude that the Mixed Armistice Commission was incompetent to deal with the complaint of Israel. The basis of this decision was that the Armistice forbade "warlike or hostile act[s]" and "aggressive action" only by armed forces against armed forces and that Egypt was taking its action through custom officials and police against neutral shipping and Israeli merchant shipping.[196] There does not appear to be any sound basis for disagreement with this construction of the language of the Agreement. The "spirit" of an agreement which runs counter to what its words say presents greater difficulty. Presumably, the traditional law had allowed the continued exercise of the right of visit and search after an armistice because there always existed the possibility that hostilities might be resumed. A belligerent should accordingly be entitled to take measures, as through trading with the enemy

---

[193] U.N. Doc. No. S/1376 (1949).

[194] U.N. SECURITY COUNCIL OFF. REC. 4th year, 433d and 434th meetings 6 (1949).

[195] See the views expressed by the representatives of Great Britain, France, the United States, Brazil, the Netherlands, Turkey, China, India, and Ecuador. U.N. SECURITY COUNCIL OFF. REC. 6th year, 552d and 553d meetings (S/PV.552 and 553) (1951).

[196] Cablegram dated 12 June 1951 from the Chief of Staff of the Truce Supervision Organization (U.N. Doc. No. S/2194) (1951). He added, however, "While I feel bound to take this technical position on the basis of the relevant provisions of the General Armistice Agreement, I must also say that the action of the Egyptian authorities in this instance is, in my view, entirely contrary to the spirit of the General Armistice Agreement . . ."

legislation and visit and search, to deny to the enemy the means of mounting a renewed attack. According to the view of the United Arab Republic, history has more than adequately borne out its contention that its restrictive measures were directed against a real threat of renewed hostilities by Israel. Furthermore, to say that the Armistice was an "important step" toward permanent peace is to go to the essence of the agreement, which is indeed a step toward peace but not the peace itself.

All possibilities of a resolution of this issue on a juridical basis disappeared with the pronouncement of Sir Gladwyn Jebb that "it is unnecessary for the Security Council to become entangled in the maze of legal arguments."[197] This view, which was apparently shared by other members of the Security Council, effectively placed the question in the domain of international politics. The resolution adopted on September 1, 1951,[198] called upon Egypt "to terminate the restrictions on the passage of international commercial shipping and goods through the Suez Canal wherever bound and to cease all interference with such shipping beyond that essential to the safety of shipping in the Canal itself and to the observance of the international conventions in force." One of the recitals stated that "since the Armistice regime which has been in existence for nearly two and a half years is of a permanent character, neither party can reasonably assert that it is actively a belligerent or required to exercise the right of visit, search, and seizure for any legitimate purpose of self-defence." The Security Council found that the practices of Egypt were "an abuse of the exercise of the right of visit, search and seizure." The other seven members of the Security Council which joined with Great Britain in voting for the resolution appeared to share the views of Sir Gladwyn Jebb that the matter should be approached from the political, rather than the legal, perspective. For this reason, the resolution must be regarded, in all probability, as a legislative or law-creating act, rather than a judicial or law-declaring one. Its value as a guide to conduct with respect to the Suez Canal or armistices in situations other than the present one is accordingly severely limited. Nevertheless, it constituted law for the purposes of the controversy with which it dealt,[199] and Egypt was under an obligation

---

[197] U.N. SECURITY COUNCIL OFF. REC. 6th year, 550th meeting 56 (S/PV.550) (1951). This statement is criticized in Note, *The Security Council and the Suez Canal*, 1 INT'L & COMP. L.Q. 85, 91-92 (1952).

[198] U.N. Doc. No. S/2298/Rev. 1 (1951); see generally regarding the discussion of this question in the Security Council, Shwadran, *Egypt before the Security Council*, 2 MIDDLE EASTERN AFFAIRS 383 (1951).

[199] But cf. Ghobashy, *Egypt's Attitude Towards International Law As Expressed in*

to comply with it. Instead of bowing to the injunction laid upon it by the resolution of September 1, Egypt alleged that the resolution was "void" and instructed its officials to maintain in force the restrictions which had hitherto been applied.[200] The Political Committee of the Arab League adopted a resolution contending that the measure was incompatible with the right of self-defense and an interference in the internal sovereignty of Egypt.[201] On these legally unjustifiable grounds, Egypt continued to visit and search and to detain vessels and to seize in prize the cargoes they carried.[202] A renewed Israeli complaint early in 1954 led to further consideration by the Security Council of the Egyptian defiance of its earlier direction.[203] Again Egypt made the same arguments that it had in 1951, as if the question were still an open one.[204] A draft resolution introduced by the representative of New Zealand, which would have called upon Egypt "in accordance with its obligations under the Charter to comply" with the 1951 resolution, encountered a Russian veto,[205] and the matter was left as before.

No test case with respect to the actual passage of an Israeli ship through the Canal arose until September 1954. On the twenty-eighth of that month, the *Bat Galim*, a vessel of 500 tons flying the flag of Israel and carrying a cargo of meat, plywood, and hides, entered the roadstead at Port Tewfik in preparation for a transit of the Canal.[206] According to the Egyptian account, the vessel opened fire with small arms on two Egyptian fishing boats. Both the vessel and the ten Israeli sailors aboard were arrested and detained

---

the United Nations; The Egyptian Israeli Dispute on the Freedom of Navigation in the Suez Canal, 11 REVUE ÉGYPTIENNE DE DROIT INTERNATIONAL 121, 131 (1955), expressing the view that the resolution was adopted under article 37 of the United Nations Charter and constituted a "political recommendation" of terms of settlement.

200 The Times (London), Sept. 5, 1951, p. 3, col. 5.

201 The Times (London), Sept. 4, 1951, p. 3, col. 3.

202 For example, 140 tons of meat consigned for Israel were confiscated from an Italian vessel in December 1953. Letter dated 18 December 1953 from the Permanent Representative of Israel Addressed to the President of the Security Council (U.N. Doc. No. S/3153) (1953).

203 U.N. SECURITY COUNCIL OFF. REC. 9th year, 658th to 664th meetings (S/PV.658 to S/PV.664) (1954).

204 U.N. SECURITY COUNCIL OFF. REC. 9th year, 659th meeting (S/PV.659) (1954); *id.* 661st meeting 4-18 (S/PV.661) (1954).

205 U.N. SECURITY COUNCIL OFF. REC. 9th year, 664th meeting (S/PV.664) (1954).

206 Letter dated 28 September 1954 from the Representative of Israel Addressed to the President of the Security Council (U.N. Doc. No. S/3296) (1954); concerning the *Bat Galim* case generally see Ereli, *The Bat Galim Case before the Security Council,* 6 MIDDLE EASTERN AFFAIRS 108 (1955).

by the Egyptian authorities.[207] The investigation which was later carried out by United Nations observers found virtually no evidence to support the Egyptian allegation of Israeli violence, which had been emphatically denied by Israel. The Military Armistice Commission found that no provision of the General Armistice Agreement had been violated.[208] Israel took advantage of this incident to call attention once more to the adverse effect of the Egyptian measures on shipping bound for Israel and declared that ninety per cent of the traffic which would normally go to and from Israel was being deterred by the Egyptian regulations.[209] The Egyptian defense, in addition to the usual argument about the illegality of the resolution of the Security Council of September 1, 1951, was that the resolution had no application to Israeli shipping and that the measures taken by Egypt were directed to the defense of the Canal against hostile measures which might be taken by Israeli vessels in transit.[210] In a conciliatory manner a number of the members of the Security Council urged Egyptian compliance with the 1951 resolution, but no new resolution was adopted.[211] The president of the Security Council merely expressed the view that "it is evident that most representatives here regard the resolution of 1 September 1951 as having continuing validity and effect," and noted that hope had been expressed for an amicable arrangement about the release of the ship and the cargo.[212]

The nationalization of the Suez Canal Company and the operation of the Canal by Egypt brought no material changes in the restrictions which that country placed on transit through the Canal. If passage through the Canal was complicated by the assumption of control over it by Egypt, the problem was one of the technical exploitation of the waterway rather than

[207] Letter dated 29 September 1954 from the Permanent Representative of Egypt to the President of the Security Council (U.N. Doc. No. S/3297) (1954).

[208] Report by the Chief of Staff of the Truce Supervision Organization to the Secretary-General Concerning the SS Bat Galim (U.N. Doc. No. S/3323) (1954).

[209] U.N. SECURITY COUNCIL OFF. REC. 9th year, 682d meeting 15 (S/PV.682) (1954).

[210] U.N. SECURITY COUNCIL OFF. REC. 9th year, 686th meeting 24 (S/PV.686) (1954); see also the views of the Egyptian senior delegate to the Mixed Armistice Commission, as reported in Report by the Chief of Staff cited *supra* note 208.

[211] U.N. SECURITY COUNCIL OFF. REC. 10th year, 687th meeting (S/PV.687) (1955); 688th meeting (S/PV.688) (1955). The representative of France, it may be noted, delivered himself of the strange statement that the measures taken by Egypt should not be allowed to interfere with "any free use of the Canal, even by warships of a Power that is an enemy of Egypt." *Id.* 687th meeting 9 (S/PV.687) (1955).

[212] U.N. SECURITY COUNCIL OFF. REC. 10th year, 688th meeting 20 (S/PV.688) (1955). Egypt had previously agreed to release the cargo and the crew of the *Bat Galim*. *Id.* 686th meeting 27 (S/PV.686) (1954).

of legal restrictions on its use. The two conferences which were held in London in August and September 1956 brought no solution to the problem of nationalization.[213] The question of free passage seems not to have been in issue, except to the extent that the contention was made by the users of the waterway that nationalization of the Company would in the future pose dangers for the use of the Canal. The resolution which the Security Council adopted on October 13, 1956,[214] noted the declarations made by the foreign ministers of Egypt, France, and the United Kingdom and agreed that any settlement of the Suez question should meet six requirements, the first two of which were: "(1) [T]here should be free and open transit through the Canal without discrimination, overt or covert—this covers both political and technical aspects; (2) the sovereignty of Egypt should be respected . . ." Statements of this generality did nothing to resolve the question of free passage which had hounded the proprietor of the waterway and its users for the previous eight years. When hostilities broke out, the General Assembly could do no more than to urge that "upon the cease-fire being effective, steps be taken to reopen the Suez Canal and restore secure freedom of navigation."[215] Upon the reopening of the Canal, the Egyptian Government made a declaration[216] that it would, among other things, "afford and maintain free and uninterrupted navigation for all nations within the limits of and in accordance with the provisions of the Constantinople Convention of 1888."[217]

Discrimination in the granting of passage was dealt with in the following terms: "In pursuance of the principles laid down in the Constantinople Convention of 1888, the Suez Canal Authority, by the terms of its Charter, can in no case grant any vessel, company or other party any advantage or favour not accorded to other vessels, companies or parties on the same conditions."[218] These statements do not indicate any modification of the policies which had hitherto been pursued by Egypt with respect to the passage of ships flying the flag of Israel or carrying goods to or from Israel, for it had been maintained throughout the controversy that the restrictions placed by Egypt on

[213] See THE SUEZ CANAL PROBLEM, JULY 26-SEPTEMBER 22, 1956 (Dep't of State Pub. 6392) (1956).

[214] U.N. Doc. No. S/3675 (1956).

[215] Resolution 997 (ES-I), Nov. 2, 1956, U.N. GEN. ASS. OFF. REC. 1st Emergency Sp. Sess., Supp. No. 1, at 2 (A/3354) (1956).

[216] Egyptian Declaration of April 24, 1957, annexed to letter to the Secretary-General of the United Nations from the Egyptian Minister for Foreign Affairs of the same date (U.N. Doc. No. A/3576, S/3818) (1957).

[217] Article 3(a).

[218] Article 7(a).

the use of the Canal were fully consistent with the Convention of 1888. If all Israeli vessels and all ships carrying contraband were to be treated on a uniform basis, there would be no conflict with the principle of nondiscrimination enunciated above. Furthermore, the principle is cast in terms of "advantage or favour," which prohibition of passage or seizure of contraband could not be considered to be. Even if "advantage or favour" were construed to apply to the right of free transit accorded vessels not so circumstanced, the limiting words "on the same conditions" would point to a distinction between innocent shipping and shipping carrying contraband or of enemy character.

That Egypt had no intention of allowing unqualified freedom of passage to Israeli ships and cargoes was clear from what happened after the filing of the Egyptian declaration. By tacit agreement, no Israeli vessels transited the Canal, but ships chartered to Israeli firms and Israeli cargoes were allowed passage, although sometimes subjected to delays.[219] The test case was the passage of the *Brigitte Toft,* a vessel under Danish flag chartered to an Israeli company with its cargo consigned to Israel. The ship was allowed to transit the Canal, but an Israeli member of the crew, alleged to be a journalist, was removed from the vessel and restored to his friends and relations only after protests by Denmark and Israel.[220] The United Arab Republic tightened up again in the spring of 1959. The *Inge Toft,* another Danish ship under Israeli charter, and other foreign-flag vessels chartered to firms alleged to be under Israeli control were detained and forced to off-load the goods of Israeli origin which they carried.[221] The prize courts of the United Arab Republic held that, under the established law of prize, the vessels lost their neutral character by reason of being chartered to the enemy and that the enemy goods aboard were therefore subject to condemnation in prize.[222]

[219] *Palestine Passions,* 192 THE ECONOMIST 87 (1959); N.Y. Times, June 5, 1959, p. 6, col. 3.

[220] Letter dated 23 July 1957 from the Acting Permanent Representative of Israel to the United Nations Addressed to the President of the Security Council (U.N. Doc. No. S/3854) (1957). The episode is described in Watts, *The Protection of Alien Seamen,* 7 INT'L & COMP. L.Q. 691-93 (1958).

[221] See letter dated 17 March 1959 from the Permanent Representative of Israel Addressed to the President of the Security Council (U.N. Doc. No. S/4173) (1959) and letter dated 31 August from the Acting Permanent Representative of Israel to the President of the Security Council (U.N. Doc. No. S/4211) (1959).

[222] S/S Captain Manoli, Prize Court of Alexandria, June 25, 1959, 15 REVUE ÉGYPTIENNE DE DROIT INTERNATIONAL 186 (1959); The Inge Toft, Prize Court of Alexandria, Sept. 10, 1960, 16 *id.* at 118 (1960). The goods in question were found to

This disposition of the matter was less drastic than the authorities might have led one to expect, since neutral ships operating under enemy control had in the past been condemned for the rendering of unneutral service.[223]

It was reported that Secretary-General Hammarskjöld had, in view of the protests made about these cases, made arrangements that goods of Israeli origin, title to which had already passed to the purchaser, would not be seized if carried aboard neutral vessels.[224] Nevertheless, there continued to be detentions of neutral vessels and seizures of their enemy cargoes in the years which followed.[225]

The vast complexity of the Suez question and the quantity of words which have been written or spoken about it cannot conceal the fact that it has established remarkably little new law about the passage of enemy and neutral ships through interoceanic canals when the territorial sovereign is at war. Before the conclusion of the General Armistice Agreement in 1949, the measures taken by Egypt were akin to the British controls of the Second World War. One must conclude that what was done in this instance is consistent with a principle having wider application, namely that the territorial sovereign is under no obligation to allow enemy shipping or contraband to pass through an interoceanic canal during a war to which it is a party. The right of Egypt to continue these restrictions upon the coming into force of the General Armistice Agreement of 1949 turns upon the proper interpretation of that instrument and not upon canal law. Whether the resolution of the Security Council of September 1, 1951, be taken as a political or a legal exercise, whether as law-constituting or law-declaring, its burden is that the Armistice had placed special obligations upon the parties. The crucial period is thus that of 1948 and 1949, when these extraneous elements had not been added to the problem. Even as to this era the further objection might be raised that Egypt's belligerency was inconsistent with its obligations under

---

be enemy-owned in both instances. See also for differing perspectives on the *Inge Toft* case Yahuda, *The* Inge Toft *Controversy,* 54 Am. J. Int'l L. 398 (1960), and El Khatib and Ghobashy, *The Suez Canal: Safe and Free Passage,* Arab Information Center Information Paper No. 9 at 26-27 (1960).

[223] See Colombos, The Law of Prize 239 (3d ed. 1949), and the cases cited therein.

[224] N.Y. Times, Feb. 19, 1960, p. 1, col. 5.

[225] The Astypalia, Prize Court of Alexandria, Nov. 2, 1960, 16 Revue égyptienne de droit international 130 (1960); N.Y. Times, April 2, 1961, p. 39, col. 5; *id.* Dec. 10, 1961, p. 24, col. 4.

the Charter of the United Nations.[226] If this contention were to be sustained, Egypt might be considered to have been without belligerent rights. This, however, is a wider question than whether Egypt or the United Arab Republic may exercise belligerent rights with respect to the Canal. If the state of the law is such that Egypt acted unlawfully in resorting to war, the principles governing free passage through the Canal in time of peace would be applicable. To deal with the Suez question in isolation would be to treat the symptom instead of the disease.

Rivers

The instances in which international rivers have been the scene of naval operations have been limited in number. American, French, and Belgian flotillas have operated on the Rhine during the two World Wars, primarily for the performance of occupation duties after the cessation of active hostilities.[227] Soviet Russia maintained a flotilla on the Danube for the purpose of attacking the enemy behind the lines and for attack on enemy shipping.[228] If straits and canals are to be closed to enemy warships, there is even less reason to suppose that rivers should be held open for passage by enemy vessels.

The practices followed by states would seem to indicate that the recognition of any right of passage through international waterways for enemy warships when the littoral state is a belligerent would be altogether unthinkable. No more can it be expected that littoral states should deny themselves the opportunity of visiting and searching merchant vessels passing through the waterway in order to determine whether such vessels have enemy character or carry contraband and of seizing and condemning in prize ships and cargoes in conformity with international law. The corresponding right of neutral warships and innocent merchant vessels to make use of the waterway must, under modern conditions, take second place to the legitimate need of the littoral state to defend itself and to derive strategic advantage from

---

[226] As seems to be asserted by Gross in *Passage through the Suez Canal of Israel-Bound Cargo and Israel Ships*, 51 AM. J. INT'L L. 530, 566, 568 (1957).

[227] Le Hagre, *Les Formations maritimes françaises du Rhin*, 18 NAVIGATION DU RHIN 181 (1946); 26 REVUE DE LA NAVIGATION 334 (1954).

[228] EMBASSY OF THE U.S.S.R., WASHINGTON, D.C., 5 INFORMATION BULLETIN, No. 14, at 4; *id.* No. 33, at 4 (1945); the activities of monitors on the Danube during the First World War are described in CHAMBERLAIN, THE REGIME OF THE INTERNATIONAL RIVERS: DANUBE AND RHINE 122 (1923).

control of the waterway. However, international law must require that the authority of the territorial sovereign be exercised reasonably and with due regard to the degree of seriousness of the danger anticipated.

CLOSING OF INTERNATIONAL WATERWAYS
AS A SANCTION

At the time when Italy was pursuing its war of aggression against Ethiopia, it was suggested that Italy's line of communications[229] could be cut by closing the Suez Canal to the warships and merchant ships of that country as a sanction under article 16 of the Covenant of the League of Nations.[230] The proposal was actually discussed with some seriousness in the House of Commons.[231] According to the position taken by the proponents of this measure, article 20 of the Covenant, abrogating all obligations or understandings which might be inconsistent with the terms of that instrument, released Great Britain from any obligation to allow to an aggressor the right of free passage in time of peace and of war which was guaranteed by the Convention of Constantinople of 1888.[232] The opponents of the proposal pointed to the *Wimbledon* case as evidence of a general right of free passage and maintained that the institution of neutrality continued to exist.[233] The main impetus to the idea of closing the Canal to Italy appears to have come from Great Britain and the United States, while lawyers in France, Germany, and Italy generally took the position that the blocking of the Canal would be a violation of the Suez Canal Convention. It is not necessary here to decide the question, now

[229] See CHURCHILL, THE GATHERING STORM 176-77 (Boston, 1948).

[230] See generally Buell, *The Suez Canal and League Sanctions,* 6 GENEVA SPECIAL STUDIES, No. 3 (1935); GUIBAL, PEUT-ON FERMER LE CANAL DE SUEZ? (1937); Hoskins, *The Suez Canal in Time of War,* 14 FOREIGN AFFAIRS 93, 101 (1935); SERUP, L'ARTICLE 16 DU PACTE ET SON INTERPRÉTATION DANS LE CONFLIT ITALO-ÉTHIOPIEN 159 (1938); Note, *The Suez Canal,* 180 LAW TIMES 54-55 (1935).

Churchill later wrote, "There is no doubt on our present knowledge that a bold decision would have cut the Italian communications with Ethiopia, and that we should have been successful in any naval battle which might have followed." *Op. cit. supra* note 229, at 176-77.

[231] 302 H.C. REP. (5th ser.) 1842 (1934-35); and see the remarks of Mr. Attlee, *id.* at 2194-95.

[232] Buell, *supra* note 230, at 11-13.

[233] GUIBAL, *op. cit. supra* note 230, at 68-101. Sir John Fischer Williams in a letter to the *Manchester Guardian* wrote: "The League has no authority to disregard the Convention," since the latter constituted "part of the public law of the world." Manchester Guardian, Oct. 23, 1935, p. 20, col. 3.

academic, whether the Suez Canal Convention was indeed incompatible with the League of Nations Covenant.[234] From the point of view of practicability, the suggestion seems to have been without merit, unless the major powers were prepared to deny the Canal to Italy and to defend it with their military, naval, and air forces. Sir Arnold Wilson, an eminent authority on the Suez Canal, remarked at the time that the closing of the Canal appeared to be "of all possible sanctions the most complicated, the most dangerous, and, quite possibly, the most ineffective."[235] The whole burden of giving effect to the decision would have fallen upon the British Government and upon the French administrators of the Canal.

There has been no real revival of these proposals since the formation of the United Nations, although a similar association of ideas appears in the suggestion that arms shipments to Egypt should have been cut off because of its actions with regard to the Suez Canal and because it refused to support the United Nations' action in Korea.[236] In theory, the closing of the Canal could fall within the interruption of "sea . . . and other means of communication" envisaged in article 41 of the United Nations Charter. However, a requirement that a state through the territory of which an international waterway runs should give effect to the decisions of the Security Council by interrupting the offending state's use of the waterway would place an unwarranted and indeed unbearable burden upon the littoral state.[237] An acci-

---

[234] A report by the Legal Sub-Committee on the Application of Sanctions and International Conventions Concerning Freedom of Communications to the Co-ordination Committee concluded that "the League is entitled to hold that no individual Member can release itself from the obligations which result from Article 16 of the Covenant by invoking obligations assumed towards a country not belonging to the League." This statement was made in response to a question whether conventions concluded with states not members of the League, which contain provisions on freedom of communication, prevent members from taking such measures of interruption or control of passage as might be necessary for the application of article 16. Annex to Proposal No. V adopted by the Co-ordination Committee, Oct. 19, 1935, in *Proposals and Resolutions of the Co-ordination Committee and the Committee of Eighteen,* LEAGUE OF NATIONS OFF. J., Sp. Supp. No. 150, at 12 (1936).

[235] 302 H.C. DEB. (5th ser.) 2202-03 (1934-35).

[236] 481 H.C. DEB. (5th ser.) 1184 (1950-51).

[237] El-Hefnaoui speaks of the "gros sacrifices" for Egypt implicit in any action by the United Nations designed to close the Suez Canal to an aggressor and expresses the view that the Egyptian people could not be persuaded of the necessity of Egypt's subjecting itself to this obligation unless the Security Council and the United Nations were "favorables à l'examen de sa cause et à la réalisation de ses aspirations nationales." LES PROBLÈMES CONTEMPORAINS POSÉS PAR LE CANAL DE SUEZ 188 (1951).

dent of geography would place the onus upon that state single-handedly to cut off the communications of a state which had committed a breach of the peace or an act of aggression. Until a more effective system of collective security is provided, those states through which waterways run may be expected to assume no less and no more responsibility for restoring peace and order than other states not so circumstanced. The denial of the use of international waterways must properly form a part of a more comprehensive scheme for giving effect to the decisions and recommendations of the principal organs of the United Nations.

Classification of Vessels

So long as customary or conventional international law makes a distinction between vessels of war and merchant ships as regards transit through international waterways in time of war, questions must inevitably arise concerning the category into which particular types of vessels fall. In some cases, vessels have somewhat arbitrarily been placed in one group or another. Thus prizes are often assimilated to warships.[238] But in other instances, the appropriate classification of vessels is not easily arrived at. Surprisingly, the administration of the Montreux Convention, which attempts the most precise definition of various types of vessels,[239] has occasioned greater difficulties in this respect than have other instruments and accordingly illustrates some of the pitfalls to be avoided.

One of the methods of circumventing the Convention which was attempted by the belligerents during the Second World War was to transfer warships to a state not at war, which would accordingly have the right, denied to the belligerents, to send warships through the Turkish Straits. When reports were received in 1941 that Italy had sold destroyers to Bulgaria with this purpose in mind, Great Britain drew this information to the attention of the Turkish Government and pointed out that Bulgaria had been engaged in hostilities with Greece and Yugoslavia. Turkey denied hav-

---

[238] E.g., United States Presidential Proclamation No. 2350, Sept. 5, 1939, para. 7, 54 Stat. 2638, 4 Fed. Reg. 3821 (1939).

[239] Annex II to the Convention concerning the Régime of the Straits, signed at Montreux, July 20, 1936, 173 L.N.T.S. 213, 7 Hudson, International Legislation 386 (1941), repeats virtually without change (see p. 241 *infra*) the provisions of the Treaty for the Limitation of Naval Armament, signed at London, March 25, 1936, 50 Stat. 1363, T.S. No. 919.

ing received any request for their passage.[240] Apparently the ships did not pass through the Straits. A further device to get Axis warships through the Straits, a stratagem which proved to be somewhat more successful than transfers to third states, was to camouflage warships as merchant ships. In 1941 an Italian vessel, the *Tarvisio,* which appeared to be a commercial tanker, was permitted through the Straits. The fact that the vessel had appeared on the prewar lists of the Italian navy as an auxiliary vessel was communicated to Turkey, which thereupon forbade further passage by this and other former auxiliary vessels of Italy.[241] A number of small German craft suitable for military purposes apparently sailed through the Straits in 1942 and 1943 without challenge from the Turkish Government.[242] With the increase of Russian strength in the Black Sea in 1944, two classes of German warships attempted to make their way from that sea into the Aegean. Those in the 800-ton class had been used for troop transport, while the smaller vessels, of some 40 to 50 tons displacement, had been used for various purposes, including submarine chasing. The German navy had disarmed both types and camouflaged them as merchant vessels loaded with timber and other products. The passage of some of these craft through the Straits elicited a sharp British protest, which resulted in the resignation of the Foreign Minister and his

[240] The matter was raised in the House of Commons on Oct. 8, 1941. 374 H.C. REP. (5th ser.) 958 (1940-41). See The Times (London), Sept. 16, 1941, p. 4, col. 4; Sept. 19, 1941, p. 4, col. 3; Sept. 20, 1941, p. 4, col. 4.

Apparently this method of attempting to gain passage for belligerent warships had some precedent. It was reported that the British admiral in command of the squadron outside the Dardanelles in the First World War refused to allow a Turkish naval craft to enter the Mediterranean and compelled it to return to the Dardanelles because he had received instructions from his Government that he was to treat all Turkish ships having German officers or sailors aboard as German ships. [1914] FOREIGN REL. U.S., Supp. 113 (1928).

[241] See note from the Turkish Ministry of Foreign Affairs to the Soviet Embassy in London, Aug. 22, 1946, in HOWARD, THE PROBLEM OF THE TURKISH STRAITS 50-52 (Dep't of State Pub. 2752) (1947), and The Times (London), Aug. 18, 1941, p. 3, col. 4; Aug. 23, 1941, p. 3, col. 5; Sept. 5, 1941, p. 3, col. 2.

[242] In July 1941 Turkey permitted the passage of the motor vessel *Seefalke* of 37 tons, flying the German commercial flag. The ship was unarmed and possessed none of the characteristics of warships as defined in Annex II to the Montreux Convention. The passage of this vessel and of a number of German fast pinnaces in 1942 and 1943 was protested by the Soviet Union in 1946. See note from the Turkish Ministry of Foreign Affairs to the Soviet Embassy in Turkey, Aug. 22, 1946, in HOWARD, *op. cit. supra* note 241, at 51, and note from the Soviet Embassy in Turkey to the Turkish Ministry of Foreign Affairs, Sept. 24-25, 1946, *id.* at 55-58; and The Times (London), Jan. 14, 1944, p. 3, col. 4.

replacement by Mr. Saracoğlu.[243] The new Foreign Minister thereafter followed a policy of strict inspection of all German ships desiring to pass through the Straits and of denying passage to vessels not bona fide merchantmen.[244] In the drafting of Annex II to the Montreux Convention, ships of under 100 tons displacement had either designedly or through an oversight been omitted from the enumeration of "vessels of war," and the Turkish Government was therefore acting in conformity with the Convention in permitting their passage.[245] It is also arguable that auxiliary vessels like these in the 800-ton class do not fall within the prohibition of passage by "vessels of war" because of the distinction made in the definitions of Annex II between "vessels of war" and "auxiliary vessels."[246] In 1946 Russia made much of the alleged failure of Turkey to give proper application to the Montreux Convention as evidence of the need for giving Russia a share in the defense of the Bosporus and Dardanelles.[247] Turkey would concede only that some precision of the Montreux Convention to take account of the new categories of warships which had been developed since 1936 would be desirable.[248] The dispatch of communications ships and minesweepers to the Yalta Conference, to which reference has already been made,[249] could be justified on the same legal basis

[243] See the remarks of Mr. Eden (as he then was) in the House of Commons in 400 H.C. REP. (5th ser.) 1986-87 (1943-44); and The Times (London), June 5, 1944, p. 3, col. 5; June 8, 1944, p. 3, col. 4; and June 16, 1944, p. 4, col. 5.

[244] The Times (London), June 17, 1944, p. 4, col. 5; June 19, 1944, p. 3, col. 5; see the answer of Mr. Eden to a question in the House of Commons on June 21, 1944 in 401 H.C. REP. (5th ser) 167 (1943-44).

[245] Article I, para. B(7) of the Treaty for the Limitation of Naval Armament, signed at London, March 25, 1936, 50 Stat. 1363, T.S. No. 919, defined a seventh category of "small craft" under 100 tons. See note from the Turkish Ministry of Foreign Affairs to the Soviet Embassy in Turkey, Aug. 22, 1946, in HOWARD, *op. cit. supra* note 241, at 52; Bilsel, *International Law in Turkey,* 38 AM. J. INT'L L. 546, 555 (1944), and *The Turkish Straits in the Light of Recent Turkish-Soviet Russian Correspondence,* 41 *id.* 727, 739 (1947).

[246] Article 19 of the Convention, *supra* note 239, speaks throughout of "vessels of war" and of "warships." Article 8 states that "for the purposes of the present Convention, the definitions of vessels of war . . . shall be as set forth in Annex II to the present Convention." Annex II defines various categories of "vessels of war" and describes "auxilliary vessels" as "naval surface vessels" of over 100 tons "which are normally employed on fleet duties or as troop transports, or in some other way than as fighting ships, and which are not specifically built as fighting ships . . ."

[247] The correspondence is collected in HOWARD, *op. cit. supra* note 241, at 47-68.

[248] Note from the Turkish Ministry of Foreign Affairs to the Soviet Embassy in Turkey, Aug. 22, 1946, in Howard, *op. cit. supra* note 241, at 50.

[249] See p. 194 *supra.*

as might be adduced for the passage of German auxiliary vessels or on the ground that Turkey was in "imminent danger of war," which, under the terms of the Convention, allows Turkey to close the Straits to warships.[250] The fact that Turkey had severed its relations with Germany and was about to enter the war undoubtedly explains the position which it took.

Similar difficulties regarding merchant ships capable of being fitted out for naval use as warships or auxiliaries appear to have arisen with far less frequency in connection with canals. It is reported that the *Derfflinger*, a German ship capable of being converted into an armed cruiser, was allowed through the Suez Canal during the First World War.[251] Precise definition of auxiliary vessels and their assimilation in general to warships, as in the regulations governing transit through the Panama Canal,[252] may have had a beneficial effect in precluding controversy in the case of that waterway.

The distinction which is made both in the practice of states and in international agreements between warships and their auxiliaries, on the one hand, and merchant vessels, on the other, is attributable to their differing function. Warships, fleet oilers, transports, communications ships, mine sweepers, landing craft, and the hosts of other vessels which participate in military and naval operations are essentially instruments of aggressive action, the presence of which in an international waterway might imperil the neutrality of a littoral state, either directly or through the possibility of attack by the enemy of the belligerent in question. While all merchant shipping possesses a special importance for the successful prosecution of war, it presents something less of a danger to a neutral state than warships or allied vessels. There is room for the view that the duties in which ships are engaged constitute the criterion by which the determination is to be made whether the vessels are to be treated as warships or as merchantmen. Thus, merchant ships which are combat-loaded and in convoy because they are to participate in an amphibious landing should be treated as if they were warships, while merchant shipping traveling singly should not be so considered even if it might be carrying military supplies. If, however, as has been suggested above,[253] neutral states have been more cautious in permitting the passage of warships through straits used as channels of international communication than the

---

[250] Article 21 of the Montreux Convention, *supra* note 239.

[251] See The Gutenfels, [1916] 2 A.C. 112 (P.C.).

[252] Presidential Proclamation of Nov. 13, 1914, 38 Stat. 2039, and 35 C.F.R. §§ 4.161-4.176, which refers to the single category of "a vessel of war or an auxiliary vessel of a belligerent," in addition to private vessels.

[253] See p. 203 *supra*.

necessities of the situation indicate, the distinction between vessels of war and merchant shipping may be an unnecessary one. When the littoral state is at war, enemy ships, whether warships or merchantmen, will be denied passage through the waterway, and the extent to which visit and search of neutral vessels will be conducted will probably be governed entirely by the law applicable to such vessels on the high seas. It would be proper to conclude that excessive emphasis may have been laid on the distinction between warships and merchant vessels, but that, if different treatment is to continue to be accorded to them, warships and all types of auxiliary vessels should be defined in a broad sense as ships operating under public control for hostile or military purposes.[254]

CONCLUSIONS

The practice of states and the equities of the matter point to the conclusion that neutral littoral states are under an obligation to keep both international canals and straits open to the vessels of even the belligerents in time of war. The success of the proprietors of international canals in keeping these waterways open to belligerent warships in these circumstances suggests that states have paid undue deference to supposed requirements of security in closing international straits to warships of the belligerents while assuring free navigation for merchant vessels. However, it is clear that in the case of either type of waterway the neutral state may impose reasonable restrictions designed to protect neutral vessels, to safeguard its own security and neutrality, and to facilitate the passage of ships. The permissible extent of these regulations must vary with the nature of the situation. It is a correlative of this principle that no hostile acts may be committed in the waterway and that it must in all respects remain neutral.

Although the obligation to keep the waterway open to neutral vessels continues when the territorial sovereign is at war, this duty must yield to

[254] As, for example, in United States Presidential Proclamation No. 2348, Sept. 5, 1939, 54 Stat. 2629, 2634, 4 Fed. Reg. 3809, 3811 (1939).

In United States v. Germany, March 25, 1924, MIXED CLAIMS COMMISSION, UNITED STATES AND GERMANY, ADMINISTRATIVE DECISIONS AND OPINIONS OF A GENERAL NATURE AND OPINIONS IN INDIVIDUAL LUSITANIA CLAIMS AND OTHER CASES TO OCTOBER 1, 1926, at 75, 99 (1928), 7 UNITED NATIONS, REPORTS OF INTERNATIONAL ARBITRAL AWARDS 73, 90 (1956), a ship was held to fall within the category of "naval and military works or materials" (and thus not to be compensated for under the Treaty of Versailles) if it had "been operated by the United States at the time of her destruction for purposes directly in furtherance of a military operation against Germany or her allies."

the legitimate needs of the belligerent through whose jurisdiction the strait or canal flows. The prohibition of acts of war within the waterway seems unrealistic and archaic in view of the defensive and offensive measures taken with regard to these waterways, and a frank recognition of the right to seize vessels in the waterway would be desirable. At the same time, having regard to the requirements of security, the restrictions imposed on peaceful passage through the waterway must be reasonable in nature and have as their principal purpose the maintenance of free passage through the waterway. So little practice has arisen with regard to international rivers that it would not be proper to draw an analogy between their standing and that of straits and canals in time of war. Keeping them open to free navigation will require consideration of a number of factors which have no application to other international waterways.

CHAPTER V

LEGAL CONTROLS
IN THE FISCAL SPHERE

A N INTERNATIONAL waterway holds the possibilities of great profit or of
a great financial burden for the state through which it flows. The
mere possession of a commercially strategic canal or strait or river offers the
temptation to the proprietor state of making that channel a source of profit at
the expense of the ships or nations which employ it. If, in the economics of
ship operation, the waterway saves the ship operator and shipper considerable
time, distance, and consequently money, the tax which can be levied without
loss of custom may be a substantial one. From the perspective of the user, however, the costs of passage must be kept to a minimum, and the charges demanded must be related to the services which the proprietor state has actually
been called upon to perform.

A balance must be struck at some point between the probable demands of
the proprietor state or operating agency for the highest possible returns from
the waterway and the users' insistence that charges be kept at a low level or
eliminated entirely. There can be little quarrel with the proposition that, to
the extent that the waterway is artificial or partially artificial or must be
improved, charges laid upon ships passing through the watercourse or revenues derived from other sources must be sufficient to meet the capital costs
and expenses incurred by the proprietor state or operating agency in the
construction, maintenance, and operation of the channel, and perhaps to provide a reasonable rate of return on invested capital as well. Even the use of a
naturally formed avenue or terminus of commerce, such as a port, may be
subjected to the payment of imposts to cover the cost of navigational facilities, dredging, pilotage, and the like.[1]

The reimbursement for such costs can be accomplished in a variety of

---

[1] See Statute on the International Régime of Maritime Ports, annexed to the Convention opened for signature at Geneva, Dec. 9, 1923, arts. 2-4, 58 L.N.T.S. 300, 2
HUDSON, INTERNATIONAL LEGISLATION 1163 (1931).

ways. The most obvious is to pass the costs to the operators of the vessels which use the waterway, through the medium of tolls based on the size or capacity of the ships. Alternatively, charges may have some relation to the extent of use of the waterway, in terms of size of the vessel and frequency of use, but be borne by the governments under the jurisdiction and protection of which the ships sail. In the case of certain aids to navigation which have been established in past years, the cost has been jointly borne by the members of the group of states which undertook the responsibility according to an agreed formula.[2] Instead of passing the cost along to using vessels and user nations, a state may find it useful to abolish charges altogether, except for special services rendered, on the assumption that the reciprocal cancellation of tolls by a group, let us say, of riparian nations will mean that the cost to one riparian of maintaining the channel in condition for navigation by vessels of other nations will be counterbalanced by the corresponding advantage to its shipping of not having to pay tolls for such use to other members of the group. In this aspect, the suppression of tolls may amount to an indirect subsidy to navigation. It should hardly be necessary to emphasize that these methods of meeting costs do not by any means exhaust the list of possibilities.

Tolls are, however, the conventional way of securing revenues to cover the costs of constructing and operating a waterway. The level at which they are established is substantially influenced by economic factors and represents,

----

[2] E.g., Convention as to Cape Spartel Light-House, signed at Tangier, May 31, 1865, 14 Stat. 679, T.S. No. 245, 1 MALLOY, TREATIES, CONVENTIONS, INTERNATIONAL ACTS, PROTOCOLS AND AGREEMENTS BETWEEN THE UNITED STATES OF AMERICA AND OTHER POWERS, 1776-1909, at 1217 (1910). Article II of the Convention provided that the expenses of maintaining the lighthouse would be borne equally by the contracting states.

On occasion, groups of states have required a nation to maintain navigational facilities without reimbursement for the cost. Concerning such an obligation laid upon Japan, see Convention Establishing Tariff of Duties between Japan and the United States, Great Britain, France, and the Netherlands, signed at Yedo, June 25, 1866, art. XI, T.S. No. 188, 1 MALLOY, *op. cit. supra* at 1012.

Light dues were collected at the Suez Canal during the British protectorate of Egypt, but only for British vessels which had passed, or were to pass through the Red Sea. See Order in Council fixing Light Dues payable by Vessels passing through the Red Sea, Aug. 8, 1932, 135 BRITISH AND FOREIGN STATE PAPERS 144 (1937). The governments of Germany, Italy, and the Netherlands made contributions to the cost of the lights until the outbreak of the Second World War. A recent convention provides that the parties will defray the cost of the Red Sea lights by contributions based on the total tonnage of the vessels of each contributing government passing through the Suez Canal. Great Britain manages and maintains the lights. International Agreement regarding the Maintenance of Certain Lights in the Red Sea, signed at London, Feb. 20, 1962, arts. 2 and 3, TRACTATENBLAD VAN HET KONINKRIJK DER NEDERLANDEN, 1962, No. 128.

to some degree at least, an implicit bargain struck between the users and the operating agency. Nevertheless, there appear to be certain limits upon tolls which are imposed by law through the gradual transformation of the customary conduct of nations into binding legal rules. This is not to say that all of these legal restraints constitute clear and firm duties universally recognized by maritime nations and by states which are proprietors of international waterways. They are of varying degrees of precision and of force, but no less worthy of characterization as law for that reason.

These legal limitations on the level of tolls derive from the basic principle, to which reference has been made throughout this study, that an international waterway, in the language of one of the most important of the governing instruments, "shall always be free and open . . . to every vessel . . . without distinction of flag."[3] To take the last element of this principle first, nondiscrimination as to transiting vessels is not assured if the rate of toll charged for transit varies according to the national character or ownership of the vessel passing through the waterway. The rule of equality of treatment may be implemented in the case of tolls only through a rate structure which prescribes a uniform charge for "every vessel . . . without distinction of flag." The further requirement that the waterway be "free and open" obviously does not imply toll-free passage in all cases; the terms in fact refer to freedom of navigation or of passage. While reasonable charges for the use of an international waterway are thus not excluded, an excessively high rate of tolls can, whether intentionally or not, bar the use of the waterway through the creation of an unreasonable condition on passage.[4] But before these principles of fiscal nondiscrimination and of reasonableness of rate may be examined, it will be necessary to dispose of the cases in which charges for passage are unnecessary and unlawful or have been eliminated for other reasons.

PROHIBITION OF THE EXPLOITATION OF A GEOGRAPHIC POSITION TO SECURE A FISCAL ADVANTAGE

A consistent practice of states permits the generalization that fiscal imposts may not be levied upon vessels transiting a natural watercourse, such as a

---

[3] Convention of Constantinople, signed Oct. 29, 1888, art. I, 15 MARTENS, N.R.G. 2d ser. 557 (1891).

[4] A transit tax of excessive amount which was imposed by the East German Government in 1958 temporarily halted barge traffic into West Berlin on the Elbe River and the Mittelland Canal. The tax was construed as a weapon to force recognition of that government by the Federal Republic of Germany. N. Y. Times, May 6, 1958, p. 2, col. 3.

river or a strait, except for services other than normal navigational aids furnished to such shipping. The law has not always been so. During the seventeenth and eighteenth centuries, the commentators conceded to the state which controlled an interoceanic strait the legal authority, as well as the power, to levy tolls upon ships passing through the strait.[5] The recognition of this right was often coupled with an injunction that the amount of the toll should be moderate and should be related to the services rendered, such as the provision of aids to navigation and protection from pirates.[6] The classic instance of such levies was the Danish Sound dues, collected from vessels passing through the Sound and the Belts.[7] So ancient were these taxes, which had first been collected in the fifteenth century,[8] and so widely deferred to, that it was still possible in the mid-nineteenth century to speak of them as "part of the International law of Europe."[9] But the burden which they imposed on the commerce of other nations and their employment as a source of revenue to Denmark made them intolerable to the United States and to the maritime powers of Europe. A treaty was concluded in 1857 whereby the parties undertook by payments to Denmark to redeem the sound dues, thus freeing their vessels from these exactions in the future.[10] Tolls or charges for passage through a strait are in this day exceptional. The most important case in which they remain is that of the Turkish Straits, passage through which may, under the Montreux Convention, be subjected to "taxes and charges" which "are not to be greater than is necessary to cover the cost of maintaining the services concerned and of allowing for the creation of a

---

[5] GROTIUS, DE JURE BELLI AC PACIS, Book II, ch. iii, § xiv (1646); PUFENDORF, DE JURE NATURAE ET GENTIUM LIBRI OCTO, Book III, ch. iii, § 7 (1688); VATTEL, LE DROIT DES GENS, Book I, ch. xxiii, § 292 (1758).

[6] PUFENDORF, *loc. cit. supra* note 5; VATTEL, *loc. cit. supra* note 5.

[7] As to the history of the Danish Sound dues, see HILL, THE DANISH SOUND DUES AND THE COMMAND OF THE BALTIC (1926).

[8] *Id.* at 11.

[9] The Earl of Clarendon to Mr. Buchanan, Nov. 5, 1855, in 46 BRITISH AND FOREIGN STATE PAPERS 661 (1865).

[10] Treaty between Great Britain, Austria, Belgium, France, Hanover, Mecklenburg-Schwerin, Oldenburg, the Netherlands, Prussia, Russia, Sweden and Norway, and the Hanse Towns, on the one part, and Denmark, on the other part, for the Redemption of the Sound Dues, signed at Copenhagen, March 14, 1857, 16 MARTENS, N.R.G. pt. 2, 345 (1858); 47 BRITISH AND FOREIGN STATE PAPERS 24 (1866). The dues were redeemed in the case of the United States by the Convention between the United States of America and Denmark for the Discontinuance of the Sound Dues, signed at Washington, April 11, 1857, 11 Stat. 719, T.S. No. 67, 1 MALLOY, *op. cit. supra* note 2, at 380 (1910). Concerning the diplomatic history of the abolition of the dues, see HILL, *op. cit. supra* note 7, at 241-86; 2 BRÜEL, INTERNATIONAL STRAITS 36-39 (1947).

reasonable reserve fund or working balance."[11] There have also been occasions on which groups of nations have banded together for the purpose of providing navigational facilities in the vicinity of interoceanic passages, such as the Cape Spartel Lighthouse,[12] but these also are the exception rather than the rule.

It is one of the paradoxes in the development of international institutions that those organizations which have come to be charged with the maintenance of freedom of navigation on international rivers have their antecedents in an agency created for an entirely different and inconsistent purpose—that of exacting taxes from the users of the waterway to meet the general financial needs of the riparian states.[13] The Central Commission for the Navigation of the Rhine of today is descended from the central administration for the collection of the Octroi, or tolls, half of the profits of which, after the deduction of certain expenses for the maintenance of the river, were to be paid as an indemnity to the Archbishop of Mayence and the German princes who had lost their territories on the left bank of the Rhine to France.[14] The power to collect tolls was returned to the individual riparian states as a consequence of the change in the régime of the river which was brought about in 1815.[15] From the 1830's onward the tolls were gradually reduced and finally eliminated by the riparians. In the negotiations among the nations bordering on the river, the promise of lower tolls or of their abolition by one party proved a useful basis for securing concessions from the other. Furthermore, tolls were inconsistent with a policy of encouraging the development of

[11] Convention concerning the Régime of the Straits, signed at Montreux, July 20, 1936, art. 2, and Annex I, para. 4, 173 L.N.T.S. 213, 7 HUDSON, INTERNATIONAL LEGISLATION 386 (1941). The "taxes and charges" are specified in Annex I of the Convention.

[12] Treaty cited *supra* note 2.

[13] "Elle [la Commission Centrale pour la Navigation du Rhin] doit son origine à une matière fiscale, l'Octroi, mais elle a trouvé la justification de son existence dans un autre domaine, celui de l'intérêt de la navigation pur et simple, et cet intérêt subsistait toujours et fut même au contraire servi par la suppression des droits de navigation." VAN EYSINGA, LA COMMISSION CENTRALE POUR LA NAVIGATION DU RHIN 95 (1935); and see pp. 98-101 *supra*.

[14] Convention between France and Germany on the Octroi of Navigation of the Rhine, signed at Paris, Aug. 15, 1804, 8 MARTENS, RECUEIL 261 (1835), 1 RHEINURKUNDEN 6 (1918), in implementation of the "Recès principal de la députation extraordinaire de l'Empire concernant les indemnités à régler d'après le traité de Luneville," Feb. 25, 1803, 7 MARTENS, RECUEIL 435 (1831), 1 RHEINURKUNDEN 1 (1918).

[15] Articles concerning the Navigation of the Rhine, arts. III, IV, and VI, annexed to the Final Act of the Congress of Vienna, signed June 9, 1815, 2 MARTENS, N.R. 416 (1818).

waterborne commerce.[16] When the Treaty of Mannheim abolished the right to collect tolls in 1868,[17] the stipulation of the treaty was in actuality no more than a reflection of the existing state of affairs on the river. A similar process was taking place in the mid-nineteenth century on the other important river systems of Europe, sometimes by outright abolition of tolls,[18] sometimes by their redemption by the user nations.[19] However, the special régime which had been established for the Danube dictated another course. The wide operational functions which had been vested in the European Commission of the Danube required that that body be permitted to levy tolls in order to defray the cost of the works it had undertaken and of the administration of the river.[20] When the arrangements concerning the administration of the Danube were revised after the First World War, the International Commission established for the fluvial section of the river was given wide discretion in determining how the maintenance of the waterway and improvements should be paid for.[21] The costs of maintenance were normally to be borne by the riparian states, but if one of these was to be put to special burdens in order to accommodate the traffic from other nations, that nation might re-

---

[16] CHAMBERLAIN, THE REGIME OF THE INTERNATIONAL RIVERS: DANUBE AND RHINE 219-36 (1923).

[17] Revised Convention for the Navigation of the Rhine, signed at Mannheim, Oct. 17, 1868, art. 3, 20 MARTENS, N.R.G. 355 (1875), 2 RHEINURKUNDEN 80 (1918).

[18] E.g., on the Elbe. Treaty between Austria and the North German Confederation for the Abolition of the Elbe Dues, signed at Vienna, June 22, 1870, 20 MARTENS, N.R.G. 345 (1875), 63 BRITISH AND FOREIGN STATE PAPERS 594 (1879). Concerning the earlier abolition of the Stade Toll, through redemption, see Treaty for Redemption of the Stade Toll, signed at Hanover, June 22, 1861, 17 MARTENS, N.R.G. 419 (1861), 51 BRITISH AND FOREIGN STATE PAPERS 27 (1868).

[19] Treaty between Belgium and the Netherlands for the Redemption of the Scheldt Toll, signed at The Hague, May 12, 1863, 53 BRITISH AND FOREIGN STATE PAPERS 15 (1868), implemented by the General Treaty for the Redemption of the Scheldt Toll, signed at Brussels, July 16, 1863, 17 MARTENS, N.R.G. pt. 2, at 223 (1869); see also GRANDGAIGNAGE, HISTOIRE DU PÉAGE DE L'ESCAUT (1868); 1 GUILLAUME, L'ESCAUT DEPUIS 1830, at 334-456 (1902). As to the termination of such dues as to the United States, see Convention between the United States of America and Belgium for the Extinguishment of the Scheldt Dues, signed at Brussels, July 20, 1863, 13 Stat. 655, T.S. No. 23, 1 MALLOY, op. cit. supra note 2, at 75. The many treaties concerning the extinction of the Scheldt dues are listed in OGILVIE, INTERNATIONAL WATERWAYS 246-47 (1920).

[20] General Treaty of Peace between Austria, France, Great Britain, Prussia, Russia, Sardinia, and the Ottoman Porte, signed at Paris, March 30, 1856, art. XVI, 15 MARTENS, N.R.G. 770 (1857), 46 BRITISH AND FOREIGN STATE PAPERS 8 (1865).

[21] Convention Instituting the Definitive Statute of the Danube, signed at Paris, July 23, 1921, arts. 15-19, 26 L.N.T.S. 173.

quest the International Commission to allocate the costs between it and the nations benefited.[22] Dues could lawfully be imposed by the riparians or by the Commission itself only for the purpose of defraying the costs of "works of improvement properly so called" and "works in respect of the maintenance of works of improvement of special importance."[23] The *de facto* termination of the international status of the river by the treaty of 1947 is reflected in authorizations to the riparians to collect dues at levels to be agreed upon with the Commission.[24] Nevertheless, it is required that the dues "must not be a source of profit."[25]

Since the interoceanic canals—Suez, Panama, Kiel, and Corinth[26]—are entirely artificial waterways employed by the shipping of a substantial number of states, it is inevitable that tolls should be charged. The alternative would be the absorption of the costs of construction, maintenance, and operation by the proprietor, which each such nation would justifiably regard as a concealed subsidy of foreign as well as domestic shipping. For similar reasons and in order to make the St. Lawrence Seaway self-liquidating, tolls are charged for transit through the waterway, which has been constructed through the canalization of the St. Lawrence River.[27]

It is possible to sum up the practice of states as pointing to a general recognition of the impropriety of charging the user of an international waterway for the mere privilege of use. However, to the extent that the proprietor of the waterway is called upon to provide special facilities or services, it may legitimately demand compensation therefor. The level at which the price of use may legitimately be fixed must now be examined.

THE PRINCIPLE OF NONDISCRIMINATION

In 1912, with the Panama Canal virtually completed, the time had come to enact a statute "to fix the method by which the canal shall be maintained

[22] Article 15.

[23] Article 16.

[24] Convention regarding the Régime of Navigation on the Danube, signed at Belgrade, Aug. 18, 1948, arts. 35 and 36, 33 U.N.T.S. 196, 213.

[25] Article 37.

[26] See NEW CORINTH CANAL COMPANY, TARIFFS OF THE CORINTH CANAL (1951).

[27] St. Lawrence Seaway Act, § 12 (b) (4) and (5), 68 Stat. 97, 33 U.S.C. § 988 (b) (4) and (5); Exchange of Notes between the United States of America and Canada regarding the Saint Lawrence Seaway Tariff of Tolls, signed at Ottawa, March 9, 1959, 10 U.S.T. 323, T.I.A.S. No. 4192.

and controlled and the zone governed."[28] The Panama Canal Act[29] dealt with the question of tolls by delegating to the President the function of pre-scribing them but stipulated that no tolls were to be levied on vessels "en-gaged in the coastwise trade of the United States,"[30] which was limited to vessels of United States registry. President Taft had expressed himself strongly on the subject of this preference to United States shipping: "We own the canal. It was our money that built it. We have the right to charge tolls for its use." The relief of such vessels from tolls would be nothing more than a subsidy to the American merchant marine, differing not at all from the practice engaged in by other states.[31] British protests were forthcoming even before the bill was passed by the Congress. It was justifiably contended that the discrimination effected by the remission of tolls as to United States vessels was in contravention of article 3, rule 1, of the Hay-Pauncefote Treaty,[32] which had opened the canal to the ships of all nations "on terms of entire equality."[33] The reply of the United States Government was that article 3 referred to the neutralization of the canal alone and that its effect was only to extend conditional most-favored-nation treatment to the users. The Canal had been constructed, so the United States argument ran, on territory subject to the control and jurisdiction of that country and at its own expense, and the charging of tolls was therefore within the sovereign preroga-tives of the United States.[34] What the contention boiled down to was that the equality of all nations under the Treaty meant all *foreign* nations.[35]

---

[28] Extract from Message of the President to Congress, Dec. 21, 1911, in [1912] FOREIGN REL. U. S. 467 (1919).

[29] 37 Stat. 560.

[30] Section 5.

[31] *Op. cit supra* note 28, at 468; and see Memorandum of the President to Congress, to accompany the Panama Canal Act, *id.* at 475.

[32] Treaty between the United States and Great Britain to Facilitate the Construction of a Ship Canal, signed at Washington, Nov. 18, 1901, 32 Stat. 1903, T.S. No. 401.

[33] The British Chargé d'Affaires to the Secretary of State, July 8, 1912, in [1912] FOREIGN REL. U. S. 469 (1919).

[34] Memorandum of the President to Congress, to accompany the Panama Canal Act, *id.* at 475.

[35] That this was not the intention of the United States negotiators of the Hay-Pauncefote Treaty is borne out by their testimony, reported in a speech by Senator Root on May 21, 1914, 51 CONG. REC. 8942 at 8945 (1914).

The question of discrimination in tolls was linked to that of the fortification of the Canal through the contention, advanced on behalf of the United States, that if "all nations" were construed to include the United States, then that country would, while at war, be under the restraints imposed by the Treaty as to passage by belligerent vessels in time of war. It was pointed out that Great Britain had, by conceding that the United

The position of the United States rested upon what were at best questionable legal grounds. That country had obtained its release from the restraints placed on the construction of a transisthmian canal in the Clayton-Bulwer Treaty[36] only by undertaking the obligations of the Treaty of 1901 and then only on the condition that the principle of article 8 of the Clayton-Bulwer Treaty, requiring that the Canal be open to British and United States vessels on a footing of entire equality, be maintained in force.[37] "Sovereignty" was an empty defense to an obligation grounded in a treaty. Having regard to the fact that the Canal tolls were intended to defray the cost of the waterway, an exemption for United States vessels would place a proportionately higher share of the burden on foreign vessels, which were no less discriminated against because a small number of United States ships engaged in foreign, as contrasted with the coasting, trade were likewise subject to the payment of tolls.[38] Tolls could not be characterized as "just and equitable"[39] if the principle of a fair share of the burden for all was not maintained. Something more was involved than a mere subsidy to the American merchant marine.

The authority to discriminate granted to the President of the United States by the Panama Canal Act was not exercised. The tolls which the President set were uniform as to all vessels.[40] In the meanwhile eloquent voices, including that of Senator Root,[41] were heard in opposition to the provisions of the Act with respect to the Panama tolls. In 1914 President

States had the right to take defensive measures with respect to the Canal, recognized that these provisions as to time of war did not apply to the ships of "all nations." See Article prepared by the Law Officer of the Isthmian Canal Commission, Mr. Feuille, regarding Tolls on the Panama Canal, H.R. Doc. No. 1313, 62d Cong., 3d Sess. 7 (1913). The British reply to this argument appears in the document cited note 38 *infra*.

[36] Convention between the United States of America and Great Britain for Facilitating and Protecting the Construction of a Ship Canal Connecting the Atlantic and Pacific Oceans, signed at Washington, April 19, 1850, 9 Stat. 995, T.S. No. 122.

[37] Preamble, Hay-Pauncefote Treaty, *supra* note 32.

[38] These views were expressed in Despatch to His Majesty's Ambassador at Washington respecting the Panama Canal Act, Nov. 14, 1912, GREAT BRITAIN, PARLIAMENTARY PAPERS, MISC. No. 12 (1912) (CD. 6451), [1912] FOREIGN REL. U. S. 481 (1919).

[39] Hay-Pauncefote Treaty, *supra* note 32, art. 3, rule 1.

[40] Presidential Proclamation, Nov. 13, 1912, 37 Stat. 1769. The British protest therefore was directed to the possibility of discrimination, not its actuality. See Secretary of State to the American Chargé d'Affaires, Jan. 17, 1913, [1913] FOREIGN REL. U. S. 540 (1920); and British Ambassador to the Secretary of State, Feb. 27, 1913, *id.* at 547.

[41] 49 CONG. REC. 1818 (1913); 51 *id.* at 8942 (1914).

Wilson appeared before a joint session of the Congress and argued for the repeal of the unfortunate provision of the Panama Canal Act in words so aptly framed that they deserve quotation here:

> Everywhere else the language of the treaty is given but one interpretation, and that interpretation precludes the exemption I am asking you to repeal. We consented to the treaty; its language we accepted, if we did not originate it; and we are too big, too powerful, too self-respecting a nation to interpret with a too srained or refined reading the words of our own promises just because we have enough power to give us leave to read them as we please. The large thing to do is the only thing we can afford to do, a voluntary withdrawal from a position everywhere questioned and misunderstood. We ought to reverse our action without raising the question whether we were right or wrong, and so once more deserve our reputation for generosity and for the redemption of every obligation without quibble or hesitation.[42]

Three months later, in the face of strong opposition,[43] the objectionable provision of the Act was repealed, with a face-saving reservation of any right which the United States might have "to discriminate in favor of its vessels by exempting the vessels of the United States or its citizens from the payment of tolls for passage" through the Canal.[44]

This episode,[45] so happily resolved in conformity with law, is an instructive illustration of a principle of nondiscrimination which is of equal application to all interoceanic canals. The basic laws governing the two other major interoceanic canals leave little room for doubt that discrimination would be unlawful. Article 1 of the Convention of Constantinople of 1888 declares that the Suez Canal "shall always be free and open . . . without distinction of flag."[46] The concession which was granted to the Suez Canal Company in

---

[42] Address of the President, March 5, 1914, [1914] FOREIGN REL. U. S. 317 (1922).

[43] See *Hearings on H.R. 14385 before the Senate Committee on Inter-Oceanic Canals*, 63d Cong., 2d Sess. (1914).

[44] Act of June 15, 1914, § 1, 38 Stat. 385.

[45] From the extensive literature on the question of the tolls controversy there may be mentioned BUTTE, GREAT BRITAIN AND THE PANAMA CANAL: A STUDY OF THE TOLLS QUESTION (1913); MILLER AND FREEHOFF, THE PANAMA CANAL TOLLS CONTROVERSY (1914); NIXON, THE CANAL TOLLS AND AMERICAN SHIPPING (1914); OPPENHEIM, THE PANAMA CANAL CONFLICT BETWEEN GREAT BRITAIN AND THE UNITED STATES OF AMERICA (1913); RICHARDS, THE PANAMA CANAL CONTROVERSY (1913); TALLEY, THE PANAMA CANAL (1915); and the papers read at the annual meeting of the American Society of International Law in 1913, on the general subject of "International Use of Straits and Canals, with Especial Reference to the Panama Canal," in [1913] PROCEEDINGS OF THE AMERICAN SOCIETY OF INTERNATIONAL LAW.

[46] Signed Oct. 29, 1888, 15 MARTENS, N.R.G. 2d ser. 557 (1891).

1856 imposed the specific obligation on the Company "de percevoir ces droits, sans aucune exception ni faveur, sur tous les navires dans des conditions identiques."[47] With respect to the Kiel Canal, the ships of all nations are, since the coming into force of the Treaty of Versailles, "in respect of charges" to be "treated on a footing of perfect equality."[48] The correctness of this conclusion as to a general principle of nondiscrimination is not called in question by the fact that Panamanian Government vessels are, pursuant to the Hay–Bunau-Varilla Treaty, exempted from the payment of tolls in the Panama Canal.[49] That right, as well as the former exemption from tolls of small vessels using the Suez Canal[50]—a privilege which redounded very largely to the advantage of Egyptian ships—may be justified as part of the compensation furnished to the territorial sovereign for the rights which the operator of the canal enjoyed.[51] So also, the circumstance that United States

[47] Firman de concession et cahier des charges pour la construction et l'exploitation du Grand Canal Maritime de Suez et dépendances, Jan. 5, 1856, art. 17, para. 1, in COMPAGNIE UNIVERSELLE, RECUEIL DES ACTES CONSTITUTIFS 6 at 10.

[48] Treaty of Peace between the Allied and Associated Powers and Germany, signed at Versailles, June 28, 1919, art. 381, 112 BRITISH AND FOREIGN STATE PAPERS 1 (1919). Prior to the First World War, and the internationalization of the waterway, the tolls assessed for passage through the "Nord-Ostsee-Kanal" discriminated in favor of German coastwise traffic. See the tariff schedules annexed to the Decree of June 4, 1895, [1895] REICHSGESETZBLATT 241, and to the Decree of August 4, 1896, [1896] id. at 681, also in ARNOLD, DIE GRUNDLAGEN DER TARIFPOLITIK FÜR DEN NORDOSTSEEKANAL (No. 20 in Kieler Studien) at 57-59 (1951); see also THE KIEL CANAL AND HELIGOLAND (Foreign Office, Peace Handbook No. 41) at 11 (1920).

[49] Convention between the United States of America and the Republic of Panama for the Construction of a Ship Canal to Connect the Waters of the Atlantic and Pacific Oceans, signed at Washington, Nov. 18, 1903, art. XIX, 33 Stat. 2234, T.S. No. 431. During the tolls controversy with Great Britain, the British Government pointed to this exemption as adding another category of vessels which would not contribute to the upkeep of the Canal. [1912] FOREIGN REL. U. S. 487 (1919). As was proper, however, the exemption was maintained in force by the Act of June 15, 1914, 38 Stat. 385.

[50] Convention of March 7, 1949, between the Suez Canal Company and the Egyptian Government, exempting vessels of less than 300 tons gross, not carrying passengers, from the payment of tolls. COMPAGNIE UNIVERSELLE, RECUEIL DES ACTES CONSTITUTIFS 241, 243.

[51] The same may be asserted of the concession to Colombia of duty-free passage for its troops, materials of war, and ships of war, accorded by the Treaty between the United States of America and the Republic of Colombia for the Settlement of their Differences Arising out of the Events which Took Place on the Isthmus of Panama in November 1903, signed at Bogotá, April 6, 1914, art. 2, para. 1, 42 Stat. 2122, T.S. No. 661, 9 L.N.T.S. 301. The right created by the Treaty, which did not enter into force until 1922, was part of the price of buying peace with the former territorial sovereign of the area occupied by the Canal Zone and should be regarded as part of

naval vessels have not paid tolls in the Panama Canal reflects the fact that such ships are being employed for, among other purposes, the defense of the Canal.[52] Finally, the discrimination between Greek and non-Greek vessels effected by the tariff of charges for the use of the Corinth Canal[53] is the exception that proves the rule, since that waterway has not been opened to general use by the ships of all nations through such a dedication as would give the waterway international status.

In those instances in which tolls are still levied for the transit of international waterways other than canals, the principle of equality of tolls prevails. Such nondiscrimination is demanded by the Montreux Convention for merchant vessels passing through the Turkish Straits,[54] but no tolls are required of light warships.[55] When tolls are demanded for transit of a river, a like requirement prevails. The Treaty of 1921 required of the International Commission of the Danube and the nations creating it that "the incidence of navigation dues may in no case involve differential treatment in respect

---

the cost of acquiring the Canal. In any event, the provision referred to does not impose an onerous obligation upon the United States, nor indirectly on the users of the Canal.

[52] The United States is authorized, but not required, to pay tolls for the transit of naval vessels, but in any event tolls must be computed and treated as revenues of the Panama Canal Company for the purpose of prescribing the rate of tolls. 2 C.Z.C. § 412(c) (1962).

In opinions by the Attorney General rendered in 1915 and 1931, the view was expressed that the exemption of vessels owned by the United States but not operated for commercial purposes from the payment of tolls did not constitute discrimination in contravention of the provisions of the Hay-Pauncefote Treaty nor did it make the charges imposed on vessels required to pay tolls other than "just and equitable." Heavy emphasis was placed on the necessity of such passages in the defense of the Canal and on the sovereign rights of the United States in regard to the Canal. The Attorney General also alluded in the first of the two opinions to the fact that Great Britain had not persevered in its protest against toll-free passage for government-owned vessels and was therefore estopped from doing so. The Secretary of State to the American Chargé d'Affaires, Jan. 17, 1913, [1913] FOREIGN REL. U. S. 540 at 542 (1920); The British Ambassador to the Secretary of State, Feb. 27, 1913, *id.* at 547; Memorandum Relating to the Question of Liability of United States Vessels to Payment of Panama Canal Tolls, June 3, 1915, and Opinion of the Attorney General, furnished to the President, July 8, 1931, reproduced in Reply Brief for the Panama Canal Company, pp. 49-71, Panama Canal Co. v. Grace Line, Inc., 356 U. S. 309 (1958).

[53] NEW CORINTH CANAL COMPANY, TARIFFS OF THE CORINTH CANAL (1951).

[54] Convention concerning the Régime of the Straits, signed at Montreux, July 20, 1936, art. 2 and Annex I, 173 L.N.T.S. 213, 7 HUDSON, INTERNATIONAL LEGISLATION 386 (1941).

[55] Article 10.

of the flag of the vessels or the nationality of persons and goods . . ."[56] No parallel provision is to be found in the Treaty of 1948,[57] which confirmed the Russian domination of the river. By the Boundary Waters Treaty of 1909, the United States and Great Britain agreed upon equality of tolls for nationals of the two nations employing canals connecting boundary waters between the United States and Canada.[58] Under the national statutes governing the waterway, the principle of nondiscrimination would appear to have equal application to tolls established for the use of the St. Lawrence Seaway, whether by agreement between the United States and Canada or unilaterally by each of the two countries.[59] The tolls established are in fact nondiscriminatory.[60]

It may not be out of place to mention that although customary international law does not appear to require nondiscrimination in port duties between national and foreign vessels or among the vessels of various foreign countries,[61] a number of treaties of friendship, commerce, and navigation contain a provision whereby the parties agree on equality of port charges for national vessels and ships under the flag of the other party.[62] Such equality

[56] Convention Instituting the Definitive Statute of the Danube, signed at Paris, July 23, 1921, art. 18, 26 L.N.T.S. 173.

[57] Convention regarding the Régime of Navigation on the Danube, signed at Belgrade, Aug. 18, 1948, 33 U.N.T.S. 196, chapter IV of which deals with fiscal matters.

[58] Treaty between the United States of America and Great Britain relating to Boundary Waters between the United States and Canada, signed at Washington, Jan. 11, 1909, art. 1, 36 Stat. 2448, T.S. No. 548. The provisions of the Boundary Waters Treaty were invoked by Great Britain during the dispute over Panama tolls. [1912] FOREIGN REL. U. S. 481 at 483-84 (1919).

[59] The alternatives envisaged by the St. Lawrence Seaway Act, § 12(a), 68 Stat. 96, 33 U.S.C. 988(a). The language "fair and equitable" applied to the level of tolls in the St. Lawrence Seaway Act resembles the phrase "just and equitable" used in article III, rule 1, of the Hay-Pauncefote Treaty of 1901. If the teaching of the tolls controversy is that the latter expression demands nondiscrimination, the same significance may justifiably be accorded the words of the St. Lawrence Seaway Act.

The Canadian The St. Lawrence Seaway Authority Act, 15 & 16 Geo. 6, c. 24, makes no reference to equality or discrimination in tolls as to Canadian and foreign vessels.

[60] St. Lawrence Seaway Tariff of Tolls, annexed to Exchange of Notes between the United States of America and Canada, signed at Ottawa, March 9, 1959, 10 U.S.T. 323, T.I.A.S. No. 4192. The view has been taken that discriminatory tolls would unfailingly have elicited protests. Cohen and Nadeau, *The Legal Framework of the St. Lawrence Seaway,* [1959] U. ILL. L.F. 29, 49.

[61] See 5 MOORE, DIGEST OF INTERNATIONAL LAW 288-302 (1906).

[62] E.g., Treaty of Commerce and Navigation between Germany and the United Kingdom, signed at London, Dec. 2, 1924, art. 20, 43 L.N.T.S. 89; Treaty of Friendship,

of treatment for "dues and charges of all kinds" is expressly required of the parties to the Statute on the International Régime of Maritime Ports of 1923.[63] It is hazardous to deduce the rule of international law in the absence of agreement from a pattern of conventional stipulations, but one may venture the conclusion that there has been during this century an awareness of the desirability of establishing a standard of nondiscrimination.

The requirement of nondiscrimination in the assessment of tolls is no more than a specific application of the more general principle of nondiscrimination and equality of treatment in passage through international canals. One rule could not exist without the other, for the general requirement of equality of treatment could be subverted by a toll structure which imposed greater or less financial obligations on users, depending on the flag of the vessel, its ownership, or its registry. It must also be borne in mind that international law offers numerous instances in which general principles must be called in question because of failure to give them expression in practice. The actual conduct of states in a matter touching so strongly upon their financial interests consequently possesses heightened importance as affording convincing evidence of the validity and force of the more general and abstract principle of nondiscrimination as to transiting vessels. Without these legal standards, the dedication of an interoceanic canal to free use by the ships of all nations on a footing of equality would be an empty gesture.

However, too much must not be claimed for a rule of nondiscrimination in the assessment of tolls as an effective restraint in the fiscal sphere upon the power of the state operating the waterway or serving as its territorial sovereign. Equality of treatment does not effectively hold tolls to a reasonable level, if the economic impact of those charges is not felt to a substantial degree by the shipping trade and commerce of the territorial sovereign or other nation operating the waterway. Foreign users of the Panama Canal see their interests defended by American shipping companies, who can make

---

Commerce and Navigation between the United States of America and the Republic of China, signed at Nanking, Nov. 4, 1946, art. XXII, 63 Stat. 1299, T.I.A.S. No. 1871; Treaty of Friendship, Commerce and Navigation between the United States of America and the Federal Republic of Germany, signed at Washington, Oct. 29, 1954, art. XX, 7 U.S.T. 1839, T.I.A.S. No. 3593; and see, for nondiscrimination as to port dues in Elbe ports, Convention Instituting the Statute of Navigation of the Elbe, signed at Dresden, Feb. 22, 1922, art. 26, 26 L.N.T.S. 221.

[63] Annexed to Convention on the International Régime of Maritime Ports, signed at Geneva, Dec. 9, 1923, 58 L.N.T.S. 285, 2 HUDSON, INTERNATIONAL LEGISLATION 1156 (1931).

their voices heard in the Congress and in the courts and who can influence the conduct of their government in the myriad ways open to United States individuals and companies and unavailable to their foreign competitors. A similar restraint of substantial national use does not hold the hand of the United Arab Republic, the users of the Suez Canal being overwhelmingly foreign. At either extreme, the proprietor of the waterway will not necessarily yield to the pressure which is exerted by national users. It therefore becomes necessary to consider what other restraints on the level of tolls have been recognized in the practice of states.

REGULATION OF THE LEVEL OF TOLLS BY TREATY AND STATUTE

If, as is normally the case, the cost of constructing and operating an artificial waterway is to be borne by the shipping employing the facility, the charges for the privilege of use should ideally be established within some legal limits. An upper level is highly desirable in order to prevent either an undue burden on shipping employing the canal or an impediment to free navigation in the form of prohibitively high tolls which would force shipping to resort to other routes. As has previously been stated, it is not beyond the realm of possibility that tolls could intentionally be raised to a level which would make it impossible for ships to use the waterway and that the resultant level of charges would thus by indirection bring about a closing of the waterway which could not, consistently with law, be brought about directly.

For the purpose of analyzing the attempts which have been made to regulate the level of tolls by international agreement or by municipal law, it is helpful to keep in mind the following elements relating to the cost of construction, operation, and maintenance of the waterway:

1. Return of capital, that is, repayments by the entity of funds furnished for the construction of the waterway.
2. Costs of capital, that is, interest paid on funds borrowed by the entity for the construction of the waterway.
3. Depreciation, as to any facilities appurtenant to the waterway, such as locks and structures.[64]

[64] As distinguished from the channel, revetments, and the like, as to which depreciation would be inappropriate. See ANNUAL REPORT OF THE PANAMA CANAL COMPANY AND CANAL ZONE GOVERNMENT, 1958, at 48 (1959), interpreting the Act of Sept. 26, 1950, § 12, 64 Stat. 1043, now 2 C.Z.C. § 411 (1962).

4. Costs of operation and maintenance.

5. Provision of reserves for the purpose of making improvements in the waterway.[65]

6. Profit.

It must be emphasized that these categories are mentioned only as indicating the range of possibilities concerning which disputes might arise and not as necessary elements in the structure of tolls. A consideration of the propriety of taking these several factors into account in the computation of charges must await some account of the formulas which have been employed in the past for determining the proper level of tolls.

Again, speaking in terms of the ideal régime for an interoceanic canal, the criteria by which tolls would be computed and kept within such limits as would not interfere with the free navigation of the waterway would be established either by the instrument dedicating the waterway to general use or by a convention to which the operator or supervisor state and states representing the users would be parties. An orderly pattern of regulation has not, however, been established in practice, as the following summary of formulas for tolls in international instruments will demonstrate.

Such a standard as the requirement of the Hay-Pauncefote Treaty that tolls be "just and equitable"[66] is so vague as to offer no real guidance in the fixing of tolls. Its sole effect may be to indicate that the question of tolls has been removed from the domestic jurisdiction of the United States. Were an objective examination to be made by an arbitral tribunal of charges for the use of the Canal, the high level of abstraction upon which the standard is cast would make any decision actually one *ex aequo et bono*. At the time of the Panama Canal tolls controversy, Sir Edward Grey interpreted these words as limiting the tolls "to the amount representing the fair value of the services rendered, i.e., to the interest on the capital expended and the cost of operation and maintenance of the Canal."[67] But these words appear to have created no firm international obligation and, even if accepted as governing, offer an

---

[65] Concerning the allocation of funds derived from tolls for passage through the Suez Canal to the improvement of the waterway, see p. 263 *infra*.

[66] Treaty between the United States and Great Britain to Facilitate the Construction of a Ship Canal, signed at Washington, Nov. 18, 1901, art. III, rule 1, 32 Stat. 1903, T.S. No. 401.

[67] Despatch to His Majesty's Ambassador at Washington respecting the Panama Canal Act, Nov. 14, 1912, GREAT BRITAIN, PARLIAMENTARY PAPERS, MISC. No. 12 (1912) (CD. 6451), [1912] FOREIGN REL. U. S. 481, 486 (1919).

insubstantial basis for attacking the method of computing tolls which the Panama Canal Company has been employing since 1950.[68]

Although the Hay-Pauncefote Treaty was in general modeled upon the Convention of Constantinople of 1888, a similar limitation upon the tolls to be assessed for passage through the Suez Canal is not to be found in the earlier treaty. However, only fifteen years before this, in 1873, an attempt had been made to deal by international agreement with the question of tolls for the use of the Canal. A new method which De Lesseps had adopted, on the recommendation of a commission established for that purpose, for the measurement of the tonnage of transiting vessels had brought about a marked increase in tolls.[69] The ensuing controversy between the shipping trade and the Company[70] induced the Ottoman Porte to convene an international commission in 1873 to inquire into the matters of tolls and the measurement of tonnage. The importance of the work of that commission lies in its recommendation of the adoption of the Moorsom or British method of computing tonnage, subject to certain modifications.[71] From 1874, when De Lesseps was forced to adopt it,[72] until this day, it has been this special system of tonnage measurement for the Suez Canal[73] which has been used as the basis for the computation of tolls.[74] Less success was experienced in obtaining De Lesseps' consent to the new tolls recommended by the Commission, which

[68] As was sought in Panama Canal Co. v. Grace Line, Inc., 356 U. S. 309 (1958). For an attempt to interpret the law of the United States in light of the Hay-Pauncefote Treaty, see Brief of the United Kingdom of Great Britain and Northern Ireland, *Amicus Curiae,* in that case.

[69] HALLBERG, THE SUEZ CANAL: ITS HISTORY AND DIPLOMATIC IMPORTANCE 222-28 (1931); WILSON, THE SUEZ CANAL: ITS PAST, PRESENT, AND FUTURE 59-60 (1933).

[70] The litigation over this matter in the French courts is discussed *supra* at p. 60.

[71] Despatch from the British Delegates on Tonnage at Constantinople, together with the Report and Recommendations of the Commission as to International Tonnage and the Suez Canal Dues, GREAT BRITAIN, PARLIAMENTARY PAPERS, COMMERCIAL No. 7 (1874) at 5 (C. 943).

[72] HALLBERG, *op. cit supra* note 69, at 227.

[73] Concerning the history of attempts to provide a uniform system of tonnage measurement for all purposes, see United Nations, Transport and Communications Commission, Third session, Unification of Maritime Tonnage Measurement: Memorandum by the Secretariat, March 4, 1949 (E/CN.2/57) (1949); and REIFF, THE UNITED STATES AND THE TREATY LAW OF THE SEA 222-23 (1959). The question has been under consideration by the International Maritime Consultative Organization. N. Y. Times, Aug. 23, 1959, sec. 5, p. 16S, col. 1.

[74] See COMPAGNIE UNIVERSELLE, RÈGLEMENT DE NAVIGATION 82-101 (1953), which has been maintained in force by the Suez Canal Authority.

had to be revised before their acceptance by De Lesseps three years later.[75] At the time the international commission rendered its report, the Turkish Government made a declaration that ". . . no modification, for the future, of the conditions for the passage through the Canal shall be permitted, whether in regard to the navigation toll or the dues for towage, anchorage, pilotage, &c., except with the consent of the Sublime Porte, which will not take any decision on this subject without previously coming to an understanding with the principal Powers interested therein."[76] The Turkish declaration was given effect during the subsequent negotiations on the level of tolls between De Lesseps and the British Government, which was acting in a representative capacity on behalf of the other nations which had participated in the tonnage conference.[77] Although the amount of tolls was on this occasion the subject of international agreement, at least to the extent of requiring consultation, the application of the Turkish declaration appears thereafter to have fallen into innocuous desuetude.

Since the régime provided for the Kiel Canal by the Treaty of Versailles was established under exceptional circumstances, it would probably be unjustifiable to suggest a historical progression from the lack of fiscal restrictions on Suez in the Convention of Constantinople, through the "just and equitable" standard of the Hay-Pauncefote Treaty, to the more precise formula of 1919 governing Kiel.[78] Article 182 of the Treaty of Versailles provided: "Only such charges may be levied on vessels using the Canal or its approaches as are intended to cover in an equitable manner the cost of maintaining in a navigable condition, or of improving, the Canal or its approaches, or to meet expenses incurred in the interests of navigation." There is apparently little danger of violation of this standard, which allows charges only for the costs of operating, maintaining, and improving the Canal, since the Canal has been operated at a loss during the period following the Second World War.[79] The latest important landmark in the fixing of tolls by an

---

[75] Agreement of Feb. 21, 1876, in Further Correspondence respecting the Suez Canal, GREAT BRITAIN, PARLIAMENTARY PAPERS, EGYPT No. 9 (1876) at 25 (C. 1525).

[76] Document cited *supra* note 71 at 11.

[77] The negotiations conducted by Great Britain are described in Society of Comparative Legislation and International Law, *The Suez Canal; A Selection of Documents* . . . , INT'L & COMP. L.Q., Sp. Supp. at 46-48 (1956).

[78] Treaty of Peace between the Allied and Associated Powers and Germany, signed at Versailles, June 28, 1919, art. 382, 112 BRITISH AND FOREIGN STATE PAPERS 1 (1919); see, as to tolls charged, NORD-OSTSEE-KANAL, VERORDNUNG ÜBER DIE BEFAHRUNGSABGABEN AUF DEM NORD-OSTSEE-KANAL VOM 13. MÄRZ 1958, NACH DEM STANDE VOM 1. APRIL 1961.

[79] *A Survey of the Development of the Kiel Canal Dues,* [1961] NORD-OSTSEE-KANAL

international instrument has been the undertaking of Egypt, contained in the Declaration of April 24, 1957, that (a) tolls would continue to be levied in accordance with the agreement which had been concluded between the Government of Egypt and the Suez Canal Company in 1936;[80] (b) any increase in tolls within any twelve-month period would be limited to 1 per cent; and (c) any greater increases would "be the result of negotiations, and, failing agreement, be settled by arbitration" according to an international arbitral procedure.[81] Although the tribunal would thus in effect be given the function of setting the tolls, it would not be entirely without guidance as to the elements to be taken into account in that determination. The Declaration specifies that the Suez Canal Authority, an autonomous organ of the Government of the United Arab Republic, is to pay over to that government 5 per cent of the gross receipts as a royalty[82] and that 25 per cent of the gross receipts are to be paid into the Suez Canal Capital and Development Fund in order to provide "adequate resources to meet the needs of development and capital expenditure."[83] It is not altogether clear whether an arbitral tribunal charged with the interpretation of this commitment, which was intended to and did create legal obligations,[84] would regard the 5 per cent royalty on the gross revenues as the upper limit on the profit to be derived by the United Arab Republic from the operation of the waterway. That such was intended

---

30 at 35-36; Lorenzen, *The Administration of the Kiel Canal*, [1953] *id.*, No. 2, 20 at 25; ARNOLD, DIE GRUNDLAGEN DER TARIFPOLITIK FÜR DEN NORDOSTSEEKANAL (No. 20 in Kieler Studien) at 33 (1951).

[80] Décret portant fixation du taux maximum des droits de navigation dans le Canal de Suez, April 28, 1936, COMPAGNIE UNIVERSELLE, RECUEIL DES ACTES CONSTITUTIFS 207.

[81] Egyptian Declaration of April 24, 1957, para. 3, annexed to letter to the Secretary-General of the United Nations from the Egyptian Minister for Foreign Affairs of the same date (U.N. Doc. No. A/3576, S/3818) (1957).

[82] These profits are apparently to be used in the development of the Aswan High Dam. 189 THE ECONOMIST 392 (1958).

[83] Egyptian Declaration of April 24, 1957, *supra* note 81, para. 5.

[84] The covering letter and the Declaration both stated that the latter constituted an "international instrument" which should be deposited and registered with the United Nations Secretariat; see, as to the legal effect of unilateral declarations, 1 OPPENHEIM, INTERNATIONAL LAW 872 (8th ed. Lauterpacht 1955) and 1 GUGGENHEIM, TRAITÉ DE DROIT INTERNATIONAL PUBLIC 147 (1953). In some particulars the Declaration went beyond and in others fell short of the terms of the Security Council resolution of Oct. 13, 1956 (U.N. Doc. No. S/3675) (1956), which gave rise to it. The resolution provided that "the manner of fixing tolls and charges should be decided by agreement between Egypt and the users," and that "a fair proportion of the dues should be allotted to development."

is indicated by the fact that the net profits of the Suez Canal Company were running at approximately 30 per cent of turnover.[85] If such net profits were the source of distributable income and working capital, that percentage bears a close relation to the 30 per cent of gross revenues which the Egyptian instrument sets aside for capital and profit—25 per cent of gross to the former and 5 per cent to the latter. Seen together, the current qualified freeze of tolls and the allocation of percentages of gross revenues to designated purposes point to an intent to limit profits generally to the royalty figure. The remainder of the gross revenues would necessarily have to cover the costs of operation and maintenance and depreciation, as well as, presumably, the return and costs of capital in the form of the amounts paid to the Compagnie Universelle du Canal Maritime de Suez (now the Compagnie Financière de Suez) in settlement of its claim for the nationalization of the Canal Company's assets in the United Arab Republic.[86] If the return of capital, the costs of capital, operation and maintenance, and depreciation were therefore to be taken as the base for the computation of tolls, the gross revenues necessary to provide the stipulated amounts for the Suez Canal Capital and Development Fund and the royalty would have to exceed the costs included in the base by 30/70ths, approximately 43 per cent of that base. This interpretation of the Egyptian Declaration of 1957 is submitted with some diffidence in view of the absence of any information bearing on the precise ambit of the expressions used in the instrument. Consideration along these lines would be requisite, it must be emphasized, only if the level of tolls established by the agreement of 1936, as augmented by an increase of 1 per cent per year, *semble* as from 1956, should be insufficient to cover the return and cost of capital plus the expenses of operation and maintenance of the Canal, while yielding the 30 per cent of gross revenues allocated to special purposes.

In sharp contrast to the imprecise standards of the treaties regulating interoceanic canals, the actual tolls to be assessed for passage through the Turkish Straits are stipulated in detail in the governing instrument, the Montreux Convention. These tolls, which cover the cost of sanitary control stations, lighthouses, light and channel buoys, and lifesaving services, may

---

[85] Compagnie Universelle, The Suez Canal Company and the Decision Taken by the Egyptian Government on 26th July 1956 (26th July-15th September 1956) 49 (1956).

[86] Agreement with the Government of the United Arab Republic in respect of Compensation to be Paid to the Company, signed at Geneva, July 13, 1958, Compagnie Financière de Suez, Bulletin, No. 1, at 21 (1958).

not be changed except with the consent of the parties to the treaty.[87] Under the former régime of the Danube, tolls might be assesed only to cover "works of improvement" and the maintenance of "works of improvement of special importance," and such receipts could be applied only to the works for which they were imposed.[88] The Treaty of 1948 authorizes navigation dues, the scale of which is to be fixed "in relation to the cost of maintenance of equipment and the cost of the works"; these tolls must not be a source of profit.[89]

The alternative to legal resolution of the level of tolls by international agreement is the establishment by municipal legislation of the charges for the use of the waterway. Because of the extent to which the tolls for the Panama Canal have been subjected to judicial and legislative scrutiny, particular interest attaches to the basis upon which these charges are computed.

The statutory formula applicable to the Panama Canal since 1950 has been: "Tolls shall be prescribed at rates calculated to cover, as nearly as practicable, all costs of maintaining and operating the Panama Canal, together with the facilities and appurtenances related thereto, including interest and depreciation, and an appropriate share of the net costs of operation of the agency known as the Canal Zone Government."[90] It is expressly provided that the collection of tolls is subject to the provisions of the treaties with Great Britain, Panama, and Colombia.[91] This basis for the assessment of tolls, which was further particularized in other provisions of the act of 1950 to which reference is made below, was designed to put the Canal upon a sounder financial footing, while, through various accounting devices, keeping tolls from rising to a level which would reflect the full cost of operating the Canal.[92] Of the factors entering into the computation of tolls, the costs of operation and maintenance, interest on capital, and depreciation are readily identifiable in the statutory formula. No express provision of the statute requires that in the fixing of tolls account be taken of reimbursement to the Treasury of capital

[87] Convention concerning the Régime of the Straits, signed at Montreux, July 20, 1936, art. 2 and Annex I, 173 L.N.T.S. 213, 7 HUDSON, INTERNATIONAL LEGISLATION 386 (1941).

[88] Convention Instituting the Definitive Statute of the Danube, signed at Paris, July 23, 1921, arts. 16 and 18, 26 L.N.T.S. 173.

[89] Convention regarding the Régime of Navigation on the Danube, signed at Belgrade, Aug. 18, 1948, arts. 35 and 37, 33 U.N.T.S. 196 at 213.

[90] Act of Sept. 26, 1950, § 12, 64 Stat. 1043, now 2 C.Z.C. § 412(b) (1962).

[91] 2 C.Z.C. § 412(d) (1962).

[92] S. REP. No. 2531, 81st Cong., 2d Sess. 4 (1950); *Hearing on H.R. 8677 Before a Subcommittee of the Senate Committee on Armed Services,* 81st Cong., 2d Sess. at 7 (1950).

supplied by the United States for the construction of the Canal. Such repayment comes, if at all, only in the form of liquidating dividends paid to the
Treasury by the Panama Canal Company out of its retained revenues.[93]
These dividends are to be paid only after the Company has set aside from its
net income sufficient funds to meet "its necessary working capital requirements, together with reasonable foreseeable requirements for authorized
plant replacement and expansion." The provision of such reserves for expansion and improvement is not obligatory and is wholly contingent upon
the availability of funds over and above those stipulated by statute as required for operation and maintenance, interest, and depreciation.

The financial history of the Panama Canal gives every indication that
the United States Government will never be in the position of recouping its
initial investment in the Canal through liquidating dividends while at the
same time securing assets equivalent to that investment. A total of $48,994,010
has been paid to the United States as liquidating dividends;[94] the net direct
investment, as to which interest must be paid, stood at $330,465,010 at the
end of fiscal year 1962.[95]

The tolls levied on vessels passing through the St. Lawrence Seaway, a
waterway which by reason of its extensive canalization merits comparison
with the interoceanic canals, are required by law to be fixed with a view to
covering "as nearly as practicable" all costs of operating and maintaining the
works administered by the Seaway Corporation "including depreciation,
payment of interest on the obligations of the Corporation, and payments in
lieu of taxes." The rates are also to be such as to permit the amortization of
the cost of the waterway and other debts of the Corporation over a maximum
of fifty years.[96]

A common characteristic of the two formulas for tolls to which reference

[93] 2 C.Z.C. § 70 (1962); concerning the dividends of $15,000,000 paid for fiscal
years 1955 and 1956, see FOURTH ANNUAL REPORTS OF THE PANAMA CANAL COMPANY
AND THE CANAL ZONE GOVERNMENT, 1955, at 5 (1956); and FIFTH ANNUAL REPORTS,
1956, at 5 (1957). The Act of Aug. 25, 1959, 73 Stat. 428, provided that $10,000,000 of
retained revenues were to be deemed to be paid into the Treasury as a dividend, and
the Canal Company was in turn empowered to borrow up to an equivalent amount from
the Treasury. 2 C.Z.C. § 71 (1962).

[94] In addition to the amounts which are mentioned in note 93 *supra*, a total of
$23,994,905 had been paid as dividends prior to June 30, 1951. ANNUAL REPORT OF THE
PANAMA CANAL COMPANY AND THE CANAL ZONE GOVERNMENT FOR THE FISCAL YEAR
1958, at 50 (1959).

[95] ANNUAL REPORT OF THE PANAMA CANAL COMPANY AND THE CANAL ZONE GOVERN-
MENT FOR THE FISCAL YEAR 1962 at 48 (1962).

[96] St Lawrence Seaway Act, § 12(b)(4) and (5), 68 Stat. 97, 33 U.S.C. § 988(b)(4)
and (5). A similar formula appears in the Canadian legislation, The St. Lawrence

has just been made is that the rate is set in terms of a minimum and that the requirement is established that the tolls be sufficient to cover at least certain stipulated charges, such as depreciation, interest, and cost of operation and maintenance. The heavy usage of both waterways by shipping of the United States and of Canada and the adverse impact of excessively high tolls upon the economies of the proprietor nations explain the legislative preoccupation with keeping tolls at a reasonable level. The fiscal regulation of the two waterways thus does not offer any real guidance as to the permissible upper limit on tolls, but there is at least a suggestion in the legislation relating to the Panama Canal and to the St. Lawrence Seaway that tolls must do no more than satisfy the statutory minimum. No "profit" in the ordinary sense is anticipated in either case, the "dividends" in the case of Panama being applied, as has been shown, to a repayment of capital costs.

No clear picture of effective legal controls upon the level of tolls emerges from this survey. The Kiel Canal has been operated at a loss. Tolls for Panama purport to reflect the actual cost of operating the Canal (together with some provision for working capital), and a number of accounting devices have also been adopted to make the load lighter for the user. Suez alone provides a profit under an instrument which permits an increase of 1 per cent per year "any increase beyond that level to be the result of negotiations." If an attempt should be made to increase the level of tolls beyond the permissible 1 per cent per year and the matter were to be put to arbitration, as provided in the Egyptian Declaration of April 24, 1957,[97] the tribunal might well decide that the tolls should be fixed so that the ratio of the royalty of 5 per cent of

---

Seaway Authority Act, 15 & 16 Geo. 6, c. 24, § 16; see Craig, *Legal Aspects of Toll Fixing on the St. Lawrence Seaway*, 38 CHICAGO BAR RECORD 351 (1957).

The Comptroller General has rendered a decision that aids to navigation are to be included amongst the costs of the Seaway under this legislation and that "all costs related to the construction, operation, and maintenance of the Seaway should be recovered from the tolls collected." Comp. Gen. Dec. No. B-215817, Nov. 2, 1955, quoted in *Report to the President on the Status and Progress of the Saint Lawrence Seaway for the Fiscal Year Ended June 30, 1955*, 34 DEP'T STATE BULL. 215, 217 (1956).

For the St. Lawrence Seaway Tariff of Tolls see Exchange of Notes between the United States of America and Canada regarding the St. Lawrence Seaway Tariff of Tolls, signed at Ottawa, March 9, 1959, 10 U.S.T. 323, T.I.A.S. No. 4192. The actual revenues of the Saint Lawrence Seaway Development Corporation have been lower than anticipated and will not be sufficient to cover interest as well as expenses of operation and maintenance. An increase in tolls may therefore be called for when tolls are reviewed in 1964. See SAINT LAWRENCE SEAWAY DEVELOPMENT CORPORATION, ANNUAL REPORT, 1961, at 6 (1962); *id.*, 1962, at 21-24; N.Y. Times, June 21, 1962, p. 50M, col. 6.

[97] Declaration cited *supra* note 81, para. 3(b).

the gross revenues to costs should be maintained at the point which prevailed at the time of the Declaration.

It is, however, safe to conclude from this limited evidence of the law that no objection, grounded in customary or conventional international law, could be raised to tolls designed to provide sufficient funds for reimbursement of costs of capital, provision of necessary reserves for improvements, depreciation, and costs of current operation and maintenance. If tolls were to be set at a level high enough to provide in addition for the return of capital invested, the question might well be raised why the operator or proprietor of the canal should receive repayment of these costs while retaining the facilities. Nevertheless, neither customary nor conventional international law appears to provide a basis for challenging tolls reflecting return of capital. It is submitted, moreover, that charges for such a purpose are proper. They would reflect the fact that the waterway is permanent in nature, has as its purpose the provision of a public service, and is not held as an asset for possible sale on the market. Once invested capital had been returned to the entity furnishing it, it would follow that no case could be made for retaining tolls at a level dictated by the necessity of securing sufficient funds to permit payment of costs of capital and return of capital. A corresponding reduction in tolls would thus be in order.

Beyond that point, no certain conclusions may be drawn. It may nevertheless be hazarded that profits of any interoceanic canal running at a rate above 5 per cent of gross revenues, the present upper limit for the Suez Canal, might justifiably be questioned as going beyond the permissible range. With the Kiel and Panama canals not fully self-sustaining, Suez alone is left as a possible source of controversy, and here the maximum degree of profit seems to be reflected in the provision relating to a royalty of 5 per cent, at least so long as improvements remain to be made in the waterway.

SPECIAL PROBLEMS OF DETERMINING COSTS

Even a precise formula which links the charges for the use of the waterway to the costs of constructing, operating, and maintaining it may leave unanswered a further troublesome question—that of determining what expenses should be regarded as costs properly allocable to the user of the waterway through the tolls which he pays. Two illustrations drawn from recent controversies concerning the Panama and Suez canals will demonstrate the dimensions of the problem.

In 1950, the Congress of the United States segregated three aspects of the operation of the Panama Canal in order to provide a more accurate picture of the economy of the Canal, which might in turn serve as a rational basis for an increase in tolls.[98] These three functions were the provision of transit, the conduct of business activities, and the civil government of the Canal Zone. The first two responsibilities were vested in the Panama Canal Company, a public corporation wholly owned by the United States Government and formed from the old Panama Railroad Company.[99] The third activity was left, as before, to the Canal Zone Government. The effect of the provision of the Canal Zone Code bearing on the computation of tolls, which is quoted above, was to require that an "appropriate share" of the cost of that government be borne by both the transit and business sides of the Panama Canal Company. In the determination of that share, "substantial weight" was to be given to "the ratio of the estimated gross revenues from tolls to the estimated total gross revenues of the Panama Canal Company exclusive of the cost of commodities resold."[100] Besides this allocation of part of the cost of the Canal Zone Government to the supporting business activities of the Canal Company, the costs upon which tolls were to be computed were reduced by lowering the rate of interest paid to the Treasury on the cost of the Canal,[101] by eliminating interest for the construction period from the capital investment on which interest was paid,[102] and by requiring either the payment of tolls or the entry of equivalent credits on the books of the Company for transits of ships owned by the United States Government.[103]

In 1955, the Comptroller General complained that the Panama Canal Company had failed to allocate the costs of the Canal Zone Government, corporate overhead, and interest payments among the various activities of the Company and had improperly charged losses from the supporting business

[98] Act of Sept. 26, 1950, 64 Stat. 1038 (codified in scattered sections of 2 C.Z.C. (1962)); S. Rep. No. 2531, 81st Cong., 2d Sess. 4-5 (1950); and see Panama Canal Co. v. Grace Line, Inc., 356 U.S. 309, 316 (1958).

[99] 2 C.Z.C. §§ 61, 62 (1962).

[100] 2 C.Z.C. § 412(b) (1962).

[101] 2 C.Z.C. § 62(e) (1962).

[102] 2 C.Z.C. § 412(e) (1962). Bills were introduced in 1956 which would have required that interest during the period of construction of the Canal be included in the investment on which interest is paid and that provision be made for the amortization of what had hitherto been considered nondepreciable assets. Both bills failed of enactment. See *Hearings on H.R. 5732 and H.R. 5733 before the Subcommittee on Panama Canal of the House Committee on Merchant Marine and Fisheries,* 84th Cong., 2d Sess. (1956).

[103] 2 C.Z.C. § 412(c) (1962).

activities of the Company against the income from the "canal activity." It was his interpretation of the act of 1950 that tolls were to be computed only with regard to the transit operations of the Company and that the existing tolls had consequently been set at too high a rate.[104] In turn, this construction of the act was made the basis for a recommendation that the statute should be amended expressly to authorize the charging of any losses sustained by the supporting activities against the income of the Canal.[105] The position taken by the Panama Canal Company was that much of the losses sustained by the supporting business activities was the result of "fringe benefits" to the employees and users of the Canal and were therefore, under the terms of the Act of 1950, properly includable in the costs of operating the Canal.[106]

The views of the Comptroller General were eagerly seized upon by a number of the shipping companies employing the Canal as the basis for a suit against the Panama Canal Company, grounded on what were contended to be excessive exactions of tolls from time of the coming into force of the 1950 legislation.[107] The relief sought was reimbursement of the excess of the tolls over what the plaintiffs maintained was their proper level and a decree directing the Company to hold hearings and to prescribe tolls in conformity with law. The stipulation of the Hay-Pauncefote Treaty regarding "just and equitable" tolls was one of the provisions of law to which the plaintiffs appealed.[108] The outcome of the litigation, which eventually made its way to the Supreme Court of the United States, was to deny any remedy to the plaintiffs.[109] The problem, in the words of Mr. Justice Douglas, speaking for a unanimous

---

[104] Report on *Audit of Panama Canal Company and the Canal Zone Government for the Fiscal Year Ended June 30, 1954,* H. Doc. No. 160, 84th Cong., 1st Sess. 2-3 (1955), quoted in Grace Line, Inc. v. Panama Canal Co., 243 F.2d 844, 847 (2d Cir. 1957).

[105] Report cited *supra* note 104; the recommendation was renewed in the audit for 1955. Grace Line, Inc. v. Panama Canal Co., 243 F.2d 844, 848 (2d Cir. 1957).

[106] Panama Canal Co. v. Grace Line, Inc., 356 U.S. 309 (1958), Brief for the Panama Canal Co., p. 21, n. 49. The report of the General Accounting Office considered that the costs of supporting business activities *should* be regarded as an element of the cost of operating the Canal but denied that the existing legislation so provided. The General Accounting Office has persisted in its efforts to have the law amended in such a way as to clarify this point. *Hearings on H.R. 8983 and H.R. 10968 before the Subcommittee on Panama Canal of the House Committee on Merchant Marine and Fisheries,* 86th Cong., 2d Sess. 99-102 (1960).

[107] Grace Line, Inc. v. Panama Canal Co., 143 F. Supp. 539 (S.D.N.Y. 1956), rev'd, 243 F.2d 844 (2d Cir. 1957), rev'd 356 U.S. 309 (1958).

[108] Panama Canal Co. v. Grace Line, Inc., 356 U.S. 309 (1958), Brief for Grace Line, *passim.*

[109] Panama Canal Co. v. Grace Line, Inc., *supra* note 108.

Court, was "in the penumbra of the law where generally the Executive and the Legislative [*sic*] are supreme."[110] So many questions of judgment were involved that the question of tolls was better left, at that stage at least, to agency discretion. The political aspect of the matter before the Court was further emphasized by the fact that the Congress appeared to have acquiesced in the accounting procedures followed by the Panama Canal Company and objected to by the Comptroller General.[111]

The issues involved in this litigation have been described at some length not because they constitute any contribution to the law relating to the international status of waterways but only as illustrative of the difficulties implicit in the bare concept of "cost." A properly drafted international instrument would require some fuller explanation of the term.

The question of the costs properly allocable to the users of a waterway has arisen in an international context, but with inconclusive results, in connection with the clearing of the Suez Canal. The cost of this operation, which exceeded $8,000,000, was initially borne by the United Nations.[112] Reimbursement of the United Nations was secured through a 3 per cent surtax on Canal tolls, in accordance with a resolution of the General Assembly of December 14, 1957.[113] This was intended to provide an amount sufficient to pay off the debt in about three years.[114] The surtax can more correctly be identified as a contribution than as a charge, since the United Nations had no power to compel the payment of the surtax and the transit of vessels was not made conditional upon its payment.

The assessment was not paid by the shipping of some countries on the

[110] *Id.* at 317.

[111] *Id.* at 319. Concerning further attempts of the shipping trade to subject the fixing of tolls to administrative and judicial review, see *Hearings on H.R. 8983 and H.R. 10968 before the Subcommittee on Panama Canal of the House Committee on Merchant Marine and Fisheries,* 86th Cong., 2d Sess. (1960).

[112] Loan advances were made by eleven states to cover the expenditures made by the United Nations; the funds had been deposited with the International Bank for Reconstruction and Development, which acted as fiscal agent for the United Nations. Clearance of the Suez Canal, Report of the Secretary-General 12-13 (A/3719) (1957).

[113] Resolution 1212 (XII), U.N. GEN. ASS. OFF. REC. 12th Sess., Supp. No. 18, at 59 (A/3805) (1957).

[114] The Reimbursement of the Cost of Clearing the Suez Canal, Report by the Secretary-General, Aug. 1, 1958 (A/3862) (1958). Thanks to the increase in traffic, the surcharges collected paid off the costs of clearance of the Canal six months earlier than had been expected. *Financial Report and Accounts for the Year Ended 31 December 1961 and Report of the Board of Auditors,* GEN. ASS. OFF. REC. 17th Sess., Supp. No. 6, at 7-8 (A/5206) (1962); N.Y. Times, March 31, 1961, p. 2, col. 4.

ground that they should not be required to bear the cost of the blocking of the Canal, which had been one of the consequences of the "aggression" of Great Britain, France, and Israel against Egypt.[115] The shipping companies of other nations maintained that they had been put to sufficient expense by the closing of the Canal without being subjected to the additional burden of clearing a canal for the blocking of which they had no responsibility.[116] Faced with this well-grounded contention and without means of forcing payment, the governments of the United States and Great Britain, among others, undertook themselves to pay the surtax on transits of vessels owned by their nationals.[117] The surtax was substantially different in nature from the surtaxes which had previously been paid on tolls by shipping companies to the former Suez Canal Company[118] in that it represented a special charge for a special service, payment for which was rendered, not to the proprietor of the Canal, but to the United Nations through its collection agent, the Banque de la Société Générale de Belgique of Brussels.

To the extent that the governments assumed an obligation to pay the cost of the surtax, according to the number of transits made by vessels owned by their nationals, the financial arrangements bear some resemblance

---

[115] E.g., the U.S.S.R. and other countries of the Soviet bloc. N.Y. Times, Nov. 1, 1958, p. 1, col. 5.

[116] On April 15, 1958, delegates to the International Chamber of Shipping decided unanimously to support the surcharge, but only on the condition that they would be reimbursed by their governments. Letter from Mr. Barco, Deputy Representative of the United States on the Security Council, to Sir Humphrey Trevelyan, Under-Secretary in Charge of Special Political Affairs, United Nations Headquarters, dated July 1, 1958, on file in the U.S. Department of State; and see 581 H.C. DEB. (5th ser.) 21 (1957-58); N.Y. Times, June 3, 1958, p. 62, col. 6.

[117] In the case of the United States, this was effected through an agreement (incorporated in the letter cited in the preceding footnote) that the United Nations would be allowed to credit against the sum of approximately $3,800,000 owed by the United Nations to the United States on the advance made by that country toward the clearance of the Canal, a sum equivalent to 3 per cent of the tolls paid by United States shipping and trade interests during the period of the surcharge. The balance of the $3,800,000 remaining after the deduction of that sum would be accepted by the United States in full settlement of the advance, which had been made from the President's Special Fund under authority of the Mutual Security Act of 1954, § 401, 68 Stat. 843, as amended. See also N.Y. Times, Aug. 2, 1958, p. 34, col. 5; Aug. 7, 1958, p. 48, col. 4.

On the British undertaking to repay shipowners for the surcharge which they had paid, see 592 H.C. DEB. (5th ser.) *219-20* (1957-58).

[118] See the agreement between M. de Lesseps and Colonel Stokes, on behalf of the British Government, signed on Feb. 21, 1876, regarding the reduction of the surtax, in Further Correspondence respecting the Suez Canal, GREAT BRITAIN, PARLIAMENTARY PAPERS, EGYPT No. 9 (1876) at 25 (C. 1525).

to the system used in financing certain air navigational facilities over the North Atlantic.[119] These cost-sharing arrangements are based on the assumption that Iceland and Denmark should not be required to bear the entire cost of aids to navigation from which foreign aircraft derive far more benefits than do the domestic. The cost formula whereby the expenses of these facilities are borne by the user nations on a voluntary basis involves prorating the costs according to the number of transits made by the aircraft of each participating nation. The amounts so assessed are not passed on to the companies whose planes benefit from these services; the same absorption of the cost by governments was true of the Suez surtax on the majority of transits. In both situations, the execution of the plan has been handicapped by the non-participation of those nations which are willing to have their ships or aircraft benefit from the services without a corresponding payment.[120] Planes of non-participants cannot be stopped from using the navigational facilities of Denmark and Iceland, but the United Arab Republic could, in agreement with the United Nations, have denied passage to vessels in cases in which no provision had been made for payment of the surtax. Presumably, this arrangement was impossible for political reasons.

POLITICAL AND ECONOMIC PRESSURES IN THE ESTABLISHMENT OF TOLLS

In the absence of any stipulations as to the level of tolls under international agreements or municipal law, the question of charges for transit must be left to negotiations between the operator of the waterway and the users. The pressures which the users of the watercourse have brought to bear in order to secure acceptable rates have been either political or economic.

While the Suez Canal was being administered by the Suez Canal Company, in which 44 per cent of the shares were held by the British Government, it was natural that British shipowners should turn to their government for assistance in the lowering of tolls. The type and measure of assistance

[119] The most recent versions of these agreements are Agreement on the Joint Financing of Certain Air Navigation Services in Greenland and the Faroe Islands, signed at Geneva, Sept. 25, 1956, art. VII, 9 U.S.T. 795, T.I.A.S. No. 4049; and Agreement on the Joint Financing of Certain Air Navigation Services in Iceland, signed at Geneva, Sept. 25, 1956, art. VII, 9 U.S.T. 711, T.I.A.S. No. 4048.

[120] See Final Act of the ICAO Conference on Air Navigation Services: Greenland and Faroe Islands, London, 1949, app. I, para. 12, in International Civil Aviation Organization, *Report on the ICAO Conference on Air Navigation Services: Greenland and the Faroes* 24, 28 (ICAO Doc. No. 7103-JS/552) (1951); and *id.* at 17, regarding negotiations with Switzerland, Iceland, and Mexico.

which that government would lend to these companies have undergone marked variation. During the controversy with De Lesseps which followed the meeting of the International Commission of Tonnage at Constantinople in 1873, it was a representative of the British Government who dealt with De Lesseps on the implementation of the decisions of the Constantinople Commission, and it was the British Government which secured the agreement of the other powers to the arrangements which its representative had been able to make with the Suez Canal Company.[121] When new negotiations were undertaken on behalf of the shipowners of Great Britain in 1883, the parties to the agreement providing for the reduction in dues were the Suez Canal Company and the Association of Steam-ship Owners Trading with the East, but the agreement was submitted to the Government for approval.[122] In later years the attitude of the British Government has been that it was more important to protect the revenues derived from dividends on the shares which it held than it was to ease the financial burden which the tolls imposed on the owners of vessels plying the Canal and on those who depended on the Canal for import and export of goods. In the words of the British Suez Canal Directors, which were approved by the Treasury and the Foreign Office: ". . . [W]e have, as the representatives of the financial interests of His Majesty's Government, to protect the large revenue which now accrues to the Exchequer; and we submit that any further reductions of the Tariff would practically amount to a subsidy to ships using the Canal, at the cost, to a great extent, of pecuniary loss to His Majesty's Government."[123] Again, during the depression of the 1930's, the British Government once more declined to intervene with the Company in order to secure a reduction in tolls

[121] Further Correspondence respecting the Suez Canal, GREAT BRITAIN, PARLIAMENTARY PAPERS, EGYPT No. 9 (1876) at 25 (C. 1525).

[122] Correspondence respecting the Suez Canal, GREAT BRITAIN, PARLIAMENTARY PAPERS, EGYPT No. 3 (1884) at 50 (C. 3850). This agreement was concluded after the withdrawal from the Parliament of the Heads of Agreement with De Lesseps, which not only dealt with tolls but also would have pledged the British Government to political and financial support of a second canal. WILSON, THE SUEZ CANAL: ITS PAST, PRESENT, AND FUTURE 67-76 (1933).

The agreements between the British Government and the Suez Canal Company and those between the British shipping companies and the Canal Company may justifiably be referred to as "transnational agreements" in the useful terminology of Professor Jessup in TRANSNATIONAL LAW (1956).

[123] British Suez Canal Directors to Sir Edward Grey, Aug. 31, 1906, in Correspondence relating to Suez Canal Dues, GREAT BRITAIN, PARLIAMENTARY PAPERS, COMMERCIAL No. 2 (1907), at 3 (CD. 3345).

favorable to the shipping trade.[124] Once the assets of the Suez Canal Company in Egypt were nationalized, the commercial stake of the British Government in the Canal lost its relevance, and that government, having been relieved of its conflict of interest, was once more ready to deal with the Egyptian Government on the question of tolls to be levied on British and other users of the waterway.

The principal economic weapon of the user is his ability to resort to other means or routes of transport. One healthy lesson of the blockage of the Suez Canal was that it was possible, although at a considerable expenditure of funds and with a great dislocation of the normal pattern of trade, to forego the use of that route over a period of months. In retrospect, economic pressure in the form of reduced use of the Canal or complete abstention from its use would have been a more effective measure to take against President Nasser's nationalization of the Suez Canal than were the protest meetings in London, which were held to so little effect. An enhanced reliance on pipelines for the shipment of oil might also serve as an economic weapon against an increase in tolls. They are, however, channels of commerce perhaps more vulnerable to interdiction in certain respects than the Suez Canal itself.

The necessity of making shipment via the Panama Canal at least competitive in cost with transcontinental rail transport has operated as a restraint on Canal tolls.[125] Rail competition also plays an important part in the determination of tolls on the St. Lawrence Seaway. The St. Lawrence Seaway Development Corporation has been required to reconcile[126] the competing demands of the railroads for high tolls, designed to discourage the use of the Seaway, and of the maritime shipping trade, which has an interest, hardly requiring explanation, in setting tolls at as low a level as possible.[127] The ports which

---

[124] 267 H.C. DEB. (5th ser.) 169 and 919 (1931-32); WILSON, *op. cit. supra* note 122, at 152-69.

[125] See the testimony of a representative of the Association of American Railroads concerning the Panama Canal legislation of 1950, *Hearings on H.R. 8677 before the Subcommittee on the Panama Canal of the House Committee on Merchant Marine and Fisheries,* 81st Cong., 2d Sess. 128-44, especially at 142 (1950).

[126] Because of the necessity of competing with rail and truck carriers, domestic package freight is treated as "bulk cargo" in order to benefit from the lower rate which is provided for such cargo, as contrasted with general cargo. Saint Lawrence Seaway Development Corporation, Report of Tolls Committee, June 12, 1958, p. 6, annexed to Press Release SLSDC-141, June 18, 1958.

[127] Hayes, *Seaway Tolls and Freight Rates: As Seen by Water Carriers,* 38 CHICAGO BAR RECORD 358 (1957); Prince, *Seaway Tolls and Freight Rates: As Seen by Rail Carriers, id.* at 363; see N.Y. Times, Aug. 7, 1958, p. 48, col. 2, regarding hearings held on tolls for the Seaway.

stand to lose or gain business from the Seaway have allied themselves with the rail and maritime groups.

In view of the relatively slight difference in sailing time which results from the use of the Kiel Canal, tolls must be fixed at such a level as not to frighten off shipping from the use of that waterway. It is not without significance that the Kiel and Panama canals, which are subject to competition from other forms and routes of transport, are not wholly self-sustaining, while the Suez Canal, which offers a great financial saving over any alternative route, has been for decades, and remains, a highly profitable enterprise.

FINANCIAL RELATIONS BETWEEN THE OPERATOR
AND THE TERRITORIAL SOVEREIGN

If the operator of an interoceanic canal is an entity separate and apart from the sovereign of the territory through which the channel passes, the territorial sovereign has considered itself justified in asserting a claim to payment by the proprietor for the privilege of operating the waterway. This compensation may, in one aspect, be regarded as rent for the use of the territory needed for the exploitation of the waterway and compensation for the other rights accorded the operator. This is the basis upon which the United States paid the Republic of Panama the sum of $10,000,000 as one of the terms of securing the Hay–Bunau-Varilla Treaty in 1903.[128] The United States has also been required to make an annual payment to Panama in amounts periodically revised by treaty and now amounting to nearly $2,000,000 a year.[129]

At the time of the nationalization of its assets in Egypt, the Suez Canal Company was paying over 7 per cent of its gross profits to the Egyptian Government under the agreement reached with that government in 1949. This "allocation" was in substitution for previous "allocations" at a flat rate in Egyptian pounds, which the Government had been able to secure from the Company.[130] From the perspective of the Company, there should be in-

[128] Convention between the United States of America and the Republic of Panama for the Construction of a Ship Canal to Connect the Waters of the Atlantic and Pacific Oceans, signed at Washington, Nov. 18, 1903, art. XIV, 33 Stat. 2234, T.S. No. 431.

[129] Treaty of Mutual Understanding and Cooperation between the United States of America and the Republic of Panama, signed at Panamá, Jan. 25, 1955, art. I, 6 U.S.T. 2273, T.I.A.S. No. 3297.

[130] Exchange of notes between the President of the Compagnie Universelle du Canal Maritime de Suez and the Egyptian Minister of Commerce and Industry, March 7, 1949, COMPAGNIE UNIVERSELLE, RECUEIL DES ACTES CONSTITUTIFS 241.

cluded within the payments to the Egyptian Government the 15 per cent of the net revenues of the Company which the Egyptian Government had been forced to sell in 1880 and which were thereafter vested in the Société civile pour le recouvrement des 15% des produits nets de la Compagnie du Canal de Suez, attribués au Gouvernement égyptien.[131] The grant of 7 per cent of the gross profits of the Company was actually a response to the demand of Egypt to participate in the profits of the Company and was secured through the power which the Egyptian Government held in fact and in law over the Company. Leaving aside for the moment the question of the power of the Egyptian Government to subject the Company to legal obligations to which it had not given its consent, it would be difficult to distinguish this "allocation" from a tax levied upon the gross profits of the Company.

The lesson to be learned from the history of these payments by the proprietor to the territorial sovereign is that, despite the existence of agreements specifying the amount of the annual levy, the matter will probably be reopened from time to time by the territorial sovereign as the canal prospers or the financial needs of the territorial sovereign increase. The annual payment of the United States to Panama was fixed at $250,000 in 1903.[132] By the Treaty of 1936, it became $430,000,[133] and by the Treaty of 1955, $1,930,000.[134] To give this last figure an air of finality, a paragraph was added to the article by which the annuity was increased, in which it was said that "Notwithstanding the provisions of the Article, the High Contracting Parties recognize the absence of any obligation on the part of either Party to alter the amount of the annuity."[135] In 1961, President Chiari of Panama made it plain that his country was dissatisfied with the present annuity and was prepared to ask for more.[136]

The payments by the Suez Canal Company to Egypt mounted in pre-

---

[131] SAINT VICTOR, LE CANAL DE SUEZ 217-20 (1934); COMPAGNIE UNIVERSELLE, RENSEIGNEMENTS CONCERNANT LES TITRES DE LA COMPAGNIE 32-33 (1953).

[132] By the Hay–Bunau-Varilla Treaty of 1903, *supra* note 128, art. XIV.

[133] General Treaty of Friendship and Cooperation between the United States of America and Panama, signed at Washington, March 2, 1936, art. VII, 53 Stat. 1807, 1818, T.S. No. 945.

[134] Treaty cited *supra* note 129, art. I.

[135] The Explanatory Statement accompanying the treaty upon its submission to the Senate by the President hopefully stated that the provision would safeguard the United States "against any assertion of a right to demand an increase in the annuity." *Hearings on the Panama Treaty before the Senate Committee on Foreign Relations,* 84th Cong., 1st Sess. 5 (1955).

[136] N.Y. Times, Sept. 12, 1961, p. 15, col. 5.

cisely the same manner. After the alienation of its 15 per cent share in the net profits of the Company in 1880, the Egyptian Government was not readmitted to participation in the profits of the enterprise until 1936, when the yearly payment was fixed at £E 200,000.[137] The amount rose to £E 300,000 in the following year.[138] In 1949, the annuity became 7 per cent of the gross profits, with a minimum of £E 350,000.[139] Moreover, these figures do not give a complete picture of the economic advantages flowing to the territorial sovereign, for the instruments mentioned accorded other benefits as well, such as the retrocession of lands under the Treaty of 1955 between Panama and the United States[140] and the obligation to construct a fishing port undertaken by the Suez Canal Company in its agreement with Egypt in 1949.[141]

While these progressively increased payments suggest that the agreements between territorial sovereign and operator are somewhat more mutable than they purport to be, there is some merit in the periodic renegotiation of the terms upon which the proprietor exercises its functions. An agreement indefinite in duration may prove to be unsuited to later conditions. The value of money changes, the fortunes of the proprietor wax and wane, functions which once seemed appropriate for discharge by the operator of the waterway later appear to be ones which might equally well be performed by the territorial sovereign. The wiser course would appear to be the recognition of change by periodic revision of certain of the arrangements between the two parties and by express provision to that effect in the agreement.

At the same time that the territorial sovereign secures certain financial benefits from the proprietor of an interoceanic canal, it normally accords certain exemptions from its fiscal laws to the operator of the waterway. One of

---

[137] Exchange of notes between the President of the Compagnie Universelle du Canal Maritime de Suez and the President of the Egyptian Council of Ministers, April 27-30/ May 4, 1936, COMPAGNIE UNIVERSELLE, RECUEIL DES ACTES CONSTITUTIFS 210.

[138] Exchange of notes between the Agent Supérieur of the Compagnie Universelle du Canal Maritime de Suez and the Egyptian Minister of Finances, June 11/14, 1937, *id.* at 213.

[139] Exchange of notes cited *supra* note 130; COMPAGNIE UNIVERSELLE, ASSEMBLÉE GÉNÉRALE DES ACTIONNAIRES, 19ÈME RÉUNION, 21 JUIN 1949, RAPPORT PRÉSENTÉ AU NOM DU CONSEIL D'ADMINISTRATION 6 (1949).

[140] Treaty cited *supra* note 129, art. V, and see the testimony of Assistant Secretary Holland, in *Hearings on the Panama Treaty before the Senate Committee on Foreign Relations,* 84th Cong., 1st Sess. 55 (1955).

[141] Convention between the Egyptian Government and the Compagnie Universelle du Canal Maritime de Suez, signed at Cairo, March 7, 1949, art. 15, COMPAGNIE UNIVERSELLE, RECUEIL DES ACTES CONSTITUTIFS 245, 251.

the most important of these relates to customs duties on goods imported for the construction, maintenance, administration, and improvement of the waterway. Under the treaty concluded in 1903 whereby Panama granted to the United States the right to construct a transisthmian canal, such an exemption was accorded as to goods and equipment imported by the United States into the Canal Zone and its "auxiliary lands" either for use in connection with the canal itself or for the maintenance of the personnel employed by the United States and their families.[142] The exemption was later extended to cover import duties and taxes on goods destined for agencies of the United States and for certain categories of authorized personnel when these agencies and persons were within the Republic of Panama.[143] The 1955 treaty, like that of 1936, restricted the categories of persons who were authorized to purchase goods imported duty-free by the United States or themselves to import goods without duty.[144] As the situation stands today, the customs exemption of the United States is as it was established in 1901, while the benefits of customs-free importation of goods destined for use or consumption by individuals now flow only to members of the armed forces and to United States citizens. The termination by the Treaty of 1955 of the right of Panamanian employees to purchase goods imported free of duty and then sold through the commissaries and sales stores was a concession to local Panamanian merchants,[145] which imposed on the United States the obligation of providing remuneration to compensate for the abolition of the fringe benefit of buying at these stores.

In the Concession of 1856, the Suez Canal Company secured the right to import all goods needed for the construction and exploitation of the Canal free of customs duties.[146] In 1869, when the Company was in financial difficulties, it surrendered its right of exemption and secured from Ismail, in return for renunciation of this and other rights and for certain immovable

---

[142] Convention between the United States of America and the Republic of Panama for the Construction of a Ship Canal to Connect the Waters of the Atlantic and Pacific Oceans, signed at Washington, Nov. 18, 1903, art. XIII, 33 Stat. 2234, T.S. No. 431.

[143] Treaty cited *supra* note 133, art. IV.

[144] Treaty cited *supra* note 129, art. XII; for the restrictions introduced by the Treaty of 1936, *supra* note 133, see article III.

[145] Explanatory Statement accompanying message from the President to the Senate, May 9, 1955, in *Hearings on the Panama Treaty before the Senate Committee on Foreign Relations,* 84th Cong., 1st Sess. 10, 15, 16 (1955).

[146] Firman de Concession et Cahier des Charges pour la construction et l'exploitation du Grand Canal Maritime de Suez et dépendances, Jan. 5, 1856, art. 13, in COMPAGNIE UNIVERSELLE, RECUEIL DES ACTES CONSTITUTIFS 6 at 9.

property of the Company, the sum of 30,000,000 francs.[147] The customs exemption was again the subject of bargaining in 1902, when the Company recovered its customs exemption in compensation for the improvement of the Ismailia-Port Said railway and the extension of the port in Port Said.[148] The exemption was given a broad scope in order to cover imports in connection with all of the varied activities carried on by the Company but did not extend to importations by the employees of the enterprise.[149]

Corresponding to the exemption from customs duties accorded the United States and its employees in the Canal Zone is an immunity from taxes recognized by the initial treaty with Panama. That immunity, which was accorded to Panamanian and non-Panamanian employees of the Canal, the railroad, and auxiliary works, extended both to property taxes and "contributions or charges of a personal character of any kind."[150] In accordance with a general policy of making privileges and immunities depend primarily on nationality rather than upon place of residence or upon employment by the United States, the exemption was in 1955 limited to members of the United States Armed Forces, citizens of the United States employed by the Canal, the railroad, or auxiliary works, and nationals of third states employed by the United States and resident in the Canal Zone.[151] The removal of exemptions for Panamanian employees of the United States was again a direct financial concession by the United States to the benefit of the Republic of Panama.

Subject to minor exceptions,[152] the Suez Canal Company was subject to taxation by Egypt on its activities in that country to the same degree as any other corporation doing business in Egypt. The tax with the heaviest impact on the company was the *impôt sur les bénéfices commerciaux et industriels* (tax on undistributed profits). The Company was also, like other Egyptian corporations, required to collect the Egyptian tax on the sums distributed by the Company to its shareholders, which might also be subjected to further taxation by the countries of residence of those shareholders.[153] But this prob-

---

[147] Première Convention du 23 avril 1869, art. I, *id.* at 48; see Hallberg, The Suez Canal: Its History and Diplomatic Importance 217, n. 3 (1931).

[148] Convention du 1er février 1902, art. 11, in Compagnie Universelle, Recueil des actes constitutifs 81 at 83.

[149] Agreement between the Egyptian Government and the Compagnie Universelle du Canal Maritime de Suez, May 2, 1936, para. 4, *id.* 214 at 215.

[150] Convention cited *supra* note 142, art. X.

[151] Treaty cited *supra* note 129, art. II.

[152] See, concerning the arrangements between Egypt and the Company in 1886 and 1887 concerning the *impôt sur la propriété bâtie,* Compagnie Universelle, Recueil des actes constitutifs 68.

[153] Information furnished by the Suez Canal Company.

lem of double taxation is not peculiar to the Suez Canal Company. The subjection of the Company to Egyptian taxation and a customs exemption secured
in compensation for other obligations undertaken by the enterprise point to
the Egyptian character of the Company for juridical purposes rather than to
any peculiar international status. The general immunity accorded by Panama
to the United States with respect to the Panama Canal is, on the other hand,
no more than a reflection of the fact that the United States is a foreign state
and that it exercises many of the powers of a territorial sovereign within the
Canal Zone. An independent operating agency created for the administration
of an interoceanic canal or other international waterway would have to be
endowed with immunities from taxation and customs resembling those of the
United States, if it were to be effectively divorced from the control of the
nation within the boundaries of which it carried on its functions.

In respect of the application of one other form of Egyptian fiscal regulation, the Suez Canal Company was able to maintain a considerable measure
of independence over an extended period of time. In connection with the
discussion of the legal status of the Suez Canal Company in Chapter II, some
words have already been said about the valuation of the franc in which the
tolls for transit were to be collected and the shareholders to be paid their
dividend.[154] In 1935, Egypt, following the example of most of the other
countries in the world, declared null and without effect gold clauses in contracts which involved international payments and specified amounts in
Egyptian pounds, pounds sterling, or any other currency which was legal
tender in Egypt.[155] In subsequent litigation between a creditor and the Company, the Mixed Court of Appeals held that the decree could not have application to the Company, because the users of the Canal paid tolls in terms of
the gold franc and payments to creditors on the same basis would therefore
constitute no drain upon the foreign-exchange resources of Egypt. The
creditors of the Company would have to be paid on the basis of the gold
franc.[156] The Mixed Court of Appeals held in 1947 that in order to determine
the amounts payable to the creditors and shareholders of the Company in
Egypt, it would be necessary to determine the price in dollars in New York

[154] See p. 61 *supra*.

[155] Décret-loi No. 45 de 1935 sur les Contrats Internationaux, Egypt, JOURNAL
OFFICIEL DU GOUVERNEMENT EGYPTIEN, No. 39, May 4, 1935, (Extraordinaire), EGYPTIAN
GOVERNMENT, MINISTRY OF JUSTICE, TABLE DES LOIS, DÉCRETS ET RESCRITS ROYAUX, 1935,
p. 138.

[156] Crédit Alexandrin v. Compagnie Universelle du Canal Maritime de Suez, Cour
d'Appel Mixte, Feb. 26, 1940, 52 BULLETIN DE LÉGISLATION ET DE JURISPRUDENCE
ÉGYPTIENNES II. 185 (1939-40).

of the amount of gold contained in the gold franc and then convert the dollar amount so computed into its equivalent in lawful Egyptian currency.[157] The effect of these two judgments of the Mixed Court of Appeals was to give the Company a special standing under the Egyptian law abrogating gold clauses in international payments.

In 1947 arrangements were made between the Egyptian Government and the Suez Canal Company—with some objection on the part of the latter to the limited exemption accorded it[158]—which relieved the Company from strict compliance with the Egyptian foreign-exchange control law, which would have required the Company to repatriate to Egypt all of its earnings in foreign currency. Under this agreement, the Company was required to collect tolls in either pounds sterling or dollars. Sterling was to be allowed to remain in an Egyptian account in London, but dollars were required to be sold to a bank in Egypt. It was arranged that payments might be made out of these accounts in order to meet the obligations of the Company and to permit the transfer of funds to the offices of the Company in Egypt, Paris, and London.[159] As the result of the view of the Egyptian authorities that too large a proportion of the Company's funds were being maintained outside Egypt, the Company was forced to conclude an agreement with the Egyptian Minister of Finances and Economy in May of 1956, under which the Company was to invest in Egypt a basic sum of £E 8,000,000, subsequently to be increased by yearly increments. The Egyptian Government promised that the amount brought into Egypt could be transferred out at the termination of the concession, subject to the payment of any obligations which the Company might have outstanding at that time.[160] While this new arrangement would

---

[157] Compagnie Universelle du Canal Maritime de Suez v. Campos, Cour d'Appel Mixte, May 17, 1947, 59 *id*. II. 219 (1946-47).

[158] Ms. letter, Agent Supérieur, Compagnie Universelle du Canal Maritime de Suez, to the Egyptian Controller of Exchange Operations, Sept. 30, 1947, on file in the headquarters of the Compagnie Universelle (now the Compagnie Financière de Suez) in Paris. One of the objections of the Company was that the requirement that shipping not paying tolls in pounds sterling pay only in dollars would lead to a discriminatory practice in violation of the concessions granted it.

[159] Ms. letters, Egyptian Controller of Exchange Operations to the Agent Supérieur, Sept. 18, 1947, and Oct. 13, 1947, on file in the headquarters of the Compagnie Universelle (now the Compagnie Financière de Suez) in Paris.

[160] Ms. letters between the Egyptian Minister of Finance and Economics and the President of the Compagnie Universelle du Canal Maritime de Suez, May 30, 1956, on file in the headquarters of the Compagnie Universelle (now the Compagnie Financière de Suez) in Paris.

have placed an unwelcome burden on the Company, which was in all likelihood in some apprehension about its ability to withdraw the funds in 1968, the special arrangements made for the Canal Company reflect some awareness of its special position and problems. With the nationalization of the Egyptian assets of the Canal Company several months after the 1956 agreement, a similar problem of the effect of foreign-exchange laws is unlikely to arise again in the administration of international waterways. However, the experience gained from the dealings of the two parties may have some relevance to the privileged financial position which must be provided for a public or international operating agency which receives payments from foreign countries and in turn makes payments abroad.

As a general principle, it would appear that no obstacle to freedom of navigation is interposed through a demand by the operator, proprietor, or territorial sovereign of the waterway that tolls be paid in the currency of that state or alternatively in a hard currency. The Suez Canal Company's practice in this respect has its counterpart in the insistence of the United Arab Republic that tolls be paid in dollars or in Swiss francs, a demand which has not gone without protest from countries in the Soviet bloc.[161] The dependence of the United Arab Republic on foreign contractors for the carrying on of maintenance and improvement of the waterway makes it of particular importance that it have available the necessary funds with which to reimburse such enterprises.[162]

[161] N.Y. Times, May 19, 1957, p. 1, col. 8.

[162] In at least one case, the facilities of a foreign government have been employed, as in the hiring for six months of the U.S. Army Corps of Engineers dredge *Essayons* for the dredging of the Canal and its roadsteads. UNITED ARAB REPUBLIC, SUEZ CANAL REPORT, MARCH 1959, under "Canal News" (unpaged).

The International Bank for Reconstruction and Development loaned the Suez Canal Authority $56,500,000 in various currencies at the end of 1959 in order to finance the "cost of goods" required in connection with the deepening of the Canal to 37 feet and other improvements under the "Nasser Program." The loan was guaranteed by the Government of the United Arab Republic. Loan Agreement (Suez Canal Development Project) between International Bank for Reconstruction and Development and Suez Canal Authority, dated Dec. 22, 1959 (Loan Number 243 UAR); and Guarantee Agreement (Suez Canal Development Project) between United Arab Republic and International Bank for Reconstruction and Development, signed in the District of Columbia, Dec. 22, 1959 (Loan Number 243 UAR).

CHAPTER VI

SOME TECHNICAL PROBLEMS OF THE ADMINISTRATION OF INTERNATIONAL WATERWAYS

ANCILLARY FACILITIES AND FUNCTIONS

FREEDOM OF transit through an artificial waterway, guaranteed by the most solemn of promises given by those with any power to obstruct the waterway, can be wholly illusory if the waterway is not maintained and operated in an efficient manner. Vessels of above a certain draft will be unable to traverse the channel if it is not properly dredged.[1] Failure to move shipping through the canal promptly, whether individually or in convoys, can add tremendously to the cost of transit by immobilizing a vessel even for a short time, for the daily operating cost may run into thousands of dollars.[2] However, the maintenance of efficiency of operation is largely a tech-

---

[1] To take the Suez Canal as an example, the Seventh Program of Improvement, the last completed by the Compagnie Universelle du Canal Maritime de Suez, had brought the maximum draft permissible in the Canal to 35 feet. *Note sur le 8ème programme des travaux d'amélioration du Canal de Suez*, Supplement to LE CANAL DE SUEZ, No. 2,306 (Feb. 15, 1955). Silting during the blockage of the Canal in 1956 and 1957 had reduced the figure to 33 feet (CLEARANCE OF THE SUEZ CANAL, REPORT OF THE SECRETARY-GENERAL, Nov. 1, 1957, at 11 (A/3719) (1957)). The Eighth Program of Improvement to be conducted by the old Canal Company looked to a maximum draft of 36 feet (*Note sur le 8ème programme, supra*). The "Nasser Project," the program of improvement undertaken by the Suez Canal Authority of the United Arab Republic, looked to an increase in draft to 37 feet. This objective was achieved in May of 1961. UNITED ARAB REPUBLIC, SUEZ CANAL AUTHORITY, SUEZ CANAL REPORT, 1961, at 37.

[2] At the time of the blocking of the Suez Canal by the collision of a tanker with the El Ferdan Bridge (N. Y. Times, Jan. 1, 1955, p. 1, col. 2), it was estimated that the delay of one liner would cost her owners £3500 per day.

284

nical and engineering problem, and in recent years[3] the proprietors of major interoceanic canals have shown little disposition to shirk their responsibilities in this respect.[4]

But the operation of an interoceanic canal may call for something more than a dredged channel, locks that work properly, and traffic control which moves ships through the canal without delay. In the past, certain other facilities have had to be provided and other functions undertaken to permit the waterway to fulfill its purpose. These activities are in one sense collateral to the running of the waterway, but in another sense they are necessary adjuncts to a route of maritime commerce where land and water meet. These ancillary functions and facilities have, moreover, proved in the past to be major sources of friction between the territorial sovereign and the operator of the waterway, and some regard must therefore be paid to them and to their bearing upon the international status of these waterways.

History and geography have had a great deal to do with the functions which the proprietors of interoceanic canals and of other waterways have had to assume during the lives of these watercourses, and even during their prenatal existence. Before the construction of the Suez Canal, the first of the three great interoceanic canals, could be undertaken, De Lesseps had to bring to the sands of the Isthmus of Suez an adequate supply of fresh water for those who would construct and operate the waterway, for the needs of the communities which would later spring up along the Canal, and for the supply of the shipping which would ply it. The fresh-water canals, which carried water from the Nile to points along the route of the Canal, were thus the first task of construction to be accomplished.[5] Even today, they are the life-

---

[3] In the earlier days of the Suez Canal Company, complaints about its service were not infrequent. The British shipping trade was particularly critical of the operation of the Canal during the 1870's and 1880's. See, e.g., Reply of M. Ferdinand de Lesseps to the "Report on the Working of the Suez Canal, addressed to the Admiralty by Captain Rice," Incl. 2 to British Suez Canal Directors to Earl Granville, Nov. 21, 1883, in Correspondence respecting the Suez Canal, GREAT BRITAIN, PARLIAMENTARY PAPERS, EGYPT No. 3 (1884), at 33 (C. 3850).

[4] Exceptional conditions called for the clearing of the Suez Canal by a United Nations salvage fleet in 1957. See CLEARANCE OF THE SUEZ CANAL, REPORT OF THE SECRETARY-GENERAL, Nov. 1, 1957 (A/3719) (1957).

[5] HALLBERG, THE SUEZ CANAL: ITS HISTORY AND DIPLOMATIC IMPORTANCE 139, 189 (1931); 1 CHARLES-ROUX, L'ISTHME ET LE CANAL DE SUEZ 358-64 (1901). The construction of the sweet-water canals was one of the requirements laid on De Lesseps by the Firman de Concession et Cahier des Charges pour la construction et l'exploitation du Grand Canal Maritime de Suez et dépendances, Jan. 5, 1856, art. I, in COMPAGNIE UNIVERSELLE, RECUEIL DES ACTES CONSTITUTIFS 6.

blood of the Canal, for without their water neither the employees of the Suez Canal Authority nor the populations of the communities along the Canal, which support, and are supported by, the commercial activity of the Canal, could endure.[6] One of the most striking sights which greets the visitor to Ismailia, which is roughly at the midpoint of the Canal, is the verdure of that city and the fields about it, divided from the desert beyond at the unmistakable line at which irrigation halts. It was this absolute essentiality of the sweet-water canals to the support of human life that necessitated their being placed under the same international protection as the Suez Canal itself under the Convention of Constantinople of 1888.[7]

The construction of a canal through the desert and wasteland of the Isthmus required the Canal Company to furnish facilities and services which would already have been in existence or would be supplied by local governments or by private individuals in more highly developed areas. At one time or another, the Company was obliged to construct roads, build a railroad, generate electricity and gas, maintain parks and gardens, and operate water-purification plants.[8] In many respects, it performed the duties of a municipal corporation by maintaining and lighting streets, supplying water, removing garbage, providing fire-fighting services, and making itself responsible for sanitation.[9] The *domaine commun,* administered for the joint benefit of the Suez Canal Company and the Egyptian Government, put the Company in the position of an estate manager and real estate dealer.[10]

[6] HOSKINS, THE MIDDLE EAST: PROBLEM AREA IN WORLD POLITICS 55 (1954).

[7] Article II, 15 MARTENS, N.R.G. 2d ser. 557 (1891); see also Convention entre le Gouvernement Egyptien et la Compagnie Universelle du Canal Maritime de Suez, pour la construction du Canal d'Eau douce du Caire à l'Ouady, March 18, 1863, art. 2, COMPAGNIE UNIVERSELLE, RECUEIL DES ACTES CONSTITUTIFS 32, concerning the construction of a portion of the canal by the Egyptian government.

[8] COMPAGNIE UNIVERSELLE, THE SUEZ CANAL: NOTES AND STATISTICS 65-68 (1952); see, by way of example of such arrangements, the Convention of Feb. 1, 1902, by which the Company agreed to widen the narrow-gauge railway between Ismailia and Port Said and to turn it over to the Egyptian Government for the duration of the concession. COMPAGNIE UNIVERSELLE, RECUEIL DES ACTES CONSTITUTIFS 81.

[9] E.g., by the Second Convention of Dec. 18, 1884, art. 11, COMPAGNIE UNIVERSELLE, RECUEIL DES ACTES CONSTITUTIFS 62 at 64; and see *id.* at 64, n. 210, concerning municipal services in the Canal area in 1950. The Company remained responsible for the lighting, sewage system, and the maintenance of the streets and gardens at Port Tewfik. The Company also continued to supply water to the four principal municipalities along the Canal.

[10] The Company was granted by Egypt certain lands over and above those needed for the actual Canal and the works necessary to its operation. These areas were those which might be required for the establishment of stores and workshops, housing for

The tropic vegetation and primitive conditions of the Isthmus of Panama presented the United States with many of the same administrative and operational problems the desert had created for the Suez Canal Company. The Panama Canal Company not only has provided quarters for its employees and officials but also finds itself in the hotel business today because of the desirability of providing comfortable living conditions for the employees of the Canal when construction was at an early stage in 1905.[11] The controversy over commissaries between the United States and Panama, which was partially resolved by the Treaty of 1955,[12] stems from difficulties encountered a half century before in providing food and other necessities to laborers at low prices. Although the United States had initially been granted by Panama the right to import goods for resale without the payment of duties to Panama,[13] the privilege of buying at the commissaries was at first denied to employees and workmen "who are natives of tropical countries."[14] Gouging by Panamanian merchants had sent prices up, and the United States had been forced to grant compensating increases of pay. The response was to reopen the commissaries to native employees and workmen, not only for their benefit but also as a means of avoiding a further drain upon the Treasury of the United States.[15] Complete control of the Panama Railroad was acquired at approximately the same time by the buying up of the small shareholdings in private hands. The inability of the Railroad to deal with the traffic which came

---

employees of the Canal, gardens, and works to prevent the shifting of sand into the Canal. Convention of Feb. 22, 1866, art. 4, COMPAGNIE UNIVERSELLE, RECUEIL DES ACTES CONSTITUTIFS 41 at 42. They were administered by the Commission d'Administration, composed of representatives of Egypt and of the Suez Canal Company, and the proceeds of the sale of the lands were shared by the Company and the Government. First Convention of April 23, 1869, art. 4, *id.* at 49; Second Convention of April 23, 1869, *id.* at 51; Second Convention of Dec. 18, 1884, *id.* at 62.

[11] DuVAL, AND THE MOUNTAINS WILL MOVE 142 (1947).

[12] Treaty of Mutual Understanding and Cooperation between the United States of America and the Republic of Panama, signed at Panamá, Jan. 25, 1955, art. XII, 6 U.S.T. 2273, T.I.A.S. No. 3297.

[13] Convention between the United States of America and the Republic of Panama for the Construction of a Ship Canal to Connect the Waters of the Atlantic and Pacific Oceans, signed at Washington, Nov. 18, 1903, art. XIII, 33 Stat. 2234, T.S. No. 431.

[14] Order of the Secretary of War, Jan. 7, 1905, in EXECUTIVE ORDERS RELATING TO THE PANAMA CANAL 33 (1921).

[15] DuVAL, *op. cit. supra* note 11, at 186-87; BISHOP, THE PANAMA GATEWAY 152-53 (1913). The order was formally revoked in 1911. Order of the Secretary of War, Jan. 5, 1911, in EXECUTIVE ORDERS RELATING TO THE PANAMA CANAL 103 (1921).

to it through the construction of the Canal necessitated complete ownership of the line by the United States.[16]

Since the United States exercised territorial jurisdiction within the Canal Zone, the performance of all the normal functions of government, including even the provision of municipal services, could be expected. So also, proprietary functions collateral to the construction and operation of the Canal bore a close relation to the successful exploitation of the Canal. None of the special activities of the Panama Canal could be regarded as alien to the responsibilities we conceive it proper for a government to discharge. It was no less natural that the European Commission of the Danube, a public international organization created by the states on whose behalf it acted, should have maintained a hospital and made contributions, in kind and in money, to the municipal services and amenities of the city of Sulina and other towns on the lower Danube.[17]

But the performance by a private company, the Suez Canal Company, of a wide range of functions ordinarily regarded as governmental became somewhat incongruous with the passage of time, notwithstanding the fact that in the early days of the Canal the Company was the sole agency which could carry on these necessary activities. Not only were Ismailia and Port Tewfik and Port Fouad[18] at one time "company towns," but they also suffered from the stigma of being company towns controlled from abroad. If justification for the broad range of functions performed by the Suez Canal Company were to be sought in the wide measure of governmental interest, both financial and political, in the Company, that concept only suggested that France and Great Britain were exercising governmental powers in Egypt. It was the omnipresence of the Company in the life of the area bordering the Canal, combined with the foreignness and evident prosperity of the enterprise, which must have had much to do with the view, widely held in Egypt after the withdrawal of the British from their bases, that the Company was the last outpost of colonialism on Egyptian soil.

This is not to say that many of the responsibilities assumed by the United States and by the Suez Canal Company could not be, or were not dropped or restored to the local authorities, Panamanian or Egyptian, as the needs which

---

[16] DuVal, *op. cit. supra* note 11, at 154-55.

[17] La Commission européenne du Danube et son oeuvre de 1856 à 1931, at 342-49 (1931).

[18] The establishment of the new community was agreed upon by the Egyptian Government and the Company in 1925; it was initially placed under the administration of the Commission du Domaine Commun. Convention of Oct. 11, 1925, Compagnie Universelle, Recueil des actes constitutifs 145.

had initially dictated their performance vanished. Pressures from Panama forced the United States to close its commissaries to Panamanian employees so that Panamanian businessmen would have the benefit of their custom.[19] The same consideration was responsible for the agreement of the United States that it would thereafter refrain from selling goods, other than fuels and lubricants, to ships passing through the Canal,[20] a lucrative commercial activity which had dated from an era in which Panamanian firms had been unable to supply the needs of such vessels.[21] The standards of health in Panama had so far improved by 1955 that the United States could safely restore to Panama the right to regulate sanitation in the cities of Panama and Colon,[22] a privilege which the United States had of necessity been granted before the construction of the Canal.[23] The Suez Canal Company gradually withdrew from its municipal functions, and the watercourses which supplied fresh water to the area of the Canal were retroceded to the Egyptian authorities.[24] One gathers that these services may have been administered with

[19] Treaty cited *supra* note 12, art. XII.

[20] Memorandum of Understandings Reached, Item 4, annexed to treaty cited *supra* note 12; see Statement by the Department of State, in *Hearings on the Panama Treaty before the Senate Committee on Foreign Relations,* 84th Cong., 1st Sess. 189 (1955).

[21] The terms upon which such goods would be sold by the United States were spelled out in a note from the Secretary of State to the Members of the Panamanian Treaty Commission, March 2, 1936, to which the Panamanian representatives replied that the limitation on sales was welcome but that they must reserve their rights under article XIII of the Hay–Bunau-Varilla Treaty, which provided that customs exemptions were provided for the United States and persons in its service. The notes are annexed to the General Treaty of Friendship and Cooperation between the United States of America and Panama, signed at Washington, March 2, 1936, 53 Stat. 1807, 1845, T.S. No. 945, at 43-47; and See PADELFORD, THE PANAMA CANAL IN PEACE AND WAR 219 (1943).

[22] Treaty cited *supra* note 12, art. IV. This did not affect the authority of the United States as to the Canal Zone and ships transiting the Canal.

The operators of international waterways and the territorial sovereigns of the land through which the waterway flows have had a strong concern in preventing the introduction of disease as the result of contact between the transiting vessel and the shore. See, as to the mouth of the Danube, *op. cit. supra* note 17, at 358-61; as to Suez, Convention amending the International Sanitary Convention of June 21, 1936, signed at Paris, Oct. 31, 1938, 198 L.N.T.S. 205; and as to the Turkish Straits, Convention concerning the Régime of the Straits, signed at Montreux, July 20, 1936, art. 3, 173 L.N.T.S. 213, 7 HUDSON, INTERNATIONAL LEGISLATION 386 (1941). The necessary controls are now generally performed by the nation having territorial jurisdiction over the waterway and surrounding area.

[23] Treaty cited *supra* note 13, art. VII.

[24] E.g., Convention between the Egyptian Government and the Suez Canal Company, dated March 7, 1949, art. 14 and Annex G, COMPAGNIE UNIVERSELLE, RECUEIL DES ACTES CONSTITUTIFS 245 at 250 and 258.

somewhat less efficiency by the Egyptian authorities than by the Canal Company, but this is not necessarily the determinative consideration. Inertia and special pressures have, on the other hand, kept in the hands of the operator of the waterway some functions which should probably be abandoned. Perhaps the clearest instance is the Panama Railroad, now suffering from obsolescence and heavy competition from highway transportation, but kept alive at the behest of the Congress.[25]

The primitive conditions which prevailed a century ago in the Isthmus of Suez and a half century ago in the Isthmus of Panama account for the remarkable agglomeration of functions assumed by the proprietors of these two waterways. Just as inevitably, the material development which has taken place in the economy of these areas and the social progress of the two countries have produced both an ability and a desire to perform activities which once had to be left to the skilled foreign proprietors of these waterways. There is no such history in the case of the Kiel Canal because there has never been any necessity in a highly developed country that the authorities who administer the Canal branch out into collateral activities. The present practice appears to be that the operator of the waterway, whether it be an arm of the government of the territorial sovereign or of a foreign government, confines its activities very largely to those closely related in a functional sense to the navigation of the waterway itself. On the other hand, not all functions relating to navigation have become the responsibility of the operator. One of the anomalies of the operation of the Suez Canal is the fact that the searchlights which are needed for night navigation and are temporarily mounted on the bows of vessels unequipped with such lights are furnished by private firms. So also are the services of the boatmen who moor vessels to the revetments in the sidings when it is necessary for convoys to pass. The nonperformance of these services can cause grave embarrassment to the operation of the Canal.[26] The only explanation why these duties were not assumed by the former Suez Canal Company lies in history.

Pilotage and towage are two of the essential services rendered to ships employing the three major interoceanic canals. Pilots are required for the transit of all but small vessels through these waterways. At Suez and Panama,

---

[25] H.R. Rep. No. 2974, 84th Cong., 2d Sess. (1957). The Panama Line has ceased commercial operations, and only one vessel is maintained on the New Orleans-Panama run. Annual Report of the Panama Canal Company and the Canal Zone Government for the Fiscal Year 1961, at 26-27 (1961).

[26] For example, in case of a strike by these workers, concerning which see p. 294 *infra*.

they are supplied without charge by the canal administrations.[27] The pilots employed on the Kiel Canal are members of two guilds under governmental supervision and are not civil servants; they differ in this respect from the other persons employed in the operation of the Canal.[28] The cost of their services is accordingly not included in the tolls. By international agreement—but only after some controversy—Canada and the United States have established requirements for pilotage services on the St. Lawrence River, have created a pool of pilots, and have fixed the fees for their services.[29]

Towage for those vessels requiring it is provided by the administrations of the three major interoceanic canals; a charge is made for this service.[30]

Special legislation exists for navigation through the canals.[31] It was characteristic of the quasi-public character of the activities of the Suez Canal Company that it promulgated its own rules for navigation of the Canal.[32] While the regulation of navigation is normally within the legislative competence of the territorial sovereign, it was appropriate that the Company, which had full responsibility for the operation of the Canal, should be conceded this law-making authority to regulate traffic within the Canal. The *règlement de navigation* for the Rhine, on the other hand, is not promulgated by the Central Commission of the Rhine but is the result of agreement by the nations represented on the Commission.[33] As has been mentioned at an

[27] Compagnie Universelle du Canal Maritime de Suez, Règlement de Navigation, arts. 3 and 28 (1953 ed.), which has been maintained in force by the Suez Canal Authority; Exec. Order No. 4314, Sept. 25, 1925, rules 26 and 27, 35 C.F.R. §§ 4.22 and 4.24 (Panama).

[28] Lorenzen, *The Administration of the Kiel Canal*, [1953] Nord-Ostsee-Kanal, No. 2, 20 at 23-24; Ehler, *The signal for the Canal Pilot is set, id.* at 44.

[29] Agreement between the United States of America and Canada concerning Pilotage Services on the Great Lakes and the St. Lawrence River, signed at Washington, May 5, 1961, 12 U.S.T. 1033, T.I.A.S. No. 4806. The President was authorized to require pilotage by the Great Lakes Pilotage Act of 1960, 74 Stat. 259, 46 U.S.C. § 216. Concerning the background of the antecedent dispute, see Statement of Mr. White, *Hearings on S. 3019 before the Merchant Marine and Fisheries Subcommittee of the Senate Committee on Interstate and Foreign Commerce*, 86th Cong., 2d Sess. 4-9 (1960); N.Y. Times, Jan. 21, 1960, p. 62, col. 5.

[30] Règlement de Navigation, *supra* note 27, arts. 5 and 30A; Canal Zone Order No. 30, 18 Fed. Reg. 281, 35 C.F.R. § 4.106 (Panama); *Tugs of the Canal Administration*, [1953] Nord-Ostsee-Kanal, No. 2, 61.

[31] Compagnie Universelle, Règlement de Navigation (1953 ed.); Annexe pour les navires à cargaisons dangereuses (1954 ed.); 35 C.F.R., subpart 4E (Panama Canal); Betriebsordnung für den Kaiser-Wilhelm-(Nord-Ostsee-) Kanal (1954).

[32] Règlements governing the Suez Canal, cited *supra* note 31.

[33] Revised Convention for the Navigation of the Rhine, signed at Mannheim, Oct. 17, 1868, art. 32, 20 Martens, N.R.G. 355 (1875), 2 Rheinurkunden 80 (1918); and

earlier stage of this study,[34] the Commission does possess certain appellate judicial functions with respect to cases arising out of the navigation of the river.[35] The existence of a central tribunal to which an appellant may in his discretion resort assists in maintaining the uniformity of the law with respect to navigation, which is otherwise subject to conflicting national interpretations.

One of the duties closely related to the navigation of the Canal itself which had fallen to the Suez Canal Company was the improvement, maintenance, and administration of the port of Port Said, including the free zone which was established there.[36] The special facilities and skills available to the Company endowed it with particular competence to carry out this type of work at one of the terminals of the Canal. The association of the port of Port Said with the Canal was not disturbed when the operation of the Suez Canal was vested in the Suez Canal Authority of the United Arab Republic.[37] Similarly, the United States has maintained wharves and facilities for transshipment within the Canal Zone, although the operation of these also is beginning to yield to the importunities of the Republic of Panama.[38]

For legal purposes, some distinction must be drawn between those ancillary services necessary to navigation through an international waterway and those which are merely conveniences. As to the first category, it seems reasonable to conclude that freedom of transit, as guaranteed in principle by

---

see COMMISSION CENTRALE POUR LA NAVIGATION DU RHIN, RÈGLEMENT DE POLICE POUR LA NAVIGATION DU RHIN DU Ier JANVIER 1955 (1955). A list of all the *règlements* appears in LES ACTES DU RHIN 55 (1957).

[34] See p. 120 *supra*.

[35] Treaty cited *supra* note 33, art. 37; see WALTHER, LA JURISPRUDENCE DE LA COMMISSION CENTRALE POUR LA NAVIGATION DU RHIN, 1832-1939 (1948).

[36] Convention of Feb. 1, 1902, arts. 10-15, COMPAGNIE UNIVERSELLE, RECUEIL DES ACTES CONSTITUTIFS 81 at 83.

[37] Statuts de l'Organisme du Canal de Suez, art. 9, annexed to Décret-loi du Président de la République No. 146 de 1957, JOURNAL OFFICIEL DU GOUVERNEMENT ÉGYPTIEN, No. 53*bis* "C," July 13, 1957.

Cf. the placing of the ports of Strasbourg and of Kehl under the supervision of the Central Commission of the Rhine under article 65 of the Treaty of Peace between the Allied and Associated Powers and Germany, signed at Versailles, June 28, 1919, 112 BRITISH AND FOREIGN STATE PAPERS 1 (1919); see 1 L.N.T.S. 367 for the text of the implementing agreement.

[38] Memorandum of Understandings Reached, Item 10, annexed to Treaty of Mutual Understanding and Cooperation between the United States of America and the Republic of Panama, signed at Panamá, Jan. 25, 1955, 6 U.S.T. 2273, 2336, T.I.A.S. No. 3297.

international instruments, cannot be assured without them and that their denial, either through unjustifiable action or inaction on the part of the proprietor or territorial sovereign, is in contravention of that principle. One has only to look to the importance attached to the recruitment of pilots for the Suez Canal to be persuaded of the necessary relation which exists between their services and freedom of transit through the waterway. Even the management of terminal ports may have a significant impact upon the use of an artificial waterway. By way of contrast, the lack of such facilities as bunkerage, free ports,[39] and shops and drydocks for the repair of transiting vessels[40] would not offer a positive impediment to passage, although their absence might make the route a less desirable one for the shipping trade. On the other hand, if transit is simply not possible without the assistance of pilots, traffic control, signaling facilities, and towage, a theoretical right of freedom of passage becomes illusory when these services are lacking. It therefore follows that if a state is under an obligation to allow freedom of transit, it has a duty not only of abstention from interference with navigation but also of affirmative action to provide the requisite services which will keep the waterway navigable in a factual sense. Practice indicates that there is no obligation to provide these services without compensation; the transiting vessel can be required to bear the cost, either through tolls or special charges.

LABOR

The history of the Suez Canal begins and ends with labor problems. The earliest of these related to the provision of laborers by the *corvée*—what would today be called forced labor—by the Egyptian Government for the construction of the Canal. According to an agreement reached in 1856, the necessary laborers for work on the Canal would be supplied to the Company by Egypt according to the needs of the enterprise.[41] The British Government,

[39] See, with regard to the blacklisting of a vessel by Egypt for the carriage of stores to the British forces in the Canal area and the delay of the ship for one month as the result of the consequent inability of the master to secure necessary repairs of the propulsion machinery, Leolga Compania de Navagacion v. John Glynn & Son Ld., [1953] 2 Q.B. 374. The legality of the blacklisting was not in issue.

[40] The Suez Canal Authority maintains a shipyard and provides repair facilities for vessels, including the use of a floating drydock. SUEZ CANAL REPORT, 1961, *supra* note 1, at 50-59.

[41] Règlement sur l'emploi des ouvriers indigènes, July 20, 1856, art. I, referred to in COMPAGNIE UNIVERSELLE, RECUEIL DES ACTES CONSTITUTIFS 29, and quoted in 55 BRITISH AND FOREIGN STATE PAPERS 1005, note (1864-65).

then opposed to the construction of the Canal, and mindful of the reliance
of the Canal Company on the 20,000 native workers which were made available to it, persuaded the Turkish and Egyptian authorities to bring an end to
these arrangements for the provision of labor.[42] France interceded on behalf of
De Lesseps; and the Viceroy of Egypt, caught in the Anglo-French cross-
fire, agreed to the submission of the dispute to the Emperor Napoleon. In his
award, the Emperor held that there had been a violation of the agreement between the Company and the Egyptian Government relating to the provision
of workers and that the Company was entitled to an indemnity of 38,000,000
francs for the termination of the agreement.[43] Thereafter the Company was
forced to rely upon labor which it secured,[44] four fifths of which was required
by the concession to be Egyptian.[45]

Labor difficulties were not again encountered in such acute form until the
last ten years of the administration of the Canal by the Suez Canal Company.[46] Over that decade there were continuing economic and political
difficulties over the employment policies of the Company. Strikes in 1951 and
early 1952 were largely attributable to the guerrilla warfare which was for a
time waged against the British forces in the Canal zone.[47] The absence of
many workers from their usual places of employment during these troubled
months served as a gesture of protest in some cases or reflected fear of unsettled conditions in others. The first workers to go out on strike were the
mooring-boat workers and the searchlight electricians, who were not, it will

[42] The diplomatic context of the abolition of the *corvée* is discussed in HALLBERG,
op. cit. supra note 5, at 197-205.

[43] Award of Napoleon II, July 6, 1864, 2 DE LAPRADELLE AND POLITIS, RECUEIL
DES ARBITRAGES INTERNATIONAUX 362 (1924).

[44] Convention of Feb. 22, 1866, art. 1, COMPAGNIE UNIVERSELLE, RECUEIL DES ACTES
CONSTITUTIFS 41.

[45] Firman de concession et cahier des charges pour la construction et l'exploitation
du Grand Canal Maritime de Suez et dépendances, Jan. 5, 1856, art. 2, *id*. at 6, and
convention cited *supra* note 44.

[46] This leap across eighty years is not intended to suggest that there were no labor
problems during the intervening years. There were British complaints as to harbor
police, the lack of English pilots, and inefficiency of the Company's harbor pilots in
1883. Reply of M. Ferdinand de Lesseps to the "Report on the Working of the Suez
Canal, addressed to the Admiralty by Captain Rice," in Correspondence respecting
the Suez Canal, GREAT BRITAIN, PARLIAMENTARY PAPERS, EGYPT No. 3 (1884) 37-41
(C. 3850). An "increase to a large extent" of the number of English-speaking officials
was one of the "Conditions for the future Administration of the Suez Canal," Nov. 30,
1883, *id*. at 50.

[47] Kirk, *Egypt: The Wafd and Great Britain, 1950-1,* in CALVOCORESSI, [1951]
SURVEY OF INTERNATIONAL AFFAIRS 260 at 282-92 (1954).

be recalled, employees of the Suez Canal Company.[48] It was estimated that there might be delays of one to two days for vessels not equipped with their own searchlights, since night navigation was out of the question for vessels not so outfitted.[49] With the defection of other workers, including stevedores and employees of the Egyptian ports and lights administration, vessels of the British Navy were sent into the Canal and its terminal ports in order to carry out the berthing and other operations normally performed by Egyptian workers.[50] Early in January of 1952, when disorder in Egypt was coming to a peak, 1500 Egyptian employees of the Company itself went out on strike for a short time, but this strike was apparently not politically motivated.[51] The Egyptian Government was prompt to respond when reports leaked out that the British Government had approached the governments of Norway and other maritime states with important interests in the Canal and had requested their assistance in the operation of the Canal. It was assumed by the Egyptian Government that this aid would, following the British practice, take the form of sending foreign warships into the terminal ports and the Canal. It warned the diplomatic representatives of the nations to whom overtures had been made that any such action would be regarded as armed aggression.[52] The period of guerrilla warfare, disorder, and rioting in Egypt, of which the strikes were but one aspect, was brought to an end with the downfall of the Wafd Government.[53] An investigation by the International Labor Organization of Egyptian complaints that Great Britain had forced Egyptian workers to work under threat of arms was quietly laid aside.[54] These disturbances in the Canal Zone had much to do with renewed suggestions that the Suez Canal should be placed under international administration.[55]

[48] The Times (London), Oct. 30, 1951, p. 4, col. 5; and see p. 290 *supra*.

[49] *Ibid.*

[50] The Times (London), Nov. 1, 1951, p. 6, col. 6; Nov. 3, 1951, p. 6, col. 5; Dec. 4, 1951, p. 6, col. 5; Jan. 10, 1952, p. 3, col. 2; 512 H.C. Deb. (5th ser.) 641-42 (1952-53); Note from the Suez Canal Company to the Egyptian Minister of Commerce and Industry, Nov. 27, 1951, reprinted in Ichtirakia (Cairo), Nov. 28, 1951.

[51] The Times (London), Jan. 8, 1952, p. 5, col. 2; Jan. 9, 1952, p. 6, col. 3.

[52] The Times (London), Jan. 14, 1952, p. 4, col. 2; Jan. 16, 1952, p. 4, col. 2; Jan. 21, 1952, p. 6, col. 2.

[53] Kirk, *The Egyptian Revolution and National Aspirations,* in CALVOCORESSI, [1952] SURVEY OF INTERNATIONAL AFFAIRS 203 at 204 (1955).

[54] International Labour Office, Report on Enquiry by the Representative of the Director-General into Conditions in the Suez Canal Area, Jan. 25, 1952 (G.B. 118/16/8) (1952); 34 INTERNATIONAL LABOUR OFFICE, OFFICIAL BULLETIN 222 (1951); 35 *id.* at 263 (1952).

[55] In a letter from the Swedish shipowners' association to the Swedish Foreign

The capacity of a labor dispute to shut down an artificial waterway was again demonstrated during a one-day strike on the St. Lawrence Seaway in 1962. As the consequence of a jurisdictional dispute between the Canadian Labor Congress and the Seafarers International Union of Canada, lock workers refused to operate the locks for ships with Seafarer crews.[56] The strike was brought to an end by the referral of the matter to a one-man industrial commission. The report of the commission recommended the enactment of legislation which would impose penalties on a labor union violating a collective bargaining agreement in such a way as to impede waterway operations and on anyone leaving a vessel unmanned so as to hinder traffic.[57]

There had been a threat of a strike by pilots two years before when the dispute between American and Canadian pilots was at its height, but no suspension of work took place.[58] The essential character of the work performed by certain categories of persons not in the employ of the operator of the waterway points to the desirability of having any prohibition against strikes applicable to employees of the operator of the waterway also govern other persons furnishing services necessary to the passage of ships.

Over the years, agitation by the Egyptian Government for the greater Egyptianization of the personnel of the Suez Canal Company was of greater consequence to the Company than the politically inspired strikes in the years immediately preceding the nationalization of the Canal. The working force of the Company had long had an international complexion, but a predominantly non-Egyptian one. In 1936, only 14 of the 547 employees of the Company falling within the category of nonlaborers were Egyptian.[59] A majority of the laborers in the employ of the Company were at that time Egyptian,[60] but these were not the real bone of contention between the Egyptian Government and the Company. The process of Egyptianization of the personnel of the Company began in that year with an agreement, negotiated at the same time as the fixing of new tolls for transit, that it

---

Ministry, reported in The Times (London), March 11, 1952, p. 3, col. 3. Politically inspired strikes, such as that of March 5, 1955 (Al Izaa Al Masria, March 12, 1955), continued to take place in later years.

[56] N.Y. Times, July 7, 1962, p. 42, col. 5; July 19, 1962, p. 55, col. 8.

[57] N.Y. Times, July 21, 1963, p. 18S, col. 7. The commission was appointed under the Canadian Industrial Relations and Disputes Investigation Act, 11 & 12 Geo. 6, c. 54.

[58] N.Y. Times, Oct. 14, 1960, p. 64, col. 6. See page 95 *supra* concerning the settlement of the dispute.

[59] Information furnished by the Compagnie Universelle du Canal Maritime de Suez.

[60] About 65 per cent, according to information furnished by the Company.

would progressively increase the number of Egyptian employees of the Company until 25 per cent of that category of personnel should by 1958 be of Egyptian birth.[61] In 1947 the strong spirit of nationalism which fired Egypt after the war and brought an end to the régime of capitulations inspired a new law directed to the Egyptianization of the control, ownership, and personnel of corporations.[62] The legislation, which was applicable to all Egyptian corporations, required, among other things, that at least 75 per cent of the employees of a corporation, earning a minimum of 65 per cent of the total payroll of the company, be of Egyptian nationality. 90 per cent of the workmen, earning at least 80 per cent of the wages paid by the company, were to be Egyptians. Three years were provided in which to adjust to these new requirements.[63] The Suez Canal Company, at first inclined to think that its special status under the concessions exempted it from the operation of the law, shortly came about to the view that legal discretion was the better part of legal valor.[64] By negotiating with the Egyptian authorities it would be possible to preserve the principle that the Company had special standing and was not governed by the ordinary Egyptian law as regarded matters of vital importance to its function of operating an interoceanic canal, but this could not be achieved without making some concessions to Egyptian aspirations that the personnel of the Company should have a somewhat less French and generally foreign character. The precise manner in which Egyptians would be introduced into the staff of employees of the Company was spelled out in great detail in a convention concluded in 1949.[65] The salient provisions of this agreement were that for every ten administrative positions created or becoming vacant, nine Egyptians[66] would be appointed; for every five technical

[61] Exchange of letters between the President of the Compagnie Universelle du Canal Maritime de Suez and the President of the Egyptian Council of Ministers, April 27-30/May 4, 1936, COMPAGNIE UNIVERSELLE, RECUEIL DES ACTES CONSTITUTIFS 210.

[62] Loi No. 138 de 1947 édictant certaines dispositions relatives aux sociétés anonymes, JOURNAL OFFICIEL DU GOUVERNEMENT EGYPTIEN, No. 74, Supplement, Aug. 11, 1947.

[63] Article 5.

[64] LE CANAL DE SUEZ, No. 2,238 (June 15/25, 1949), at 9281-83, quoted in EL-HEFNAOUI, LES PROBLÈMES CONTEMPORAINS POSÉS PAR LE CANAL DE SUEZ 269-73 (1951).

[65] Convention, signed March 7, 1949, COMPAGNIE UNIVERSELLE, RECUEIL DES ACTES CONSTITUTIFS 245; concerning the ratification of the Convention, see EL-HEFNAOUI, *op. cit. supra* note 64, at 265-69; see also Weinberger, *The Suez Canal in Anglo-Egyptian Relations,* 1 MIDDLE EASTERN AFFAIRS 347, 349-50 (1950).

[66] Under article 6 of the Convention, an Egyptian for these purposes was an individual born of a father considered Egyptian at the time of the birth of the individual in question; see, concerning the interpretation of this article, Boutros Jaouiche v.

positions, four Egyptians; that 18 Egyptians would be promoted or appointed in the middle and upper grades of the Company's service;[67] that 20 new Egyptian pilots would be appointed; and that thereafter one of every two pilots appointed would be Egyptian.[68] Other categories of personnel were dealt with in similar fashion, and the Company was allowed some exceptions as to highly qualified technicians. The Company carried out its obligations under this agreement to the extent that by 1955 50 per cent of the employees and 15 per cent of the pilots were Egyptian.[69] These appointments did not mean that the administration of the Canal was entrusted to a management which had acquired a strong Egyptian flavor. By and large, the higher positions in the administration of the Canal in Egypt were still filled by Frenchmen at the time that the Canal was nationalized. The training of qualified Egyptian administrative and technical personnel who would be in a position to operate the Canal upon the termination of the concession in 1968 was already becoming a matter of importance to the Egyptian Government. Egyptianization of the personnel of the Company was dictated not only by economic conditions and national pride but also by the desire to provide an orderly transition to Egyptian control,[70] which, prior to President Nasser's sudden action, appeared to the Government and the Company to be at least a decade distant.

The really troublesome point about the implementation of the agreement of 1949 turned out to be the hiring of pilots to be placed aboard the steadily increasing number of ships daily making their way through the Canal. After 1949, candidates of Egyptian nationality were presented to the Company by the Egyptian Corporations Department,[71] which the Company complained was more interested in seeing Egyptians hired than it was in providing qualified pilots to the Company. It proved difficult to maintain the 50-50 ratio on Egyptian and non-Egyptian pilots when qualified Egyptians with a master's ticket and a substantial amount of sea service were not available. The Egyptian merchant marine and navy were not large enough to provide

---

Ministère du Commerce, Conseil d'Etat, Jan. 27, 1953, 9 REVUE ÉGYPTIENNE DE DROIT INTERNATIONAL 142 (1953).

[67] Article 2.

[68] Article 4.

[69] Information furnished by the Compagnie Universelle du Canal Maritime de Suez.

[70] See the views expressed to the same effect in 1951 by Dr. El-Hefnaoui, later to become one of the directors of the Suez Canal Authority, *op. cit. supra* note 64, at 289.

[71] See article 11 of the Convention of 1949, *supra* note 65, and accompanying note.

a reservoir of individuals prepared to enter the Company's service with the same qualifications as were demanded of foreign pilots. The whip hand was held by the Egyptian Government, which was in a position to deny visas to foreign pilots unless the stipulated number of Egyptians were hired.[72] Prolonged negotiations were necessary before a compromise could be worked out concerning the relative number of Egyptian and non-Egyptian pilots hired, and while these negotiations were proceeding, the Company was hard pressed to supply the demands laid upon it by the increase in transits through the Canal.[73] The importance of this category of technicians to the operation of the Canal was again to be demonstrated during the period immediately after the nationalization of the Canal and the departure of the great majority of pilots formerly in the service of the Company.

Egyptian sensitivity about discrimination in wages between the foreign and Egyptian employees of the Company is also to be discerned in various dealings with the Government on this subject in the late 1940's. Commissions of conciliation between the Company and its employees and workmen directed that there should be absolute equality in all respects, including pay, leave, family allowance, and other benefits, between Egyptian and foreign personnel with the same aptitudes, occupying the same posts, and with the same length of service.[74] One of the commissions was even willing to go so far as to say that an Egyptian employee should have the same length of leave as, for example, an American employee of the Company and that if an Egyptian wanted to go to the United States on leave he should have the same allowance as the American, since it would be inconsistent with the equality of the two that the Egyptian should only be paid travel allowance to his home in Egypt.[75] The principle of nondiscrimination in conditions of employment was written

[72] Under article 10 of the Convention of 1949.

[73] The foregoing statements are based on correspondence between the Compagnie Universelle du Canal Maritime de Suez and the Egyptian authorities over the period from Feb. 8, 1954 to March 21, 1955, on file in the headquarters of the Company (now the Compagnie Financière de Suez) in Paris; see also 520 H.C. DEB. (5th ser.) 373 (1953-54).

[74] Ms. arrêté of the Ministre des Questions Sociales, dated Nov. 11, 1947, approving the decision of the "Commission de Conciliation entre la Compagnie du Canal de Suez et Ses Employés" of Oct. 19, 1947; Ms. decision of the "Commission de Conciliation entre la Compagnie du Canal et Ses Ouvriers," dated March 27, 1948, on file in the headquarters of the Company (now the Compagnie Financière de Suez) in Paris.

[75] Decision of the Commission de Conciliation, dated Oct. 19, 1947, *supra* note 74. As the result of a later compromise the travel allowance was reduced to a flat sum for all indigenous employees.

into the Convention of 1949, subject to the right of the Company to pay special allowances for the first trip to Egypt and the last trip from Egypt, an allowance for the schooling of children, mortgage loans, and an expatriation allowance not to exceed 25 per cent.[76] It was estimated at the time of the nationalization of the Canal that Egyptian employees of the Company were receiving pay three to five times that received by comparable persons employed by other Egyptian enterprises.[77]

The same principle of nondiscrimination in the compensation of locally recruited personnel and those brought in from abroad has professedly been applied by the United States with respect to the employment of citizens of the United States and of Panama in the Canal Zone.[78] However, in practice, there have since the building of the Canal been two categories of employees, the one described as "U.S. rate" or "gold" employees, from the currency in which such persons were originally paid, and the other as "local rate" or "silver" employees. The distinction is not made on ground of nationality but according to the amount of the pay of the employee, those above a certain level being regarded as "U.S. rate" and entitled to a wider range of privileges than the "local rate" employees.[79] Panamanian nationals are eligible for positions paying wages qualifying as "U.S. rate" and if appointed to these positions earn the same pay as United States citizens, less the special differential allowed the latter for overseas service in the Canal Zone. In actuality, the "local rate" employees are largely Panamanian nationals or nationals of other

---

[76] Convention of 1949, *supra* note 65, art. 1.

[77] COMPAGNIE UNIVERSELLE, THE SUEZ CANAL COMPANY AND THE DECISION TAKEN BY THE EGYPTIAN GOVERNMENT ON 26TH JULY 1956 (26TH JULY-15TH SEPTEMBER 1956) 49 (1956).

Both Egyptian and foreign pilots complained in 1959 when wages and fringe benefits were reduced under the terms of new contracts concluded by the Suez Canal Authority. The matter was taken into the courts. Manchester Guardian, June 25, 1959, p. 9, col. 5.

[78] Exec. Order No. 1888, Feb. 2, 1914, paras. 1 and 6, in EXECUTIVE ORDERS RELATING TO THE PANAMA CANAL 158 (1921); Exchange of notes between the Secretary of State and the Members of the Panamanian Treaty Commission, March 2, 1936, annexed to the General Treaty of Friendship and Cooperation between the United States of America and Panama, signed at Washington, March 2, 1936, 53 Stat. 1807, 1856-1858, T.S. No. 945 at 56-58.

[79] PADELFORD, THE PANAMA CANAL IN PEACE AND WAR 240-43 (1943); *Hearings on the Panama Treaty before the Senate Committee on Foreign Relations*, 84th Cong., 1st Sess. 11-13 (1955). The employees of the Armed Forces were an exception. In accordance with the general policy of the Army, Navy, and Air Force, two wage scales existed, one for United States citizens and the other for foreign nationals.

tropical countries.[80] The appearance of discrimination is thus created and has been resented by Panama.

When the Treaty of 1936 was concluded, an exchange of notes accompanying the Treaty stated that the United States would "maintain as its public policy the principle of equality of opportunity and treatment" set down in the executive orders pertaining to the Canal and would favor, subject to the efficient administration of the Canal and the Canal Zone, such implementation of that principle as would assure equality of treatment between nationals of the two countries.[81] Supposed discrimination nevertheless remained a sore spot between the two countries.[82] It was hoped that the Treaty of 1955 would bring an end to the quarrel through arrangements satisfactory to Panama. These called for the establishment of a basic wage level for a given grade which would be the same for citizens of the United States and of the Republic of Panama. The United States citizen would be entitled in addition to an overseas differential, an allowance for extra costs to which he might be put, and more liberal leave benefits and travel allowances.[83] The United States also undertook to seek legislation which would extend the Civil Service Retirement Act[84] to employees who were Panamanian citizens. These persons had previously been protected by the so-called Cash Relief Act,[85] which was looked upon by Panamanians as a second-class re-

[80] PADELFORD, *op. cit. supra* note 79, at 240.

[81] Exchange of notes cited *supra* note 78; see 40 OPS. ATT'Y GEN. 515 (1947), expressing the view that the Civil Service Commission is required to admit citizens of the Republic of Panama to civil service examinations for employment by the Panama Canal and the Panama Railroad Company.

[82] The provision in the act authorizing new construction in the Canal Zone (Act of Aug. 11, 1939, 53 Stat. 1409) which required that skilled jobs be filled by Americans was regarded by Panama as being in contravention of the agreement of 1936. See Wright, *Defense Sites Negotiations Between the United States and Panama, 1936-1948*, 27 DEP'T STATE BULL. 212, 217 (1952).

For a later reference to concern about the matter of equality in employment, see the Joint Statement issued on the occasion of President Rémon's visit to the United States in 1953, 29 DEP'T STATE BULL. 487, 488 (1953).

[83] Memorandum of Understandings Reached, Item 1, Annexed to Treaty of Mutual Understanding and Cooperation between the United States of America and Panama, signed at Panamá, Jan. 25, 1955, 6 U.S.T. 2273 at 2329, T.I.A.S. No. 3297.

[84] 46 Stat. 468, as amended, 5 U.S.C. §§ 2251-68 *et seq*. Panamanian employees came under the coverage of the Act pursuant to section 13 of the Act of July 25, 1958, 72 Stat. 405, 410, which was enacted in order to give effect to the Treaty of 1955. Panamanians previously retired continue to receive retirement benefits under the "Cash Relief Act." 2 C.Z.C. § 181 (1962).

[85] Act of July 8, 1937, 50 Stat. 478, now 2 C.Z.C. § 181 (1962), authorizing the

tirement act because of the lower benefits it provided.[86] The United States pledged itself to provide equality of opportunity to Panamanian citizens for all employment, except where considerations of security dictated otherwise. To the disappointment of Panama[87] the United States Government made it clear at the time of the consideration of the Treaty by the Senate and on subsequent occasions that the agreement would in fact produce no real change in the employment practices of the United States.[88]

President Eisenhower responded to further Panamanian complaints[89] by the nine-point program of 1960, calling for, among other things, new housing for Panamanian employees, improved employee benefits, and wider access by Panamanians to "skilled and supervisory positions."[90] When President Chiari visited the United States roughly two years later, the employment problem again came up, and President Kennedy agreed that labor questions, including equal employment opportunities, would be among those discussed by representatives of the two countries.[91] The end of the problem has so often been just in sight that it would be hazardous to speak with any degree of confidence about its successful resolution in the immediate future.

Superficially, the application of the general principle of nondiscrimination in the operations of the Panama Canal Company and of the Suez Canal Company might appear to be the same. The reality of the situation was, and remains, that the United States, because it is a sovereign nation and because it has more than a mere proprietary interest in the Canal, continues to keep United States citizens in the more important posts and has never acceded to the gradual "Panamanianization" of the staff required to operate the Canal and its associated enterprises. Its position is quite different from that of

---

Governor of the Panama Canal to pay "cash relief" to persons unfit for "further useful service by reason of mental or physical disability resulting from age or disease."

[86] *Hearings on the Panama Treaty, supra* note 79, at 12.

[87] N.Y. Times, Aug. 26, 1956, p. 37, col. 3.

[88] As put by an Assistant Secretary of the Army to the Senate Foreign Relations Committee: "Those categories for which employees must be recruited from the continental United States will necessarily continue to be paid at the rates paid for comparable employment in the United States; and the occupational categories filled by recruitment on the Isthmus of Panama will be continued to be paid at rates based on the local prevailing rates for comparable employment in the Panama area." *Hearings on the Panama Treaty, supra* note 79, at 70.

[89] N.Y. Times, March 20, 1959, p. 8, col. 6; and see Secretary Herter's statement at the news conference of Dec. 10, 1959, 41 Dep't State Bull. 936, 937 (1959).

[90] 42 Dep't State Bull. 798 (1960).

[91] Joint Communiqué, June 13, 1962, 47 Dep't State Bull. 81 (1962).

a corporation holding a concession which was to terminate before many years had elapsed.

The nationalization of the Suez Canal brought home to Egypt labor problems which differed from those which had been encountered by the Suez Canal Company and by the United States Government as operator of the Panama Canal. The matter was no longer one of compelling the operator of the waterway to afford larger employment opportunities to local persons and of eliminating discrimination between domestic and foreign employees but rather of the operator's being hard put to supply adequate staff to run the canal. At the time of the nationalization of the Suez Canal in July of 1956 a majority of its staff, excluding laborers, was still foreign.[92] "These foreign workers include all the key men, the technicians and engineers," a "leading authority on canal management" was quoted as saying in Cairo. "Without them, the Egyptians couldn't run the canal for more than a week."[93] The key problem as to staffing was still, as it had been in the past, the pilots. It was estimated that 250 were needed for the operation of the Canal. There were 207 on the Company's payrolls, of whom 14 were in August absent on home leave; only 40 of the total number were Egyptians.[94] Although the Company had directed its staff to stay on the job pending the outcome of the London Conference in August and September 1956,[95] the Egyptian Government insisted that the British and French governments and the Canal Company were attempting to induce defections and to persuade pilots on home leave not to return to Egypt.[96] At the conclusion of the London Conference, the Company notified its non-Egyptian employees, including the pilots, that they could now end their duties.[97] Despite the fact that the law nationalizing the Suez Canal Company had directed its employees to remain at work under pain of punishment,[98] the Egyptian Government allowed the foreign pilots

[92] *Op. cit. supra* note 77, at 39.

[93] Wall Street Journal, Aug. 17, 1956, p. 1, col. 4.

[94] N.Y. Times, Aug. 23, 1956, p. 1, col. 2; Aug. 21, 1956, p. 1, col. 2.

[95] Statement issued by the Company on Aug. 6, 1956, *op. cit. supra* note 77, at 31.

[96] Letter dated 17 September 1956 from the Representative of Egypt Addressed to the President of the Security Council (S/3650) (1956); N.Y. Times, Aug. 27, 1956, p. 3, col. 1.

[97] Message from the Company to its personnel, Sept. 11, 1956, and Extracts from the statement made by the French Prime Minister after the Cabinet Meeting on 12th September 1956, *op. cit. supra* note 77, at 33.

[98] Decree Law No. 285 of 1956, July 26, 1956, arts. 4 and 5, in REPUBLIC OF EGYPT, WHITE PAPER ON THE NATIONALISATION OF THE SUEZ MARITIME CANAL COMPANY 3 (1956).

and other staff who wished to leave Egypt to do so.[99] A tremendous burden descended upon the small corps of pilots, most of them Egyptian, who were left behind. Although the number of daily transits fell off somewhat from the average before nationalization of the Canal and some delays were encountered, ships were kept moving. This was accomplished only through nearly continuous labor by the Egyptian pilots.[100] In the meanwhile, the Egyptian Government had been recruiting pilots in foreign countries—in Germany from among the Kiel Canal pilots and river pilots, in the United States, and in Soviet Russia.[101] The suggestion that the Users Association, the formation of which had been agreed upon at the London Conference, could alleviate this situation by furnishing pilots when the Suez Canal Authority was unable to do so could not,[102] under the circumstances prevailing at the time, have provided a realistic solution to Egypt's problem. The proposal was too heavily freighted with politics and Egypt's pride was too much bound up in the operation of the Canal to make this a workable expedient. By October of 1956, just before the Anglo-French military expedition, the Suez Canal Authority had 233 pilots from 17 countries in its employ, of whom about 100 were Egyptian.[103] The number of transits was roughly at the same level as for the comparable period of the previous year, and all this had been accomplished with a minimum of accidents. In this remarkably short period of time—barely three months—the principal personnel problem of the Canal had been solved and freedom of transit was being maintained,[104] despite the measures taken by the British and French governments to hamper the operation of the Canal.

What has been said in the preceding pages about the relationship between the territorial sovereign and the proprietor of an international waterway as regards labor matters is not intended to suggest that the practice of nations has been productive of a group of legal principles which may be of application to such controversies in the future. Nevertheless, some conclusions may

[99] N.Y. Times, Sept. 13, 1956, p. 1, col. 7.

[100] N.Y. Times, Sept. 18, 1956, p. 4, col. 3; Sept. 24, 1956, p. 1, col. 7.

[101] N.Y. Times, Aug. 25, 1956, p. 3, col. 5; Aug. 26, 1956, p. 3, col. 4; Aug. 29, 1956, p. 3, col. 1; Aug. 31, 1956, p. 3, col. 1; Sept. 18, 1956, p. 4, col. 3; Sept. 19, 1956, p. 1, col. 6.

[102] Statement by Prime Minister Eden, Sept. 12, 1956, 558 H.C. DEB. (5th ser.) 10-11 (1956); Statement by Mr. Dulles at the Second Suez Conference at London, Sept. 19, 1956, in THE SUEZ CANAL PROBLEM, JULY 26-SEPTEMBER 22, 1956, at 355 (Dep't of State Pub. 6392) (1956).

[103] N.Y. Times, Oct. 26, 1956, p. 5, col. 3.

[104] SUEZ CANAL AUTHORITY, MONTHLY BULLETIN, Sept. 1956, at 4.

be drawn from the experiences of the operating agencies for the Suez and Panama canals.

1. The unhappy experiences of the Suez Canal Company in securing foreign pilots and of the Suez Canal Authority in having to provide staff to replace the non-Egyptian personnel of the Company demonstrate in a forceful manner that the provision of a staff adequate in quantity and quality to maintain and operate the waterway is a necessary condition of the provision of free navigation.

2. Either a foreign government or a private operating company will be subjected to strong pressure from the territorial sovereign to employ a large number of nationals of that country and to pay them at a wage scale which will avoid discrimination between such persons and nationals of foreign countries. It seems improbable that even an international organization would be immune from such pressures.

3. The importance of the task performed by the staff of an operating agency is such that the right to strike should probably be required to yield to the needs of those who rely upon the waterway as a channel of communication and of commerce. In this respect, such persons would be under the same restraints as are applied to civil servants in a number of jurisdictions.

4. A user nation or the territorial sovereign is not privileged to interfere with the staffing of the operating agency, and a user state would be estopped from challenging the resultant delays or obstructions to its shipping were it to do so.

CHAPTER VII

INTERNATIONAL ADMINISTRATION OF INTERNATIONAL CANALS

A N ATTEMPT has been made in the preceding chapters to indicate the general and the limited analogies which exist between interoceanic canals and other forms of international waterways and to bring experience derived from all three types of waterways to bear upon the legal and organizational problems to which the three major interoceanic canals of Suez, Panama, and Kiel give rise. There can be little doubt about the extent to which the international community is concerned with the free and efficient navigation of these waterways. The important question is how this concern can be translated into principles of law and administration which will give due regard to the possibly conflicting interests of the users of the waterway, the territorial sovereign, and the operator or supervisor of the waterway. The existence of these principles, even if they be translated into specific rules of application, may not be enough. While the three major canals are regulated by rules of law of both customary and conventional origin, all three are currently in fact administered by the governments of sovereign states. There is room for the view that if these waterways are to be efficiently and economically administered with a minimum of interference with free passage, their operation and supervision may be too important a function to be left entirely to the discretion of their proprietors, who may not be uniformly heedful of the legal obligations laid upon them.

THE EXISTENCE OF A GENERAL BODY OF LAW GOVERNING INTEROCEANIC CANALS

Before turning to the question of the institutional means of promoting order and stability in the operation of interoceanic canals, it is necessary to take account of the fact that any proposal for one international waterway, such as Suez, must be capable of application to the other interoceanic canals. Reference has already been made in a number of contexts to the occasions on which analogies have been drawn between the respective legal positions of Suez, Panama, and Kiel. The assertions of the United States that the legal situations of Suez and Panama are actually quite different[1] are without foundation, for the very argument which could be used to demonstrate that the United States retains almost complete autonomy with respect to the Panama Canal, subject only to its treaty obligations to Panama and to Great Britain, could be applied with equal force to establish a corresponding autonomy on the part of the United Arab Republic with respect to the Suez Canal. The very reluctance of the United States to have the question of Suez dealt with through the organs of the United Nations and its enthusiasm for consultation outside of that body can largely be attributed to a desire to keep the question of Panama out of the discussions.[2] The contention that Suez had been "internationalized"[3] but that Panama had not is untenable, since its effect would be to deny to third parties as to the treaties regulating Panama the very rights which are recognized to flow to the nonsignatories of the Convention of Constantinople of 1888. Had Egypt taken the bold and imaginative step of suggesting that it would be prepared to place the Suez Canal under an international régime if the United States were prepared to place the Panama Canal in a similar position, it is difficult to see how the United States could have avoided having the two questions linked. From both the legal and the political points of view, the situation of the two waterways is virtually identical. The same principles of free passage govern both waterways, both share a strategic significance, both have been the stakes of political and military contention, and both serve the same important fundamental purpose of providing communication between two great oceans. One would therefore have to ask as to any proposal to "internationalize" the Suez Canal: Would this proposed régime be acceptable and viable for Panama and Kiel as well?

---

[1] As stated by President Eisenhower at his news conference of August 8, 1956. THE SUEZ CANAL PROBLEM, JULY 26–SEPTEMBER 22, 1956, at 45 (Dep't of State Pub. 6392) (1956).

[2] Bloomfield, *The U.N. and National Security*, 36 FOREIGN AFFAIRS 597, 602 (1958).

[3] In a news conference on August 28, 1956, Secretary of State Dulles spoke of the Suez Canal in these terms and said of Panama that ". . . there is no international treaty giving other countries any rights at all in the Panama Canal except for a treaty with the United Kingdom which provides that it has the right to have the same tolls for its vessels as for ours." *Op. cit. supra* note 1, at 301. As previously discussed (*supra* pp. 70-71), a right of free passage for the vessels of all nations is granted in the treaties of the United States with Great Britain and Panama.

The proprietors of interoceanic canals rest under certain obligations to those nations whose vessels employ these waterways. These duties are the consequence of the hardening into customary international law of usages which have received the acquiescence of nations and have in turn justifiably induced reliance by those nations.[4] This body of law shares many common features with the principles which govern the navigation of international straits and international rivers. However, the law of nations as a body of law applicable *ex proprio vigore* does not prescribe a legal régime for canals as it does for interoceanic straits. The law of canals acquires its force with respect to a particular canal only if that waterway has been dedicated to international usage—a dedication which has in the practice of states taken the form of treaties constituting a grant of rights to the other contracting parties and to third states as well.[5] The extent to which the rules of customary international law apply may be limited by the terms of the dedication. In this respect, the origin of the law governing interoceanic canals resembles that of the law of international rivers, for both presuppose a dedication followed by the entry into force of norms of customary international law bearing upon waterways so dedicated.

The three great artificial waterways which link the oceans have been opened on a nondiscriminatory basis to free navigation by the warships and merchant vessels of all nations in time of peace. This freedom of navigation cannot be secured merely through a prohibition on direct interferences with free passage, such as prohibitions on transit directed against certain vessels or certain nations. The waterway must also be kept free of illegitimate interference with passage effected through the application of local law dealing with such matters as defense, police, safety, navigation, customs duties, sanitation, and the like.[6] While the power of a territorial sovereign to subject

---

[4] The force of acquiescence in asserted rights as establishing principles of international law received its most important confirmation in the Fisheries Case (United Kingdom v. Norway), [1951] I.C.J. Rep. 116. Concerning the role of acquiescence in the creation of customary international law, see MacGibbon, *The Scope of Acquiescence in International Law,* 31 Brit. Y.B. Int'l L. 143 (1954), and *Customary International Law and Acquiescence,* 33 *id.* at 115 (1957).

[5] See pp. 182-84 *supra.*

[6] An impediment of this nature may either exclude a vessel entirely from the use of the waterway or subject it to delay, the impact of which is measurable in terms of the cost to the shipowner of having the vessel inoperative for that period. Reference has already been made in another context (p. 221 *supra*) to the use which has been made of customs inspections as an instrument of the closing of the Suez Canal to vessels carrying contraband. Concerning unnecessary delays in transit through the

both the operator of the waterway and the users of the waterway to regulations of this nature cannot be doubted, these powers must be exercised in a reasonable manner, which is no more than to say that their exercise must result in a minimal interference with freedom of transit. What a state through the territory of which an interoceanic canal runs is precluded by law from doing directly, it cannot be allowed to accomplish indirectly through impediments which find their purported justification in the necessity of protecting the functioning of the canal and the public order of the state.

If, while the littoral state is neutral, the waterway is to be employed by warships and merchant vessels of the belligerents and by vessels carrying supplies to and from the belligerents, the right of free passage must be maintained for this shipping as well as for vessels having no part in the hostilities. If action is necessary to protect the neutrality of the state through which the interoceanic canal flows, reasonable measures may be taken to achieve that purpose, even as navigation through interoceanic straits may be restricted and controlled by riparian states in the interest of their defense. Finally, it must be recognized that a state through the territory of which an interoceanic canal passes has no obligation to allow entry by warships of its adversaries, by merchant vessels of its enemies, or by vessels belonging to neutral nations which, if intercepted upon the high seas or in the ports or territorial sea of the belligerent, would be subject to control by that state under the rules currently applicable to the conduct of an economic blockade.[7]

Freedom of navigation conceived solely in terms of freedom from legal restraints imposed by the operator or supervisor of the waterway or by the territorial sovereign is not enough. The operator or supervisor must be under a corresponding affirmative obligation to maintain the waterway in navigable condition, to improve it in order to accommodate an increased volume of traffic or larger vessels, and to repair it as the occasion arises. Whether the canal be a sea-level canal like Suez or one through which transit must be accomplished with the aid of locks, as is characteristic of the Kiel and Panama canals, the canal cannot be the same passive channel of communication that a strait is. Passages must be scheduled, locks filled and emptied, signals operated, convoys organized in a manner which bears a striking analogy to the

---

Turkish Straits caused by the making of sanitary inspections and of the condemnation of this practice by the Straits Commission, see RAPPORT DE LA COMMISSION DES DÉTROITS À LA SOCIÉTÉ DES NATIONS, 1925, at 23-26 (1926).

[7] See 1 MEDLICOTT, THE ECONOMIC BLOCKADE (1952); Fitzmaurice, *Some Aspects of Modern Contraband Control and the Law of Prize,* 22 BRIT. Y.B. INT'L L. 73 (1945).

operation of a railroad. If further facilities, such as pilotage and tugs, are required in order to permit transit by shipping generally or by special types of vessels, these too must be provided. None of these services can be supplied without adequate tolls or other means of financial support for the operator of the canal. Tolls have very largely been eliminated on international rivers because the riparians are by and large both users and administrators of the waterway, so that the financial burden of maintaining the waterway in a navigable condition bears some rough correlation to the benefit derived from the use of the river as a channel of waterborne commerce and as a means of access to the markets of other nations. A nation's right to have the portion of the waterway within foreign territory navigated by its vessels without cost to them is compensated for by the grant of a corresponding right to the vessels of the other riparians. Maintenance and improvement of the river then becomes an indirect subsidy to waterborne commerce, just as the provision of highways without cost to persons employing them for the transport of goods constitutes a form of financial assistance to such carriers. The situation with respect to canals is altogether different, for in this case no such reciprocity of maintenance and use within a group of states exists. Given that fact, the assessment of tolls becomes inevitable, and these tolls must be adequate to supply, as a minimum, the necessary costs of maintenance and operation.

The financial régime of interoceanic canals furnishes persuasive evidence that negative prohibitions on interference with freedom of navigation and affirmative obligations with respect to the running of the waterway are not mutually exclusive. While the floor under tolls must without doubt be the necessary cost of operating the waterway prorated according to the quantum of use made by the individual vessels navigating it, the absence of any recognized ceiling could result in a blocking of the canal by pricing it out of the market. To raise tolls to an unreasonably high limit would therefore assuredly constitute an interference with freedom of navigation in as material a sense as a blockade by warships of the operator or territorial sovereign.

THE FUNCTIONS OF AN INTERNATIONAL AGENCY CHARGED WITH THE
ADMINISTRATION OF INTEROCEANIC CANALS

These principles, which have been stated only with the utmost generality, are indisputably difficult to apply and enforce under troubled economic or political conditions. The nationalization of the Suez Canal Company inspired those adversely affected by the action of the Egyptian Government to propose the panacea of "internationalization" of the Canal, of which the establishment of an international operating or supervisory body was to be an essential element. These proposals owed some indebtedness to the precedents furnished by international river commissions and by the Straits Commission which formerly guarded the Turkish Straits. The failure of a substantial proportion of these agencies to withstand bad political weather, as previously discussed in these pages, was conveniently overlooked in the desire to wrest control of Suez from the Egyptian authorities without actually ousting Egypt from all jurisdiction over the waterway. What was first suggested as the International Authority for the Suez Canal[8] to take over the operation of the Canal became in turn the proposed Suez Canal Board,[9] in which Egypt would participate, and ultimately degenerated, thanks to Egyptian intransigence, into the Suez Canal Users Association.[10] This last, denied any jurisdiction over the Canal itself, became no more than an organ of consultation among the nations whose ships employed the Canal and the instrument of what was hoped to be a financial boycott of the new Egyptian authority operating the Canal.[11] In both these respects, it must be reckoned a failure—a consummation which its composition quite plainly indicated from the outset.

Despite the failure of the proposals for internationalization of the operation and control of Suez, the establishment of an international agency for one or all of the interoceanic canals remains at least a theoretical possibility.[12] Such an international body could have jurisdiction ranging from complete control

[8] Tripartite Proposal for the Establishment of an International Authority for the Suez Canal, Aug. 5, 1956, *op. cit. supra* note 1, at 44.

[9] Attachment to the Aide Mémoire Delivered by the Suez Committee at Meeting with President Nasser, Sept. 3, 1956, *id*. at 306, 307.

[10] Declaration Providing for the Establishment of a Suez Canal Users Association, Sept. 21, 1956, *id*. at 365.

[11] The prospect held out by Secretary of State Dulles that a vessel which had paid its tolls to the Association and had been furnished a pilot by the Association might present itself at the entrance to the Canal and be placed in one of the transiting convoys by the Egyptian authorities (Extemporaneous Statement by Secretary of State Dulles at the Second Suez Conference, Sept. 19, 1956, *id*. at 359-61) was virtually the remotest of possibilities that could have been envisaged at that time.

[12] Upon his return from a study mission which took him to Panama, Senator Aiken proposed that one possible way out of the continuing conflict between the United States and Panama over the Panama Canal would be the "internationalization" of the Canal under the auspices of the United Nations or the Organization of American States. *Report to the Senate Foreign Relations Committee,* 86th Cong., 2d Sess. 15 (1960).

of the operation of the waterway and passage through it, including all of the attributes of a sovereign state over its own territory, to a merely advisory and consultative role without power to direct or to adjudicate.[13] The agency might find its competence limited to preventing prohibition or impediment of free passage through the operation of laws and regulations promulgated by the operator of the waterway or by the sovereign of the territory through which it flows. In that event, the function of the agency would be in the main adjudicative. If, on the other hand, the international organ were to be charged with the responsibility of maintaining freedom of passage in its operational sense, it might be expected that the agency would be a technical one devoted to the administration, maintenance, and operation of the Canal. In either event—whether the agency were to be adjudicative or operational—it is not unlikely that these powers would have to be supplemented by the function of making rules. It is assumed that for these purposes the agency would have more than an advisory and consultative function and that it would be capable of ordering as well as of recommending.

Should an international body be entrusted with the task of ensuring that passage through the canal was not impeded or precluded by legal restraints upon transit, that agency would presumably remain independent of the operator of the waterway and of the territorial sovereign. The disputes which would be submitted to it would be those arising between the proprietor and sovereign on the one hand (who admittedly might be one and the same) and the users of the waterway on the other.[14] The differences with which it dealt might include such questions as the following: Are customs inspections of transiting vessels conducted with sufficient speed? Is the sovereign of the territory justified in forbidding transit of the waterway to vessels which had been blacklisted as the result of trade with enemies of that sovereign? Are the safety regulations promulgated by the agency operating the waterway excessively stringent and consequently an unreasonable burden upon freedom of navigation?

[13] It must be remembered that "the Agents in Egypt of the Signatory Powers" of the Convention of Constantinople, Oct. 29, 1888, 15 MARTENS, N.R.G. 2d ser. 557 (1891), were by article VIII of that convention "charged to see that it is carried out." In the event of any threat to "the security and free passage of the Canal," they were to meet, "make the necessary verifications," and report to the Khedival Government. With this essentially fact-finding function must be contrasted the requirement of the last paragraph of article VIII that they were to "demand" ("réclameront") the removal of "any work" or the "dispersion of any assemblage" which would interfere with the freedom and safety of navigation.

[14] That is, differences of the nature referred to in Chapters III and IV.

The subject matter upon which such a tribunal would operate would be so specialized and so much removed from the normal concerns of international courts that special procedures and remedies would have to be applied. The International Court of Justice or a specially constituted tribunal could grant injunctive relief,[15] but, depending on the procedures of the court, a substantial period of time might elapse between the inception of the interference and the order of the court. For the losses suffered by the vessel or vessels prior to the injunction (or after the injunction, in the event of noncompliance by the offending state), damages might be awarded.

However, the harm caused by an unlawful interference with freedom of navigation might be such as not to be compensable in damages. A delay in the transit of a warship or some other public vessel or a closing of the canal to such forces could lead to the loss of a battle or even of a war. Interference with the passage of merchant ships might, unless promptly restrained, cause severe dislocations in the economy of a user country. In such instances, the tribunal would not be able to deal with the matter in the leisurely manner of some international courts of the past.[16] Occasions would inevitably arise on which a prompt order that a vessel be allowed immediate transit would be the only way in which justice might be done.[17] Such an order compelling the operator of the canal or the sovereign of the territory to allow passage forthwith would permit only the most summary of proceedings or no formal proceedings at all if the matter of hours was crucial. While the number of cases to be decided could not be expected to be large, the need to grant prompt relief in certain cases would point to the desirability of a body able to act on extremely short notice in urgent cases. The function of granting permanent injunctions and of providing reparations for delay or exclusion might be left to a judicial tribunal, while the power temporarily to enjoin interferences with navigation in cases demanding expeditious treatment might be vested in a body of a political character. To draw upon the precedent furnished by the organ of consultation established by the Convention of Constantinople,[18]

---

[15] For example, Trail Smelter Case (United States v. Canada), Decision of March 11, 1941, 3 UNITED NATIONS, REPORTS OF INTERNATIONAL ARBITRAL AWARDS 1905, at 1966 (1949).

[16] See in this connection the proposal of Sir Cecil Hurst in *Wanted! An International Court of Piepowder,* 6 BRIT. Y.B. INT'L L. 61 (1925).

[17] It might be observed at this point that the "provisional measures" which may be indicated by the International Court of Justice under article 41 of its Statute (59 Stat. 1055, T.S. No. 993) are not binding upon the parties.

[18] Convention of Constantinople, *supra* note 13, art. VIII.

the diplomatic representatives of certain designated states might justifiably
be given this authority by reason of their availability for prompt action.
Whether particular acts of the operator of the waterway or the territorial
sovereign were lawful could, of course, turn out to present difficult and complex issues of law. In that event, the political organ would probably feel constrained to take no action and would be called upon to turn the matter over
to the court designated for the purpose.

Some, but not all, of the problems of adjudication in particular cases
might be resolved by giving the tribunal the power to establish detailed rules
concerning free passage in application of the more general principles to be
derived from customary international law and the treaties governing the
waterway. Such rules would not be dissimilar to the *règlements de navigation*
which have been framed by the Central Commission of the Rhine[19] and by
other river commissions and the rules promulgated by the Straits Commission
for the Turkish Straits.[20] Regulations of this nature might permit certain
questions to be decided on the basis of a full and deliberate examination of
the problem before the hard case of a vessel's demanding and being denied
immediate transit should be presented. In this respect, the agency charged
with responsibility for regulating impediments to free navigation would
combine legislative and judicial functions in the manner of the Central
Commission of the Rhine or an administrative agency of the United States
Government. It should hardly be necessary to add that the rule-making
function might supplement but could not replace the essentially adjudicative
role of the international organ.

To confine the jurisdiction of such an international agency to impediments to freedom of transit in the legal sense would, however, leave the way
open to the impairment of navigation through the indirect means of failure
to supply facilities necessary for transit of the canal. A warship could be de-
layed for a substantial period of time by refusing to grant it space in a convoy or by the closing of a set of locks for repair. Large and heavily laden
tankers could be denied the use of the canal through the device of failure to
do the requisite dredging or deepening. To push tolls to an exceptionally
high level or to assess tolls in a discriminatory fashion could also result in

---

[19] See p. 117 *supra*.

[20] See Renseignements sur les conditions de passage des navires et aéronefs à
travers les détroits reliant la Mer Noire à la Mer Egée, published annually by the
Straits Commission; for those of a typical year see Rapport de la Commission des
Détroits à la Société des Nations, 1929, at 11 (1930).

denying use of the waterway to shipping generally or to particular classes of vessels. Admittedly, a number of these obstacles to navigation would operate in a less selective fashion than the legal restraints to which reference has just been made, in that the latter are more susceptible of precise application to a particular vessel for a particular purpose for a particular period of time. Fault might likewise be somewhat easier to assess in the case of these legal restraints than in the case of failures to maintain the waterway in proper operating order, which may arise from intent, negligence, inadvertence, or causes beyond the control of the operator of the waterway.

The fact that no hard and fast line can be drawn between legal and technical restraints on freedom of navigation suggests that a tribunal having jurisdiction over disputes between users on the one hand and the operator or territorial sovereign on the other might very soon find itself involved in questions of a technical, operational, and financial order. If there should be complaints from the world shipping community that tolls had been raised to an unwarrantedly high level in order to afford a larger profit to the operator of the waterway,[21] an examination of the accounts and of the operations of the authority administering the canal would be necessary before any intelligent decision could be made on the matter. The blocking of the waterway by the closing of a bridge at inconvenient hours would call for a survey of the scheduling of transits through the canal as well as of the use made of the bridge. A complaint that vessels were being subjected to undue delay before being formed into convoys to pass through the canal could be answered only on the basis of a close inquiry into the capacity of the canal, the efficiency of its operation, and the practicability of increasing the speed of transits through proper scheduling. If these questions are to be answered in a meaningful way, the supervisory agency would not find it enough to deal with the individual case. It would instead have to establish rules, requirements, and plans for the future. With the assumption of that function would come entry into the day-to-day administration of the canal.

Once an international agency had been charged with the actual operation and maintenance of the waterway, it would be inevitable that it would lose some of that impartiality and detachment characteristic of a judicial body. Separation of the managerial and judicial functions would promote the in-

---

[21] As in the case of the controversy over Suez Canal tolls which culminated in the agreement of Nov. 30, 1883, between De Lesseps and representatives of the British shipowners. See GREAT BRITAIN, PARLIAMENTARY PAPERS, EGYPT No. 3 (1884) 50 (C. 3850).

dependence of the judicial arm and enhance respect for its judgments. But the descent into the market place to operate the waterway would place the international agency in a position in which it would have to act as a mediator between the users and the territorial sovereign and would be subject to pressures and demands from both.

In its relations with the territorial sovereign, the agency would have to enjoy a particularly wide range of immunity from the legal restraints which the territorial sovereign would undoubtedly attempt to place upon it. Freedom from the economic burden of taxes on the activities, the profits, the property, and even perhaps on the employees of the operating agency could be secured only by the working out of detailed financial arrangements with the sovereign. These arrangements would likewise govern the entry, conditions of employment, and freedom of movement of persons, other than nationals of the territorial sovereign, who might be brought in as technicians, as well as the circumstances under which local labor requirements might be met. The extent to which the international body operating the waterway was to be freed from the restraints of local law and the immunities to be accorded to the officials and employees of the agency would have to be agreed upon. The agreements negotiated by the old Suez Canal Company and the Egyptian Government and, to a lesser extent, the international agreements between the Republic of Panama and the United States would be valuable sources of guidance for matters which the international agency and the territorial sovereign would have to resolve. Although the precise terms of the arrangements might not be appropriate to the operation by an international agency of an interoceanic waterway, agreements which grant foreign governments the right to maintain military facilities on the territory of a state and define the immunities to be accorded the forces manning those facilities[22]

[22] See Agreement between the Parties to the North Atlantic Treaty regarding the Status of Their Forces, signed at London, June 19, 1951, 4 U.S.T. 1792, T.I.A.S. No. 2846. Agreements relating to the uses of bases and other facilities exist in great numbers and variety; see Agreement between the United States of America and the Republic of the Philippines concerning Military Bases, signed at Manila, March 14, 1947, 61 Stat. 4019, T.I.A.S. No. 1775, as extended and amended; Agreement between the Government of the United States of America and the Government of the United Kingdom of Great Britain and Northern Ireland concerning a Long Range Proving Ground for Guided Missiles To Be Known as "The Bahamas Long Range Proving Ground," signed at Washington, July 21, 1950, 1 U.S.T. 545, T.I.A.S. No. 2099; Defense Agreement between the United States of America and Spain, signed at Madrid, Sept. 26, 1953, 4 U.S.T. 1895, T.I.A.S. No. 2850; Agreement between the United States of America and the Kingdom of Greece concerning Military Facilities, signed at Athens, Oct. 12, 1953, 4 U.S.T. 2189, T.I.A.S. No. 2868.

would offer some guidance as to the matters with which the operator and the sovereign would have to deal.

Although it is thus possible to conceive the functions which an international agency for the administration of an international canal might be called upon to fulfill and the privileges and immunities with which it might be necessary to endow it, the hard reality remains that such an organization would in greater or lesser degree do no more than give an institutional framework to conflicts of interest in terms of commerce, economics, politics, and strategy. The arena would cease to be the diplomatic one and would become the agency itself, as a means through which conflicting demands might be asserted and perhaps reconciled.[23] Denial of passage to a vessel would no longer be a matter for resolution on the diplomatic level between the user nation and the sovereign of the waterway. Adjustment would instead be sought within the organs of consultation, adjudication, or decision established as parts of the organization.

The pressures to which such an international organization would be subjected would be far greater than those to which international river commissions have been exposed. In the first place, the strategic and economic importance of interoceanic waterways is immeasurably greater. Secondly, the river commissions frequently give expression to an interdependence and to a mutuality and reciprocity of interest which is lacking in the case of interoceanic canals. A riparian represented in an international river commission would in all likelihood, in addition to its concern with the river as part of its territory or the boundary of its domains, have sound economic reasons for maintaining the portion of the river subject to its control in navigable condition and would desire to lend its support to its shipowners who conducted commerce on the waterway. Failure to maintain the river in navigable condition would have an adverse impact not only on foreign vessels but on the shipping of the delinquent state. Lack of adequate maintenance might also lead to retaliation by another riparian with ill effects for the shipping of the state which had been the first to let the state of the waterway deteriorate. And quite independently of nationalities, a certain community of interest would exist among the various ship operators on the waterway. It should hardly need repetition at this point that such an identity or reciprocity of interests is lacking in as strong a degree in the case of the three major interoceanic canals. Admittedly, the fact that the vessels of the United States and of Germany are

[23] It will be recalled that the Central Commission of the Rhine was in its infancy little more than a permanent diplomatic conference.

respectively the largest users of the Panama and Kiel canals must be taken into account. But it must not be forgotten that these are pluralities and not majorities. In the case of the most important of these waterways, the interest of the United Arab Republic in the use by its shipping of the Suez Canal is comparatively limited. Nor do the United Arab Republic, the United States,[24] and Germany occupy important positions as users of the waterways of which they are not the proprietors. And finally, the river commissions have by no means shown themselves to be invulnerable to political pressures and to the strategic demands of the riparians and of powerful nonriparians. The Central Commission of the Rhine still thrives, but many of the other commissions have passed into history.

The significance of an international regulatory or operating agency would not necessarily be limited to its role as an area for the assertion of conflicting demands by the interests represented within it. To the extent that the agency assumed an autonomous existence, it could be expected that it might acquire interests and functions and rights of its own, the protection or assertion of which might place it in conflict with user states, with shipping interests, with other international organizations, or with the territorial sovereign within the domain of which it might operate. This presupposes that the agency had been so constituted as to be more than a deliberative body within which the territorial sovereign, the users, and the actual operator of the waterway would attempt to harmonize their policies and reconcile their conflicting demands.

As we have had occasion to see, if an international agency were to be given more than a judicial function and were to engage in policymaking and day-to-day operation of the waterway, the conflicting demands of states would be asserted within the organization itself. So similarly, the agency would become increasingly vulnerable to conflict with individual nations, even with those whose interests were most strongly represented within it, if it were given wide responsibility for the operation and maintenance of the canal. If the international agency were to fix tolls and to determine the revenues to be derived by the territorial sovereign, it would be forced to defend itself

[24] The existence of large flag of convenience fleets actually owned by United States companies make the statistics on passages through interoceanic canals somewhat misleading, based as they are on the flags of the transiting vessels. More sophisticated statistics would have to add to the total of transits of the Suez Canal by United States vessels many of the passages of Liberian vessels, which as a class are the second largest users of the canal.

against the accusations of shipping companies and of governments of maritime states that the tolls were too high. At the same time, the territorial sovereign would be in a position to assert that the operator was guilty of extravagance in its operations, that it did not charge enough for the services which it rendered, and that the resulting revenues to the sovereign were consequently kept at an unreasonably low level. In turn, the international agency might find itself in the position of asserting that its operating costs were being inflated through taxes or customs duties or minimum wage scales which were demanded by the sovereign of the territory through which the waterway ran. Conceived as an independent operating agency, an international body charged with the running of one or more of the interoceanic canals would fall heir to all of the problems which have been faced in the past by independent operating agencies. The most notable of these agencies, now of solely historical interest, has been the Suez Canal Company. With the acquisition by an operating agency of autonomous and supranational status, the internationalization of the waterway would have brought its administration full circle. Admittedly, the creation of such a body by the United Nations or by a group of nations would in all probability endow it with more influence and authority than that possessed by a private company or a single government operating the waterway. In terms, however, of its legal status and of possible areas of dispute, an international operating agency would be no more and no less than the operator of the waterway. The possible forms of conflict in which it might become involved would not differ from those to which, as indicated in this study, operating agencies have been subject in the past.

The Organization of a Possible International Agency

It now becomes necessary to consider what might be the composition of an international agency playing one of the possible roles which have been hypothecated. In view of the many questions of law and of policy which would necessarily have to be decided in the course of operating an interoceanic canal, the mere appointment of a sole arbitrator or of a general manager would not be enough. The responsibility for supervision and for policy would undoubtedly have to be placed in an institution regulated by a body of persons representative of the membership of the organization.

If "internationalization" should prove politically acceptable only if it were adopted for all three of the major interoceanic canals, it would then be necessary to decide whether the institution established to carry out this purpose should be a separate one for each waterway[25] or a single one exercising supervision over all three. Should the functions of the body be primarily those of adjudication, the question is easily answered. Economy would be the most important practical consideration in calling for only one. Disputes of sufficient moment to be referred to an international tribunal would be infrequent enough, at least, not to warrant three standing tribunals for the purpose of dealing with them. From the standpoint of the development of the law, uniformity of decision could not be secured through the separate operation of a number of tribunals. Furthermore, the referral of a substantial number of cases to one tribunal would both facilitate the work of the tribunal and enhance the acceptability of its decisions. A substantial volume of business produces a sophisticated jurisprudence by widening the categories of problems which the tribunal must face. At the same time, the fact that one state may in one case appear as a claimant, qua user of a waterway, and as a respondent, qua operator or territorial sovereign, means that it will be forced to see itself in a number of capacities before the court, as one who will benefit in one case and incur some burden in another.

On the other hand, there is no reason why the maintenance and operation of three widely dispersed waterways—one in Europe, a second in Africa, and the third in America—should be subjected to a detailed centralized control which would be geographically removed from at least two and possibly three of these principal waterways. It would be expected that the routine function of passing ships through the waterways and of keeping the watercourse in good order would rest with local managers, while policies on such matters as denial of passage, the level of tolls, or major improvements would be left to a single commission, exercising supervisory powers over the three waterways.

The range of possibilities concerning the institution which might be established for the operation of the three major interoceanic canals would be a wide one. Such an organization might or might not be within the structure of the United Nations.[26] An agency which was no more than an instrumen-

---

[25] The Treaty of Versailles, 112 BRITISH AND FOREIGN STATE PAPERS 1 (1919), provided in article 386 that in the event of a dispute about the interpretation of the provisions relating to Kiel or a violation of these articles, an appeal could be had "to the jurisdiction instituted for the purpose by the League of Nations." To avoid the reference of small questions to the League, Germany was to establish a local authority at Kiel to deal with disputes in the first instance.

[26] See Summary of Internal Secretariat Studies of Constitutional Questions relating to Agencies within the Framework of the United Nations 2 (A/C.1/758) (1954) for a

tality of several of the great powers might be a viable one, but an organization so conceived would be incompatible with internationalization. An independent international organization could fulfill the minimal requirements of effectiveness only if it were to be virtually universal in its membership, since trade and transport make almost all the nations of the world users of these waterways. It seems improbable that an agency outside the United Nations would command the influence and power which would enable it to survive in the hard world of power politics.

An operating agency constituted within the general structure of the United Nations might take the form of a "specialized agency" within the meaning of articles 57 and 63 of the Charter. The relation of the organization to the United Nations would then be determined by the terms of the agreement concluded between it and the United Nations. Establishment of an interoceanic canal agency in this form would mean that its principal dealings within the United Nations would be with the Economic and Social Council.[27] Since the agency would be fundamentally unlike existing specialized agencies, the organization established for the administration of interoceanic canals would be subjected to political and military pressures which would call for a link with the United Nations different from that provided by a relationship with the Economic and Social Council. In the case of the International Atomic Energy Agency, this problem has been solved by an agreement between it and the United Nations that the Agency is an "autonomous international organization in . . . working relationship with the United Nations . . ."[28] What is theoretically a less intimate relationship with the United Nations offers the possibility of reporting directly to and of calling matters to the attention of the General Assembly and the Security Council, rather than the Economic and Social Council.[29]

A stronger bond might be forged by the establishment of the organiza-

---

consideration of the ways in which specialized international organizations may be brought into relationship with the United Nations.

[27] By reason of the requirements of articles 63 and 64 of the United Nations Charter.

[28] Agreement Governing the Relationship between the United Nations and the International Atomic Energy Agency, Annex to U.N. Gen. Ass. Res. 1145 (XII), Nov. 14, 1957, U.N. GEN. ASS. OFF. REC. 12th Sess., Supp. No. 18, at 56 (A/3805) (1957). Such an agency is a "special body," in the terminology employed by the United Nations Secretariat; see document cited *supra* note 26, at 4.

[29] Statute of the International Atomic Energy Agency, done at the Headquarters of the United Nations, Oct. 26, 1956, art. III, para. B.4. and B.5., 8 U.S.T. 1093, T.I.A.S. No. 3873.

tion as a "subsidiary organ" of the General Assembly under the authorization contained in article 22 of the Charter. The consequences of this relationship might paradoxically be a greater involvement in international politics and at the same time some protection against the buffeting of international politics. Once the agency were subordinated to the General Assembly the policies which might be established for it would be determined in what is essentially a political arena. It is not inconceivable that proposals like those made during the Italo-Ethiopian War would again be heard.[30] The closing of the three canals to vessels in the service of a belligerent state or carrying goods to or from that state might be called for by the General Assembly or by the Security Council as a sanction for an unlawful resort to the use of force. With this power of control would likewise come responsibility for the defense of the waterway against attack upon it by one of the belligerents. The level of tolls and the scale of expenditure for the operation, maintenance, and improvement of the canals would likewise be subjected to political influences. On the one hand, an increase in revenues might be sought in order to support other activities of the United Nations, such as programs of technical assistance. On the other, reductions in tolls might be sought as an indirect subsidy to an ailing transport industry or to stimulate the flow of goods between certain areas. The economic and strategic importance of the three great canals is such that they must inevitably become the focus of political controversy.

The suggestion has been made by Clark and Sohn[31] that important straits and interoceanic canals might be directly administered as "trust territories" under articles 77 and 81 of the United Nations Charter. The instrument, as presently drafted, would of course permit such an arrangement, but it must be remembered that this proposal is made in the context of a general revision of the Charter. From the perspective of the present day, the difficulties of convincing a nation which is now the sole proprietor of an interoceanic canal that it should place the waterway under international administration would be great enough without adding the requirement that the surrounding territory be placed under trusteeship. If the proprietor could be so persuaded, then the type and machinery of control which the General Assembly would have to establish would give rise to much the same sort of problems as would the creation of a controlling body in the form of an "organ" of the General Assembly.

An operating organization established as other than an "organ" of the General Assembly would in all likelihood be constituted in the bicameral

[30] See p. 237 *supra*.
[31] WORLD PEACE THROUGH WORLD LAW 158 (2d ed. 1960).

form normally found in international organizations, that is, a general representative body or assembly and a council. In accordance with the usual pattern, wider authority and heavier responsibilities would be vested in the latter of these bodies. In view of the limited size of a council, the necessity would arise of securing within its membership a proper balancing of those interests to which reference has been made.

The provision of representation for special interests is by no means a new concept of international organization. Within the Conference of the International Labor Organization, seats are provided for those representatives of labor, of employers, and of governments who make up each national delegation.[32] One half of the governmental representatives in the Governing Body of that organization must be appointed by the "members of chief industrial importance," as decided by the Governing Body, subject to an appeal to the Conference.[33] The constitutional instruments of several international organizations concerned with transportation have provided spaces in their councils for those states which have particular and identifiable interests in the matters with which the organizations deal. In electing members to the Council of the International Civil Aviation Organization, the Assembly is required to give adequate representation to "the States of chief importance in air transport . . . the States not otherwise included which make the largest contribution to the provision of facilities for international civil air navigation," and other states to give representation to major geographic areas.[34] Within the Council of the Intergovernmental Maritime Consultative Organization, balanced representation is given to the nations with the largest interest and with substantial interest in "international seaborne trade" and in "providing international shipping services."[35] Such devices as these for taking account of special national interests are a more sophisticated response to special needs than the commissions constituted by groups of states, aligned on a political basis, which have assumed special responsibilities for transportation facilities in the past.[36]

[32] Constitution of the International Labour Organisation, adopted at Montreal, Oct. 9, 1946, art. 3, 62 Stat. 3485, T.I.A.S. No. 1868, 15 U.N.T.S. 35.

[33] Article 7, paras. 2 and 3.

[34] Convention on International Civil Aviation, opened for signature at Chicago, Dec. 7, 1944, art. 50, para. (b), 61 Stat. 1180, T.I.A.S. No. 1591, 15 U.N.T.S. 295.

[35] Convention of the Intergovernmental Maritime Consultative Organization, done at Geneva, March 6, 1948, art. 17 and Appendix I, 9 U.S.T. 621, T.I.A.S. No. 4044, 289 U.N.T.S. 48.

[36] E.g., the bodies created by the Convention as to Cape Spartel Light-House, signed at Tangier, May 31, 1865, art. IV, 1 MALLOY, TREATIES, CONVENTIONS, INTERNATIONAL ACTS, PROTOCOLS AND AGREEMENTS BETWEEN THE UNITED STATES OF AMERICA AND OTHER

The councils of the modern specialized agencies may also include representatives of states which do not have a concern in the work of the organization based on special interests.[37] Some "neutral" element within a commission charged with responsibility for the operation of interoceanic canals could provide an important stabilizing element, as well as give representation to a wider public interest.

A decision in favor of a body charged with third-party determination would necessarily entail a body staffed in part at least by persons sitting in a nonrepresentative capacity as impartial judges. However, it must be recognized that under current procedures of judicial settlement, whether by the International Court of Justice[38] or by an arbitral tribunal,[39] the litigant states are entitled to have judges of their nationality sitting upon the tribunal. It is unlikely that states would be willing to have legal questions relating to the use of international waterways referred to a tribunal which would lack the national representation presently demanded of other types of international courts.

The paramount procedural obstacle to the creation of any form of international organization which would have authority to vary the terms upon which Suez, Panama, and Kiel are now open to free use by the ships of all nations would be the probable necessity of almost universal acceptance of any new régime to be established for the three waterways. If the decisions

---

POWERS, 1776-1909, at 1217 (1910) (now terminated by the Protocol relating to the Management of the Cape Spartel Light, signed at Tangier, March 31, 1958, 9 U.S.T. 527, T.I.A.S. No. 4029, transferring the management of the Light to Morocco); and the Convention relating to the Régime of the Straits, signed at Lausanne, July 24, 1923, art. 10, 28 L.N.T.S. 115, 131.

For a fuller discussion of functional organizations and forms of representation therein see JESSUP AND TAUBENFELD, CONTROLS FOR OUTER SPACE AND THE ANTARCTIC ANALOGY 87-116, 131-33 (1959).

[37] Convention cited *supra* note 34, art. 50, para. (b)(3).

[38] Statute of the International Court of Justice, art. 31, para. 2, providing for the presence of judges, whether regular members of the court or *ad hoc* judges, of the nationality of the parties.

[39] As reflected in the practice of each party's appointing one or two judges, the third or fifth member, as the case may be, being chosen by the other members or by an impartial third party, such as the President of the International Court of Justice. Under article 45 of Convention No. I of The Hague concerning the Pacific Settlement of International Disputes, signed Oct. 18, 1907, 36 Stat. 2199, T.S. No. 536, it is provided that, if the parties have not made other arrangements, each is to appoint two arbitrators, only one of whom may be its national. The neutral element in the tribunal is thus not confined to the fifth member.

of a tribunal were to be binding upon the disputants, the contending parties would have had to have accepted the jurisdiction of the court so constituted. If an operating agency were to claim the power of discriminating between vessels in time of war, that exercise of authority would be in conflict with the existing safeguards surrounding use of the waterways. There is room for doubt that a direction by the General Assembly to an agency which it had established as its subsidiary organ would be sufficient warrant for impeding the passage of a ship under the protection of a state which had not given its consent to this exercise of control.[40] As in other respects, force, rather than

[40] Those states which had consented to the jurisdiction of the agency, whether by participation in its establishment or otherwise, would be bound by its decisions. Other states would maintain that their rights were to be determined according to previous treaties rather than by resolutions of the General Assembly. The validity of the latter position would turn upon (a) the binding effect of resolutions of the General Assembly under article 13 of the Charter as to matters other than the administration of the United Nations itself, and (b) the impact of article 103 of the Charter, which provides that in the event of a conflict between obligations under the Charter and obligations under any other international agreement, the former shall prevail. The fact that the General Assembly would cast its resolution in the form of a direction to an "organ" could not endow that resolution with greater authority vis-à-vis the nations affected than a resolution speaking directly to those nations.

The more widely held view appears to be that recommendations of the General Assembly do not have the force of law in the sense of creating legal duties. See the authorities collected in Johnson, *The Effect of Resolutions of the General Assembly of the United Nations,* 32 BRIT. Y.B. INT'L L. 97, 97-111, 122 (1955-56); STONE, AGGRESSION AND WORLD ORDER 153 (1958); but cf. Sloan, *The Binding Force of a 'Recommendation' of the General Assembly of the United Nations,* 25 BRIT. Y.B. INT'L L. 1, 29, 31-33 (1948). In his separate opinion in the Advisory Opinion on Voting Procedure on Questions Relating to Reports and Petitions Concerning the Territory of South-West Africa, [1955] I.C.J. Rep. 66, 90 at 115, Judge Lauterpacht pointed out that "it is in the nature of recommendations that, although on proper occasions they provide a legal authorization for Members determined to act upon them individually or collectively, they do not create a legal obligation to comply with them."

If a resolution of the General Assembly were "to provide a legal authorization" for restrictions on passage through a waterway to be implemented by the operator or supervisor of the waterway, that *authorization* would come into conflict with the *right* of a belligerent, under the treaty regulating the waterway, to free and unimpeded passage. Such an authorization could not create an "obligation" within the meaning of article 103 of the Charter, for to do so would be to endow peace-keeping resolutions with an obligatory force for which support cannot be found in the Charter. The resolution which authorized the blocking of the waterway would thus not oblige the billigerent to honor the closing. The limitations placed on the right of self-help by the Corfu Channel Case, [1949] I.C.J. Rep. 1 at 34-35, would call in question the legal authority of the belligerent to force passage until all other remedies had been exhausted.

The Security Council could, under articles 41 and 42 of the Charter, take action

law, might dictate the outcome of such an attempt. As to those matters which do not constitute terms of the dedication to international use which has been made with respect to the waterways, universal consent would not be necessary. If there is no present firm legal limitation on the level of tolls levied by the nation which is both proprietor and operator of the waterway, there is no reason why an international organization would not have equal power unilaterally to fix the tolls demanded for transit.

THE DEFENSE OF THE WATERWAY

Each major interoceanic canal and the territory through which it flows have been of major geopolitical and strategic importance throughout their history. The words of Lawrence, written some eighty years ago with respect to the Suez Canal, have an oddly contemporary ring: "In fact, under present circumstances, there is no guarantee for the security of the canal. The Eastern Question is a gunpowder magazine, which may explode at any moment."[41] The vital strategic position which these waterways continue to occupy would, in the event of their internationalization and subjection to the control of an international agency, call for arrangements which would protect their physical security and the integrity of their international control. To discharge this function, an international agency would probably have at its immediate disposal no more divisions than the Pope. Its military strength could hardly be greater than that of any of its members. What security could, under such circumstances, be provided for each of the waterways must be considered against the background of previous attempts to isolate or defend the channel from hostile acts.

Historically, the first approach to the problem during the nineteenth and early twentieth centuries was an attempt to secure the "neutralization" of the waterway. In its narrower sense, the concept refers to the doctrine that in time of war the passage of belligerent vessels and of merchant vessels carrying supplies to the belligerents does not compromise the neutrality of the

---

binding upon members of the United Nations, even if they had not consented to the establishment and powers of the organ having responsibility for the waterway. Under article 2, paragraph 6, of the Charter its action might also be binding upon non-members. It is deserving of note that article 41 envisages the interruption of sea communications as one of the sanctions falling short of the use of force.

[41] *The Suez Canal in International Law,* in ESSAYS ON SOME DISPUTED QUESTIONS IN MODERN INTERNATIONAL LAW 41 at 62 (2d rev. ed. 1885).

territorial sovereign. It is this meaning which is probably to be attributed to the expression as it is employed in the treaties relating to the Panama Canal.[42] In its second and more important sense, "neutralization" looks to the status of territory or of a watercourse which by international agreement (and possibly also by a process of self-neutralization) is not to be employed for military purposes by the territorial sovereign and is not to be the subject of hostile military operations by other nations.[43] This form of neutralization, which normally includes the demilitarization of the area, was at one time conceived to be the best manner in which to insulate waterways from hostilities.

Even before the completion of the Suez Canal, De Lesseps had had it in mind that the principal powers should guarantee the neutrality of the waterway and should agree not to commit hostile acts in the Canal or in the adjacent land and sea areas.[44] During the period from 1856, when De Lesseps first unsuccessfully broached this proposal, to 1882, the neutralization of the Canal was frequently mentioned[45] but was not the subject of any international agreement.[46] The success of an Egyptian nationalist movement in 1882 under the leadership of Colonel Arabi Pasha and the threat which this was thought to pose to the security of the Canal brought a British military expedition, culminating in the occupation of a substantial amount of Egyptian territory, including the area of the Canal, and the temporary control of

---

[42] Treaty between the United States and Great Britain to Facilitate the Construction of a Ship Canal, signed at Washington, Nov. 18, 1901, art. III, 32 Stat. 1903, T.S. No. 401; Convention between the United States of America and the Republic of Panama for the Construction of a Ship Canal to Connect the Waters of the Atlantic and Pacific Oceans, signed at Washington, Nov. 18, 1903, art. XVIII, 33 Stat. 2234, T.S. No. 431 ("The Canal . . . shall be neutral in perpetuity . . .").

[43] See, concerning the demilitarization and neutralization of territory, *Report of the International Committee of Jurists entrusted by the Council of the League of Nations with the task of giving an advisory opinion upon the legal aspects of the Aaland Islands question,* LEAGUE OF NATIONS OFF. J., Sp. Supp. No. 3, at 15-19 (1920); REID, INTERNATIONAL SERVITUDES IN LAW AND PRACTICE 196-203 (1932); VÁLI, SERVITUDES OF INTERNATIONAL LAW 263-72 (2d ed. 1958); Graham, *Neutralization as a Movement in International Law,* 21 AM. J. INT'L L. 79 (1927). Graham observes that the institution of neutralization was on the way to extinction after the peace treaties of the First World War (at 94).

[44] WILSON, THE SUEZ CANAL: ITS PAST, PRESENT, AND FUTURE 89 (1933); EDGAR-BONNET, FERDINAND DE LESSEPS: LE DIPLOMATE, LE CRÉATEUR DE SUEZ 272-75 (1951).

[45] *Id.* at 89-90.

[46] Other than the reference in article 14 of the second Firman of Concession of Jan. 5, 1856, COMPAGNIE UNIVERSELLE, RECUEIL DES ACTES CONSTITUTIFS 6 at 14, to the Canal as "passage neutre."

the waterway by the British army.[47] Once established as the guardian of the Canal, Great Britain proposed to the principal European powers that no hostilities should take place in the Canal or its approaches, that no troops or munitions of war should be disembarked in the Canal, and that no fortifications be erected on the Canal or in its vicinity.[48] The diplomatic negotiations leading to the drafting of a treaty by a conference of the major powers[49] held in 1885 and to the definitive Convention of Constantinople, concluded in 1888, have been fully described elsewhere[50] and need not be dwelt upon here. The Canal was not demilitarized. "Permanent fortifications" were forbidden to the extent that they might "interfere with the freedom and complete safety of navigation."[51] The right of the Sultan and of the Khedive of Egypt to take measures "to assure by their own forces the defense of Egypt and the maintenance of public order" was recognized, notwithstanding the provisions of the Convention relating to free passage in time of war and to certain restrictions on the activities of the signatory powers.[52] On the other hand, the contracting parties agreed to refrain from the blockade of the Canal and from the commission of any hostile act in the Canal or its approaches, and affirmatively undertook to respect the equipment and establishment of the Canal.[53] What this amounted to was a recognition of a right to defend the Canal, subject to a right of free passage except as to enemies of the territorial sovereign, and of an obligation upon the part of the high contracting parties to refrain from belligerent activity in the Canal and its vicinity.[54]

[47] HALLBERG, THE SUEZ CANAL: ITS HISTORY AND DIPLOMATIC IMPORTANCE 254-56 (1931); 2 CHARLES-ROUX, L'ISTHME ET LE CANAL DE SUEZ 46-82 (1901).

[48] Extract from a Despatch from Earl Granville to Her Majesty's Representatives at Paris, Berlin, Vienna, Rome, and St. Petersburgh, dated January 3, 1883, respecting the Suez Canal, etc., GREAT BRITAIN, PARLIAMENTARY PAPERS, EGYPT No. 10 (1885) (C. 4355).

[49] The Draft Treaty for insuring the Free Use of the Suez Canal drawn up at this conference is reprinted in Correspondence respecting the Suez International Commission, with the Protocols and Procès-Verbaux of the Meetings, GREAT BRITAIN, PARLIAMENTARY PAPERS, EGYPT No. 19 (1885) 309 (C. 4599).

[50] HALLBERG, *op. cit. supra* note 47, at 282-309; 2 CHARLES-ROUX, *op. cit. supra* note 47, at 82-118.

[51] Convention of Constantinople, *supra* note 13, art. XI. It is significant that the paragraph of article 4 of the draft treaty of 1885 which provided: "Aucun point en commandant ou en menaçant le parcours ou l'accès ne pourra être occupé militairement" (FRANCE, DOCUMENTS DIPLOMATIQUES, COMMISSION INTERNATIONALE POUR LE LIBRE USAGE DU CANAL DE SUEZ 213 (1885)) was not included in the Convention of 1888.

[52] Article X.

[53] Articles II-IV.

[54] Sir Arnold Wilson felicitously put it that the Canal was not "neutralised" but

That the Suez Canal was not in fact neutralized is made all the clearer by the relation of Great Britain to the Canal over a period of 70 years following its military intervention in Egypt in 1882. During these decades, it was Great Britain, and not Egypt or Turkey, which guaranteed the security of the Canal through the presence of its military forces in Egypt and the protectorate which it maintained over that country.[55] The measures which were taken by the British authorities during the First World War in order to prevent the waterway from being of service to the enemy have already been described.[56] The years of that conflict saw the area of the Canal the scene of hostilities, against which the Canal was successfully defended.[57] The Suez area assumed enhanced strategic significance during the Second World War. The British base which was established there supported the equivalent of 28 infantry divisions, 13 armored divisions, and naval and air force units totaling about 65 squadrons.[58] The Suez Canal Company cooperated in these activities by making its land, its workshops, and its buildings available to the defending forces, in a number of instances at reduced cost or without charge.[59] The Canal was repeatedly attacked from the air, but with only comparatively slight effect on the movement of vessels; the Canal was blocked for only a limited time by the bombings which took place.[60] To the Egyptian nationalist, this involvement of the Canal and of the operating company in hostilities seemed both inconsistent with the asserted neutrality of the Canal and a painful consequence of the ability of Great Britain,

---

"universalised." *Some International and Legal Aspects of the Suez Canal,* 21 TRANSACTIONS OF THE GROTIUS SOCIETY 127, 140 (1935). For contemporary studies of the "neutralization" of the Canal, see TWISS, ON INTERNATIONAL CONVENTIONS FOR THE NEUTRALISATION OF TERRITORY AND THEIR APPLICATION TO THE SUEZ CANAL (1887) and CONTUZZI, LA NEUTRALIZZAZIONE DEL CANALE DI SUEZ E LA DIPLOMAZIA EUROPEA (1888).

[55] HALLBERG, *op. cit. supra* note 47, at 295; HOSKINS, THE MIDDLE EAST: PROBLEM AREA IN WORLD POLITICS 64 (1954).

[56] See p. 218 *supra.*

[57] DOUIN, L'ATTAQUE DU CANAL DE SUEZ (3 FÉVRIER 1915) (1922); HALLBERG, *op. cit. supra* note 47, at 325-49.

[58] Statement by Mr. Selwyn Lloyd on May 12, 1953, 515 H.C. DEB. (5th ser.) 1082 (1952-53).

[59] Rapport sur le fonctionnement du Service du Transit au cours de la guerre de 1939-1940, at 44-54, Manuscript dated 1944, in the files of the Compagnie Universelle du Canal Maritime de Suez (now Compagnie Financière de Suez) in Paris.

[60] COMPAGNIE UNIVERSELLE, THE SUEZ CANAL: NOTES AND STATISTICS 53-62 (1952); REPORT, SUEZ CANAL COMPANY, 1945, at 2, 5, 8.

through the power of its military strength, to remove Egypt from the neutral status which that country would more willingly have assumed.[61]

At the conclusion of the Second World War, negotiations were begun by Great Britain and Egypt looking to the conclusion of a treaty providing for the termination of the Treaty of 1936,[62] which constituted the legal basis for the continued presence of British forces in the Canal area. A draft treaty was agreed upon *ad referendum*,[63] but failed of conclusion because of outstanding issues about the Sudan and because Egypt was not satisfied with anything short of "immediate, complete" evacuation of British troops, "not conditioned by a treaty."[64] Although the Treaty of 1936 contained no provision authorizing its termination at the will of one of the parties,[65] it was denounced by Egypt on October 15, 1951, after further negotiations between the two countries had failed to provide a basis for agreement.[66] A period of conflict, during which fighting broke out between Egyptian civilians and the British forces, was brought to an end by the conclusion in October 1954 of the Agreement regarding the Suez Canal Base, whereby Great Britain agreed to the withdrawal of her forces, subject to a right to maintain certain bases on a standby basis under civilian administration.[67] Egypt, for its part, undertook that in

[61] See EL-HEFNAOUI, LES PROBLÈMES CONTEMPORAINS POSÈS PAR LE CANAL DE SUEZ 172-77, 182-86 (1952), maintaining that the activities of Great Britain during the two World Wars were inconsistent with the Convention of 1888.

Egypt was technically neutral until war was declared on Germany and Japan on February 26, 1945. Decree of Feb. 26, 1945, Egypt, Ministère de la Justice, [1945] RECUEIL DES LOIS, DÉCRETS ET RESCRITS ROYAUX 23.

[62] Treaty of Alliance between His Majesty, in respect of the United Kingdom, and His Majesty the King of Egypt, signed at London, Aug. 26, 1936, GREAT BRITAIN T.S. No. 6 (1937).

[63] GREAT BRITAIN, PAPERS REGARDING THE NEGOTIATIONS FOR A REVISION OF THE ANGLO-EGYPTIAN TREATY OF 1936, EGYPT No. 2 (1947) (CMD. 7179).

[64] Statement by the Prime Minister, March 11, 1947, 434 H.C. DEB. (5th ser.) 1141-42 (1946-47). Egypt then requested the Security Council to direct "the total and immediate evacuation of British troops from Egypt . . ." Letter from the Prime Minister and Minister for Foreign Affairs of Egypt addressed to the Secretary-General, dated 8 July 1947 (S/410) (1947). Because it proved impossible to secure the adoption of any of the resolutions submitted to the Security Council, no action was taken. U.N. SECURITY COUNCIL OFF. REC. 2d year, 198th meeting 2303, 200th meeting 2339-40, 201st meeting 2362 (1947).

[65] See GREAT BRITAIN, ANGLO-EGYPTIAN CONVERSATIONS ON THE DEFENCE OF THE SUEZ CANAL AND ON THE SUDAN, DECEMBER 1950-NOVEMBER 1951, EGYPT No. 2 (1951) 47 (CMD. 8419); 481 H.C. DEB. (5th ser.) 35 (1950-51).

[66] The Egyptian Minister for Foreign Affairs to His Majesty's Ambassador in Cairo, Oct. 27, 1951, EGYPT No. 2 (1951), *supra* note 65, at 46.

[67] Agreement between the Government of the United Kingdom of Great Britain and

the event of an armed attack by an outside power on one of the states of the Arab League, Egypt would "afford to the United Kingdom such facilities as may be necessary in order to place the Base on a war footing and to operate it effectively." [68]

Within two years after the conclusion of this treaty, Israel had mounted an attack upon Egypt in retaliation against what were alleged to be widespread violations of the truce agreement, and British forces and French forces were in the Canal area after a short war [69] with the Egyptian armed forces. The combination of air attack and the deliberate sinking of block ships in the Canal closed down the Canal completely from October 1956 to April 1957, when the clearance of the channel was completed under United Nations supervision. [70] During the months that the Canal was unusable a startled world came to the conclusion, as the result of the shifting of vessels to the Cape route and of the utilization of alternative sources of supply, that the Canal was somewhat less indispensable than it had been thought to be. Anglo-French intervention was sought to be justified on the ground that it was necessary to the defense of the Canal [71] and that a brisk skirmish with the Egyptian forces was not too high a price to pay for protecting the Canal from the consequences of hostilities between Israel and Egypt. In retrospect, the cure appears to have been worse for the patient than the disease it was designed to prevent. The short-lived war of 1956 affords the final proof that the Canal, far from being "neutralized," has been a strategic prize of great importance in three important wars of the twentieth century.

Less than a half a century ago, the right of the United States to fortify the Panama Canal, consistently with the treaties establishing the régime of the

---

Northern Ireland and the Egyptian Government regarding the Suez Canal Base, signed at Cairo, Oct. 19, 1954, GREAT BRITAIN T.S. No. 67 (1955).

[68] Article 4. "Immediate consultation" was to take place in the event of a "threat of an armed attack" (art. 6).

[69] The author, while mindful of the great weight to be attached to the view of Professor Jessup (as he then was) that the conflict did not constitute "war," has yielded to the judicial authority of Navios Corp. v. The Ulysses II, 161 F. Supp. 932 (D. Md. 1958), aff'd per curiam, 260 F.2d 959 (4th Cir. 1958), in which case Professor Jessup was an expert witness. That case held the conflict to be "declared war" within the meaning of a charter party. See also Baxter, *The Definition of War*, 16 REVUE ÉGYPTIENNE DE DROIT INTERNATIONAL 1 (1960).

[70] *Clearance of the Suez Canal, Report of the Secretary-General*, Nov. 1, 1957, particularly at 7-9 (U.N. Doc. No. A/3719) (1957).

[71] Statement by Sir Pierson Dixon, U.N. SECURITY COUNCIL OFF. REC. 10th year, 749th meeting 1-5 (S/PV. 749) (1956).

canal, was much in dispute. The legal case against fortification was based on the fact that the Hay-Pauncefote Treaty[72] used the word "neutralization" at three points, that the United States had taken as "the basis of the neutralization of such ship canal" rules substantially like those in the Convention of Constantinople of 1888,[73] and that the preamble of the Treaty stated that that agreement did not impair "the 'general principle' of neutralization" established in article VIII of the Clayton-Bulwer Treaty.[74] As in the case of the Convention of Constantinople, it was stipulated in the Hay-Pauncefote Treaty that the Canal would never be made the scene of any act of hostility,[75] but the agreement between the United States and Great Britain departed from the precedent of 1888 in according to the United States the right "to maintain such military police along the canal as may be necessary to protect it against lawlessness and disorder."[76] The unsoundness of the contention that the United States was without power to fortify the Canal may be demonstrated by reference to the history of the drafting of the Hay-Pauncefote Treaty. Of major significance is the fact that the rules established for the Panama Canal, although assertedly based on the principles of the Convention of Constantinople of 1888, contained no such limitations on fortification as are to be found in the earlier treaty. Moreover, the express prohibition of fortification which was incorporated in the Clayton-Bulwer Treaty[77] was omitted from the Hay-Pauncefote Treaty in what was quite clearly a deliberate change of principle. The accuracy of this conclusion is borne out by a memorandum of the British Secretary of State for Foreign Affairs, which was communicated to the President of the United States, to the effect that the Hay-Pauncefote Treaty contained no stipulation prohibiting the erection of fortifications.[78] As thus conceived, "neutralization" did not mean that the

---

[72] Treaty between the United States and Great Britain to Facilitate the Construction of a Ship Canal, signed at Washington, Nov. 18, 1901, 32 Stat. 1903, T.S. No. 401.

[73] Article III of the Hay-Pauncefote Treaty.

[74] Convention between the United States of America and Great Britain for Facilitating and Protecting the Construction of a Ship Canal Connecting the Atlantic and Pacific Oceans, signed at Washington, April 19, 1850, 9 Stat. 995, T.S. No. 122. Article VIII provided, *inter alia,* that any canal to be constructed would be open on a basis of equality to citizens of the United States and citizens and subjects of every other state.

[75] Hay-Pauncefote Treaty, *supra* note 72, art. III, rule 2.

[76] *Ibid.*

[77] Clayton-Bulwer Treaty, *supra* note 74, art. I.

[78] Memorandum, accompanying a dispatch of Lord Lansdowne, Foreign Secretary, to Mr. Lowther, chargé, Aug. 3, 1901, 3 MOORE, DIGEST OF INTERNATIONAL LAW 212, 215 (1906).

Canal could not be defended by military forces nor that it could not be fortified. The term was used to identify much the same concept as had been written into the Convention of 1888—that the Canal should remain open to the vessels of belligerents while the United States should be a neutral, that no warlike acts were to be committed by the transiting vessels, and that the state occupying the territory through which the waterway flowed had the right to defend it against use or attacks by its enemy. The agreement with Panama, the Hay–Bunau-Varilla Treaty,[79] left no doubt that the "neutrality" of the Canal did not exclude defensive measures. Article XXIII of the Treaty accorded the United States the right to use its land and naval forces and to establish fortifications to assure the safety and protection of the Canal and transiting vessels, when considered necessary by the United States. Despite the lively debate among international lawyers,[80] the Canal was fortified, substantial sums being expended for that purpose from 1911 onward.[81]

The authority of the United States to take measures for the defense of the Panama Canal was again spelled out in the treaty which was concluded with the Republic of Panama in 1936.[82] Article X of that treaty provided: "In case of an international conflagration or the existence of any threat of aggression which would endanger the security of the Republic of Panama or the neutrality or security of the Panama Canal, the Governments of the United States of America and the Republic of Panama will take such measures of prevention and defense as they may consider necessary for the protection of their common interests ..."

These measures were to be the subject of consultation between the two governments. The declaration of the two nations that they had "common interests" to be defended reflected the realities of the situation. So narrow is

[79] Convention between the United States of America and the Republic of Panama for the Construction of a Ship Canal to Connect the Waters of the Atlantic and Pacific Oceans, signed at Washington, Nov. 18, 1903, 33 Stat. 2234, T.S. No. 431.

[80] For the view that the Panama Canal was neutralized and could not be fortified, see Olney, *Fortification of the Panama Canal*, 5 Am. J. Int'l L. 298 (1911); Kennedy, *The Canal Fortifications and the Treaty*, 5 *id.* at 620 (1911); Hains, *Neutralization of the Panama Canal*, 3 *id.* at 354 (1909). The contrary opinion is advanced in Davis, *Fortifications at Panama*, 3 *id.* at 885 (1909); Knapp, *The Real Status of the Panama Canal as Regards Neutralization*, 4 *id.* at 314 (1910); Wambaugh, *The Right to Fortify the Panama Canal*, 5 *id.* at 615 (1911); Nixon, *Does the Expression "All Nations" in Article 3 of the Hay-Pauncefote Treaty Include the United States?*, [1913] Proceedings of the American Society of International Law 101, 120-21.

[81] Padelford, The Panama Canal in Peace and War 306-07 (1943).

[82] General Treaty of Friendship and Cooperation between the United States of America and Panama, signed at Washington, March 2, 1936, 53 Stat. 1807, T.S. No. 945.

the strip of land under the control of the United States and so intertwined
the respective jurisdictions of the two countries that one of them could not be
expected to remain neutral while the other was at war.[83] After the outbreak
of the Second World War in Europe, the United States initiated negotiations
with Panama to secure additional defense sites needed for the protection of the
Canal.[84] The government of President Arias, which was justifiably suspected of pro-Axis leanings, was unwilling to provide these facilities. It was
not until the régime of Arias had ended and the United States was a par-
ticipant in the war that an agreement was signed providing for utilization
by the United States of the defense sites and for the construction of needed
highway facilities.[85] A year after the termination of hostilities, Panama demanded that the lands which had been used by the United States be returned on the ground that the 1942 agreement had expired.[86] Strong public
sentiment against any prolongation of the rights of the United States within
Panamanian territory brought about the defeat of an agreement embodying
a negotiated compromise on the issue, and the United States had moved
out of the last sites by February 1958.[87] The difficulties encountered by the
United States in securing the sites and the refusal of Panama to extend their
occupancy after the war illustrate the political obstacles which lie in the way
of defending an interoceanic canal if the territorial sovereign of the nation
through which it flows and the operator of the waterway find themselves
in disagreement. In this instance two states, sharing jurisdiction over the
territory through which the Canal runs, could coordinate their activities
only while active hostilities were in progress.

As for the Kiel Canal, there has never been even a hint of neutralization.
The waterway was originally constructed for military purposes. When it
was first subjected to international control through the relevant clauses of

[83] Garcia-Mora, *International Law Applicable to the Defense of the Panama Canal*, 12
U. Detroit L.J. 63, 71 (1949).

[84] Wright, *Defense Sites Negotiations Between the United States and Panama, 1936-
1948*, 27 Dep't State Bull. 212, 214 (1952).

[85] Agreement for the Lease of Defense Sites in the Republic of Panama, signed at
Panamá, May 18, 1942, 57 Stat. 1232, E.A.S. No. 359; the negotiations preceding the
conclusion of the agreement are described in Wright, *supra* note 84, at 214-16.

[86] Article I of the Defense Sites Agreement had provided that it would terminate
"one year after the date on which the definitive treaty of peace which brings about the
end of the present war shall have entered into effect." The surrender of Japan was
elevated by Panama to the dignity of a definitive treaty of peace for these purposes.
See Wright, *supra* note 84, at 218.

[87] Wright, *supra* note 84, at 219; 18 Dep't State Bull. 31 (1948).

the Treaty of Versailles,[88] no mention was made of defense, or fortification, or neutralization, an omission perhaps explicable in terms of the general demilitarization of Germany. The Second World War witnessed extensive use of the Canal and of the port of Kiel for the deployment of German naval forces, particularly submarines. Germany was, until the end of the war, very largely successful in protecting the waterway from the aerial attacks mounted against it by the British and United States air forces.[89]

Neutralization—that slippery concept—has in times gone by been prescribed for other types of waterways as well. To secure the free navigation of the Lower Danube, for example, there was agreement by the parties to the Treaty of Berlin of 1878 that all of the fortifications from the Iron Gates to the mouth of the river were to be razed and no new fortifications erected. The presence of warships on the river was prohibited.[90] But these limitations were not to be found in the Treaty of Saint-Germain, which governed the administration of the river after the First World War.[91] To draw a further example from the decade of the Convention of Constantinople, with its pseudo neutralization of the Suez Canal, the Boundary Treaty which Argentina and Chile concluded in 1881[92] stipulated that the Strait of Magellan was to be neutralized and that no fortifications were to be erected on the shores.[93] The Strait of Gibraltar, like Suez a major gateway to the Mediterranean, was also subjected to a régime[94] politically linked with the Suez question,[95] which

[88] Signed June 28, 1919, arts. 380-386, 112 BRITISH AND FOREIGN STATE PAPERS 1 (1919).

[89] CAI SCHAFFALITZKY DE MUCKADELL, THE KIEL CANAL 12 (1947); see Prime Minister to General Ismay, July 15, 1940, in CHURCHILL, THEIR FINEST HOUR 646 (Boston, 1949), concerning the military importance of the Canal.

[90] Treaty between Germany, Austria-Hungary, France, Great Britain, Italy, Russia, and Turkey, signed at Berlin, July 13, 1878, art. 52, 3 MARTENS, N.R.G. 2d ser. 449 (1878-79).

[91] Treaty of Peace between the Allied and Associated Powers and Austria, signed at Saint-Germain-en-Laye, Sept. 10, 1919, arts. 301-08, 11 MARTENS, N.R.G. 3d ser. 692 (1923).

[92] Signed at Buenos Aires, July 23, 1881, art. V, 12 MARTENS, N.R.G. 2d ser. 491 (1887).

[93] It was realized that the content given to the concept of neutralization as applied to this strait would have an important bearing upon the creation of a similar régime for the Suez Canal. 2 BRÜEL, INTERNATIONAL STRAITS 235 (1947).

[94] Declaration between Great Britain and France respecting Egypt and Morocco, together with the Secret Articles, signed at London, April 8, 1904, art. VII, 101 BRITISH AND FOREIGN STATE PAPERS 1053 (1912).

[95] It was by article VI of the same instrument that Great Britain finally adhered to the stipulations of the Convention of Constantinople of 1888, 15 MARTENS, N.R.G. 2d

forbade the fortification of a portion of the shore. But these arrangements neither placed any limitations upon the passage of warships of belligerents or of merchantmen carrying military supplies and other contraband, nor gave them a standing differing in any respect from that of other straits.

The most recent experiment in the neutralization of a strait proved unsuccessful. By the Treaty of Lausanne the Turkish Straits were both neutralized and demilitarized.[96] However, an undefended route of access to the Black Sea through Turkish territory caused Turkey increasing apprehension as the major powers of Europe prepared for the war to come.[97] At the request of Turkey, a new convention was concluded at Montreux in 1936, which substituted Turkey for the Straits Commission as the custodian of the Bosporus and Dardanelles and authorized Turkey to "remilitarize" the Straits.[98] The part which the Straits played in the Second World War has been described elsewhere in this study.[99] During the last two decades, there has been intermittent pressure by the Soviet Union for a revision of the Montreux Convention which would give that country a role in the control and defense of the Straits.[100] That the Convention should be revised to

---

ser. 557 (1891), subject to the holding in abeyance of the execution of those provisions of the Convention (art. VIII) concerning consultation with the Khedive in the event of a threat to the security or free passage of the Canal.

[96] Convention relating to the Régime of the Straits, signed at Lausanne, July 24, 1923, arts. 4-9, 28 L.N.T.S. 115.

[97] Hoskins, *op. cit. supra* note 55, at 27; Routh, *The Montreux Convention Regarding the Régime of the Black Sea Straits (20th July, 1936),* in [1936] Survey of International Affairs 584 (1937); the writings on the new convention are collected in 7 Hudson, International Legislation 386 (1941). As put by the representative of Turkey at the Montreux Conference: "La démilitarisation des Détroits a été acceptée par la Turquie dans des circonstances qui diffèrent totalement de celles qui existent aujourd'hui. La garantie collective accentuée par une garantie régionale envisagée par l'article 18 de la Convention de Lausanne était en réalité le support sur lequel toute la structure du régime des Détroits avait été basée; or, il est avéré aujourd'hui que ce support est incapable de jouer le rôle qui lui est dévolu." Actes de la Conférence de Montreux, 22 juin-20 juillet 1936, Compte rendu des séances plénières et procès-verbal des débats du comité technique 22 (1936).

[98] Protocol of Signature of the Convention concerning the Régime of the Straits, signed at Montreux, July 20, 1936, 173 L.N.T.S. 241, 7 Hudson, International Legislation 404 (1941).

[99] See pp. 194-95 and 199-200 *supra.*

[100] Concerning the conversations between Hitler and Molotov on this question in 1940, see United States Department of State, Nazi-Soviet Relations, 1939-1941: Documents from the Archives of the German Foreign Office 244-46 (Dep't of State Pub. 3023) (1948). The postwar position of the Soviet Union is reflected in a

"meet present-day conditions" was agreed at Potsdam,[101] but after a preliminary exchange of views, Russian enthusiasm for a conference to revise the Convention waned in the face of British, Turkish, and American opposition to any Russian military presence in the area of the Straits.[102] And so the matter remains today, although the Montreux Convention is now subject to denunciation at the demand of one of the parties.[103] An armed Turkey still guards the gates to the Black Sea.

The term "neutralization" has also been applied—it is submitted, with doubtful accuracy—to the action taken in 1950 by President Truman with respect to the activities of the Seventh Fleet in the Formosa area. According to the statement made at that time, the Chinese Nationalist Government was called upon not to engage in air and sea operations against the mainland, while United States forces were directed to prevent any attack on Formosa.[104] These restraints were later lifted to the extent of freeing the hand of the Nationalist Government to deal with the communist forces.[105] This policy, far from "neutralizing" the Formosa Strait, called for possible military activ-

---

Russian identic note of Aug. 7, 1946, in HOWARD, THE PROBLEM OF THE TURKISH STRAITS 47 (Dep't of State Pub. 2752) (1947).

[101] The Berlin (Potsdam) Conference, Protocol of the Proceedings, Aug. 1, 1945, sec. XVI, in *A Decade of American Foreign Policy; Basic Documents, 1941-49*, S. Doc. No. 123, 81st Cong., 1st Sess. 46 (1950); statements by Mr. Bevin in February and June of 1946, 419 H.C. DEB. (5th ser.) 1356 (1945-46); and 423 *id.* 1828-29 (1945-46). The position of the United States is set forth in a note from the Acting Secretary of State to the Soviet Chargé at Washington, Aug. 19, 1946, in HOWARD, *op. cit. supra* note 100, at 49-50, which suggested that "the regime of the Straits should be brought into appropriate relationship with the United Nations." Ahmed Sükrü Esmer has characterized the Montreux régime as "the best system that can be devised for safeguarding peace in this part of the world." *The Straits: Crux of World Politics*, 25 FOREIGN AFFAIRS 290, 302 (1946-47).

[102] In October 1946, Russia indicated that it was premature to consider holding a conference on the subject. Howard, *The Development of United States Policy in the Near East, 1945-1951, Part I*, 25 DEP'T STATE BULL. 809, 811 (1951).

A Soviet note to Turkey on May 30, 1953, renounced any territorial claims against Turkey, and the latter country replied by emphasizing that the question of the Black Sea Straits is governed by the Montreux Convention. Howard, *The Development of United States Policy in the Near East, South Asia, and Africa During 1953: Part I*, 30 DEP'T STATE BULL. 274, 277-78 (1954).

[103] Convention concerning the Régime of the Straits, signed at Montreux, July 20, 1936, art. 28, 173 L.N.T.S. 213; 7 HUDSON, INTERNATIONAL LEGISLATION 386 (1941).

[104] Statement by the President, June 27, 1950, in 2 AMERICAN FOREIGN POLICY 1950-1955: BASIC DOCUMENTS 2468 (Dep't of State Pub. 6446) (1957).

[105] *Id.* at 2475.

ities by the United States for the purpose of defending Formosa. Whatever is made of the position taken by the United States, the episode constitutes no real contribution to the law of international straits, since the area of sea involved did not constitute a strait in the legal sense of the word, that is to say, a narrow waterway in which the territorial seas of the littoral state or states meet and overlap.

In terms of the historical development of legal institutions, this brief summary of neutralization as applied to international waterways can only lead to the conclusion that there has been little confidence in this device as a means of preserving the integrity of the waterway.[106] The high-water mark of the institution occurred during the second half of the nineteenth century. The one major experiment of the twentieth century, the demilitarization and neutralization of the Turkish Straits, led to an abandonment of both neutralization and internationalization (in the sense of supervision by an international commission) after only thirteen years. The purported neutralization of the Suez and Panama canals goes only to freedom of passage in time of war, a prohibition on hostilities by transiting vessels while the territorial sovereign remained neutral, and, in the former case, a prohibition of fortification.[107] With the obsolescence of the fixed fortification in the late nineteenth and early twentieth centuries, this restriction upon the defensive measures taken by Great Britain and by Egypt became of no practical significance. Today, the three great interoceanic passages of Suez, Panama, and Kiel are under the protection of the three powers which control them and operate them, defended by their forces and subjected to the protective measures which the proprietor states may take in time of war. One cannot accurately speak of them as "neutralized"; at best, they can be characterized as "neutral."

If affirmative defensive measures, backed by the armed might of the proprietor nation, are necessary to the security of these waterways, who will

---

[106] But cf. Tchirkovitch, *La Question de la révision de la Convention de Montreux concernant le régime des Détroits Turcs: Bosphore et Dardanelles,* 56 REVUE GÉNÉRALE DE DROIT INTERNATIONAL PUBLIC 189, 220 (1952), who advocates the internationalization, neutralization, and demilitarization of the Straits, as under the Treaty of Lausanne.

When the United States urged in 1946 that the régime of the Straits be brought into relationship with the United Nations, it also stated that "Turkey should continue to be primarily responsible for the defense of the Straits." The Acting Secretary of State to the Soviet Chargé at Washington, Aug. 19, 1946, in HOWARD, THE PROBLEM OF THE TURKISH STRAITS 49-50 (Dep't of State Pub. 2752) (1947).

[107] See Siegfried, *Les Canaux internationaux et les grandes routes maritimes mondiales,* 74 HAGUE RECUEIL 5, 36, 60 (1949).

guarantee the security of the interoceanic canals when they pass under international administration? Their strategic significance[108] will be no less great because they are operated by an international agency. Undefended, they will constitute an attractive prize to any power which might desire to take advantage of the commanding positions they occupy on the world's sea routes. They cannot be left as power vacuums, to be filled by the first nation which finds it possible to replace international control with its own administration in time of war.

A collective guarantee by a group of nations of the security of the waterway and the territory surrounding it would be effective only if it could be determined in advance whether the guarantee would be carried out through energetic military measures in the event of an attack upon the waterway. The existing regional defense agreements, or alliances, do not contain firm commitments of military assistance by the member nations, since the members are desirous of protecting their right to take military measures in their own discretion and on their own terms. It is difficult to conceive the United States, as one of the guarantors of the Suez Canal, binding itself to take all measures within its power to defend the Suez Canal from a possible attack by Israel or one of the Arab states. For all of the importance of these waterways, it is probable that even those states most vitally concerned in their affairs would make the defensive measures taken with respect to the waterway turn upon their national interests at the time of the actual attack. To create uncertainty about the possibility of defense is itself to remove one defense.

Were a United Nations force ever to be constituted, the attractive prospect could be held out of basing these forces within three canal zones spanning the three major canals.[109] While this deployment might solve the problem of where to put the United Nations forces, it would not necessarily guarantee the military security of the canals, which require forces of considerable size for their effective defense. Before the evacuation of the British forces from the Suez area, over 50,000 British troops were based there alone. Although the political and military effectiveness of United Nations forces could not be assessed in terms of numbers alone, it must still be borne in mind that strong military, naval, and air forces, resting upon a sound logistical base, would be

[108] Eliot, *The World's Strategic Waterways,* 1 UNITED NATIONS WORLD, Sept. 1947, 30, 31.

[109] The Commission to Study the Organization of Peace has suggested that a permanent United Nations force "could guard sites, such as international waterways, for the mutual benefit of the nations concerned." STRENGTHENING THE UNITED NATIONS 87 (1957).

requisite to the successful defense of the canals. With the waning of enthusiasm for a United Nations force, the possibility that such troops would be
available for the defense of the canals becomes correspondingly more remote.

If the two possibilities to which reference has been made are not real or
useful ones, it would be necessary to fall back on national military power for
the defense of the passages. None of the existing proprietors—the United
Arab Republic, the United States, or Germany—would be willing to place its
armed forces at the call of an international agency operating or supervising
the canal. The imposition of responsibility for defense would call for some
corresponding authority over the canal and a strong voice in decisions which
might hold a possibility of inducing an aggrieved state to resort to the use
of force. If power must be coextensive with responsibility, the greater the
responsibility placed upon the littoral state for the protection of the waterway
the greater would have to be the weight given to its participation in the
affairs of the international agency. Failure of coordination between the territorial sovereign and the international body would allow the wedge of the
political attack to be inserted between them. Indeed, it is on this very rock
of defense that the placing of the canals under an international administration
endowed with wide powers might well founder.

Conclusions

Given the probable obstacles, which have been outlined above, to the
placing of interoceanic canals under international administration, the pan-
aceas of "internationalization" and of "neutralization" appear in a far less
attractive light than they might at first impression. For the present, the
gradual development of a body of customary international law applicable
to these waterways may hold out greater hope of maintaining free passage
through these important channels of commerce, even in time of international
stress and of conflict. The financial stake of the proprietors, the revenues
which they derive from the canals, and the importance of the waterways to
the naval and commercial interests of the states through the territory of which
they flow may ultimately act as a more important stimulus to the efficient and
orderly operation of these waterways than legal norms or institutions. It must
not be forgotten, however, that this very stability and order is itself law-
creating.

There is ample room for the view that interoceanic canals are already governed by a common body of law, which is the product of state practice, of

treaties, and of adjudication. This law does not confine itself to restrictions on the authority of the territorial sovereign or proprietor but also imposes restrictions and obligations upon the users of these waterways. What these legal rules are, the sources of law which have created them, and their present standing have been alluded to from time to time in the course of this study. Having regard to the continuing importance of the subjection of these waterways to a régime founded on law, the codification of these rules would be a useful step in giving them greater precision and in securing general acceptance of the existing customary law. The writer has attempted to give such a codified form to the law in the Articles on the Navigation of International Canals, which appear in the Appendix.

ARTICLES ON THE NAVIGATION OF INTERNATIONAL CANALS

ARTICLE 1

An interoceanic canal is an artificial waterway forming a necessary or convenient passage between two portions of the high seas.

ARTICLE 2

An interoceanic canal becomes an international canal within the meaning of these Articles when it has been dedicated by treaty, unilateral declaration, or otherwise to use by ships of nations other than those of the State through the territory of which such a canal flows, and there has been substantial reliance on the dedication by the shipping of such nations.

ARTICLE 3

Except under the circumstances set forth in Articles 4 and 5, an international canal shall be free and open, at all times and on a basis of nondiscrimination, to navigation by the ships of all nations, subject to:

(a) payment of tolls fixed in conformity with paragraph (d) of Article 6;

(b) compliance with the laws and regulations relating to sanitation, customs, immigration, navigation, safety, and the like of the State through the territory of which the canal flows and of the operator of the canal, provided such laws and regulations are applied in conformity with paragraph (e) of Article 6; and

(c) such conditions and restrictions as may have been established by the instrument dedicating the canal to international use.

ARTICLE 4

1. In time of war or armed conflict to which the State through the territory of which an international canal flows is not a party, merchant vessels and ships of war of the belligerent States shall be entitled to use the canal, provided they commit no acts of hostility therein. The passage of such vessels does not compromise the neutrality of the State through the territory of which the canal flows.

2. During the period referred to in paragraph 1, the State through the territory of which an international canal flows is under an obligation to ensure that no acts of hostility or other acts inconsistent with its neutrality are committed therein. In particular, both that State and the State the vessels of which pass through the canal are under an obligation to conform to the rules of neutrality in naval warfare, as derived, *mutatis mutandis,* from the Convention of The Hague of 1907 concerning the Rights and Duties of Neutral Powers in Naval War.

ARTICLE 5

1. In time of war or armed conflict to which the State through the territory of which an international canal flows is a party, that State is authorized:

(a) to exclude from the canal or to seize or attack therein vessels of war and merchant ships of the nation with which it is engaged in war or armed conflict, and

(b) to take necessary measures to assure that the canal is not used as a means of transporting contraband, including the exercise of the right of visit and search and the seizure of contraband so found.

2. The enjoyment of the rights referred to in paragraph 1 of this Article is subject to the treaties and other international obligations of the State through the territory of which the international canal flows. Such rights must likewise be exercised, in conformity with the general principle set forth in Article 3, with due regard for the right of neutral shipping to use the canal.

ARTICLE 6

In order to assure free and unimpeded navigation through an international canal, the State through the territory of which the canal flows and the operator of the canal are under an obligation:

(a) to operate the canal in an efficient fashion;

(b) to maintain the canal in a proper state of repair;

(c) to the extent feasible, to make necessary improvements to permit use of the canal by all ships desiring to transit it;

(d) to maintain tolls at a reasonable level, without discrimination between States, such tolls to be fixed with regard to the need of:

(1) covering the costs of current operation and maintenance and depreciation,

(2) meeting the costs of capital,

(3) provision of necessary reserves for improvements,

(4) repayment of the capital investment, and

(5) provision of a reasonable return on the investment represented by the canal.

(e) so to apply local laws relating to such matters as sanitation, customs, immigration, navigation, safety, and the like as not to impair the right of free navigation; and

(f) to provide, or assure the provision of, necessary ancillary services, such as towage and pilotage.

ARTICLE 7

1. In the absence of treaty stipulations to the contrary, the State through the territory of which an international canal flows has full discretion to determine from time to time by what agency the canal shall be operated in conformity with these Articles.

2. The principle set forth in paragraph 1 is without prejudice to such arrangements as may be made to place an international canal under international administration and control.

TABLE OF TREATIES

TABLE OF CASES

INDEX

(Countries employing international waterways or serving on international commissions are generally not indexed. See specific waterway, commission, legal doctrine, or question.)